Artisanal Enlightenment

Artisanal Enlightenment

*Science and the Mechanical Arts
in Old Regime France*

Paola Bertucci

Yale
UNIVERSITY
PRESS
New Haven & London

Published with the assistance of the Frederick W. Hilles Publication Fund
of Yale University, the Annie Burr Lewis Fund, and the Kingsley Trust Association
Publication Fund established by the Scroll and Key Society of Yale College.
Copyright © 2017 by Paola Bertucci.
All rights reserved.

This book may not be reproduced, in whole or in part, including illustrations,
in any form (beyond that copying permitted by Sections 107 and 108 of the
U.S. Copyright Law and except by reviewers for the public press),
without written permission from the publishers.

Yale University Press books may be purchased in quantity for educational, business,
or promotional use. For information, please e-mail sales.press@yale.edu (U.S. office)
or sales@yaleup.co.uk (U.K. office).

Set in PostScript Electra type by Newgen North America.
Printed in the United States of America.

ISBN 978-0-300-22741-3 (hardcover : alk. paper)

Library of Congress Control Number: 2017940032

A catalogue record for this book is available from the British Library.

This paper meets the requirements of ANSI/NISO Z39.48-1992 (Permanence of Paper).

10 9 8 7 6 5 4 3 2 1

To Ivano and Damian

Contents

Acknowledgments   ix

Introduction: Savants, Artisans, *Artistes*   1

PART 1   THE NATURAL HISTORY OF THE ARTS

ONE Lost Knowledge and the History of the Arts   31

TWO. Réaumur and the Science of the Arts   51

PART 2   THE SOCIÉTÉ DES ARTS

THREE Theory, Practice, Improvement   79

FOUR Society of Arts   104

PART 3   WRITING AND MAKING

FIVE The Politics of Writing About Making   145

SIX *L'Esprit* in the Machine   176

Epilogue   207

Appendix: Members of the Société des Arts (1728–1740s)   221

Notes   235

Bibliography   269

Index   291

Acknowledgments

This is a book of the digital age, and my first acknowledgments go to people whom I have only interacted with via e-mail. First and foremost, I am grateful to Olivier Courcelle, who approached me several years ago with the proposal of co-authoring an article on the Société des Arts, based on new documents that he and the late Roger Hahn had discovered. It was while working on that article that the idea of this book first emerged.

I am also profoundly grateful to my research assistants in Paris, who visited many of the archives listed in the bibliography, where they took thousands of photos of primary sources. With a newborn (then a toddler) to take care of, I would have been unable to study these documents without their help.

For the same reason, I am grateful to all the libraries and archives that make it possible to take photographic reproductions of their materials; to the various online databases and resources, such as Gallica, Google Books, the ARTFL Project, and archive.org, that offer free and easy access to rare research materials; and to Eve Watson at the Royal Society in London, Marie-Noëlle Vivier-Gallardo at the Bibliothèque de Caen, and John Nash at the Whipple Museum in Cambridge, who promptly sent me image files of the materials I needed.

In real life, I have benefited immensely from conversations and exchanges with colleagues and friends in Europe and the United States. I discussed this project over numerous glasses of wine with Mary Terrall, in Pasadena, New Haven, Los Angeles, Chicago, and Florence. Her warm support and generous feedback have been essential to the book's development. Ann Blair, Dena Goodman, Joy Rankin, Sophia Rosenfeld, and Courtney Thompson offered invaluable comments on earlier versions of the manuscript. I have been lucky to receive useful suggestions and friendly criticism from Pierre-Yves Beaurepaire, Bruno Belhoste, Marco Beretta, Sven Dupré, Dan Edelstein, Paula Findlen, Robert Fox, Jan Golinski, Anna Guagnini, Pamela Long, Isaac Nakhimovski,

Liliane Pérez, Jessica Riskin, Lissa Roberts, Simon Schaffer, Emma Spary, Larry Stewart, Marie Thébaud-Sorger, Koen Vermeir, and Spence Weinreich. Laurel Waycott has been the most resourceful assistant I could hope for. For sharing research materials or insights, I wish to thank Michael Bycroft, Marie Leca, Giandomenico Piluso, Larry Principe, and Giorgio Strano.

I am grateful to my colleagues at Yale for being strong supporters of this project. Naomi Lamoreaux, Francesca Trivellato, and Steve Pincus read the manuscript in various phases and offered insightful advice and warm encouragement. Interacting with them has been one of the great pleasures of writing the book. Students and faculty in the History of Science and Medicine Program have contributed directly or indirectly to the book's life, and I feel fortunate for being part of this stimulating community. I wish to thank also Glenda Gilmore for her timely advice on how to interact with publishers, and Jean Thomson Black and the staff at Yale University Press for endorsing my project with heartwarming enthusiasm.

I wrote most of the book overlooking Florence from the hills of the European University Institute or Villa I Tatti, and the Arno River from the library of the Museo Galileo or our apartment in the Oltrarno. My deepest thanks to Stephan Van Damme, Luca Molà, Ann Thompson, and Regina Grafe for their friendly hospitality and inspiring conversations. Thank you also to all relatives and friends who made our Italian year an unforgettable experience.

My yoga teachers also deserve warm thanks for helping me take care of my physical and mental well-being, especially when deadlines were approaching and balance was easier to find on the mat than in front of the screen.

Dulcis in fundo: My gratitude to Ivano, my traveling companion, is beyond words. This book is as much a part of his life as it is of mine. Between its lines, there is the story of our American adventure and its amazing surprises. One of these is Damian. As this book thickened, he appeared into the world, learned to crawl, walk, run, jump, climb, swim, and talk fluently in two languages. I don't feel I accomplished as much in the same amount of time. But I am excited to write these final lines and to dedicate this book to them. *Grazie* for your patience, your impatience, and the never-ending laughter.

Artisanal Enlightenment

Introduction: Savants, Artisans, *Artistes*

According to Greek mythology, the god Hephaestus was born so ugly that his own mother, Hera, could not stand the sight of him, and eventually threw him out of Mount Olympus. Rescued by nymphs, the rejected child—described as "the deformed" or "the one with crippled feet"—learned the art of metallurgy, becoming a skilled artisan who made the most beautiful ornaments. When word of his mastery reached Olympus, Hera became eager to see something made by her son. Hephaestus sent her a marvelous throne made of noble metals and precious stones whose beauty lured her to sit in it, only to find herself trapped. No other god could free Hera from the throne's grip, so she turned to her forsaken son, begging him to come back to Olympus. She offered him the most beautiful of the goddesses, Aphrodite, as a wife. Later, Hephaestus forged the weapons that Achilles used in the Trojan War, an act of metallurgical creation that seventeenth-century commentators regarded as an allegory of the origin of the world.[1]

The myth of Hephaestus illuminates the anxieties associated with the creative powers of artisans. There was something godlike in their ability to create useful and beautiful things from mere matter, as exemplified in Plato's notion of the Demiurge, the divine artisan-creator. Yet Plato's ideal city-state, as presented in his *Laws*, excluded technicians from civic life. In this respect, Plato's ideal polis differed little from the ancient Athens he discussed in *Critias*, where artisans lived at the fringes, never taking part in decision-making processes. The creative power of artisans was something the city-state needed, and at the same time feared. As has been recently argued, philosophical discussions on artisans' political presence in classical Athens reflected concerns about the emergence of a new breed of political men described by Aristophanes as technicians.[2] To what extent could one trust artisans? They could flee to another state and put their manual arts at the enemy's service. Should they be allowed to take part in the political life of the polis? Greek authors justified the political exclusion of artisans by claiming that the work of the hand limited the capabilities of the artisan's mind. The ugliness or deformity of Hephaestus symbolized such intellectual difference. His exceptional skills, however, earned him readmission among the gods.

When in 1732 the members of the Société des Arts in Paris chose to portray Vulcan (the Roman version of Hephaestus) on the medal they would award to the best essay on the mechanical arts, they were making a strong statement about the Société's mission and about their own ambitions.[3] At the time, the myth of Hephaestus-Vulcan was common knowledge for the educated reading public in France. Among other sources, Madame Anne Le Fèvre Dacier's widely read annotated translation of Homer's *Iliad* popularized the story of the deformed god.[4] In the four years of its existence, the Société had obtained the patronage of the Count of Clermont, a prince of the blood, and had gathered more than a hundred members, most of them among the best Parisian artisans. In choosing the effigy of Vulcan for their medal, the members of the Société des Arts asserted their understanding of the role that artisans should play in French society, initiating a process of appropriation and transfiguration of the Greek myth: they claimed the story of rejection and vindication but transfigured the deformed Vulcan into the muscular maker of the weapons of Achilles. Artisans in Paris, especially the members of the Société, did not live at the fringe of civic life. They organized themselves into powerful guilds or worked under royal protection, at the Louvre or in the faubourg Saint-Antoine. Yet, with few exceptions, they lived at the fringe of the so-called Republic of Letters, the virtual community of the learned in which citizenship was obtained through intellectual networks created by participation in salon conversations, membership in academies, epistolary exchange, or publication.[5] They also lived at the fringe of state-funded institutions, such as the Académie Royale des Sciences, which, despite its commitment to evaluating technical innovation, was notoriously disdainful of artisans.[6]

The constitution of the Société des Arts was a vindication of the strategic relevance of artisanal knowledge for the French state. Operated de facto by a group of clockmakers, cartographers, and surgeons, whose specific knowledge and expertise were not represented in any academy, the Société des Arts was dedicated to the improvement of the mechanical arts. Its members believed that the improvement of the mechanical arts was essential to the economic prosperity and the colonial expansion of the French state.[7] The choice of Vulcan as the

Fig. I.1. Guillaume Coustou the Younger (1716–1777), *Vulcan* (1742). [Musée du Louvre, Paris, France; © RMN-Grand-Palais/Art Resource, New York]

Société's emblem reflected the members' appropriation and transfiguration of the Greek myth: as Vulcan's skills earned him readmission to Olympus, so they hoped that the king would grant them royal support to establish an academy of the mechanical arts that would contribute to the economic advancement of the French state. However, it was not the deformed, rejected god that the members of the Société had in mind. Their Vulcan was akin to the god of Herculean strength that the sculptor Guillaume Coustou the Younger would carve out of white marble and present in 1746 for his admission to the Académie de Peinture et de Sculpture (fig. I.1).

Coustou's transfigured Vulcan materialized the ambitions and self-understanding of a historical figure that was neither properly artisan nor savant: the *artiste*. The *artiste* is the main actor of this book, and the Société des Arts is the focusing lens through which I explore his political epistemology, by which I mean a political project based on the notion that his knowledge and expertise were uniquely suited to advance France's economic and imperial development. I argue that key Enlightenment concepts—such as improvement, progress, and useful knowledge—emerged from the contested territory of the mechanical arts, where *artistes* strove to differentiate themselves from artisans as well as from savants.

L'artiste was not an artist in the modern sense of the word. He could be a painter, a sculptor, or an architect, but he could also be a chemist, a clockmaker, or a forger. Indeed, the term *artiste* first emerged in the context of practical alchemy and chemistry. Antoine Furetière, who compiled a dictionary of the French language in the 1650s, defined *artiste* thus: "Term of chemistry. It is he who knows how to carry out successfully chemical operations. One needs be a great *artiste* to prepare minerals so as to prevent them from being noxious. Raymond Lull, Paracelsus, and Arnold the Villeneuve were knowledgeable Artistes."[8]

A few decades later, around 1700, the lexicographer Pierre Richelet embraced a broader definition in his new dictionary of the French language: the *artiste* was an artisan who, in addition to manual skills, could also boast of that ineffable essence of French intellectualism, *esprit*. This is the sense in which the term *artiste* is employed in this book. As an artisan with *esprit*, the *artiste* consistently presented himself as superior to other craftsmen because his work did not consist of rote practice but resulted from his ingenuity, an ability to combine practical skill, creative design, and inventive intelligence. The *artiste* was learned and polite, like a savant. Yet, unlike the savant, his *esprit* manifested itself not simply in the world of words but above all in the world of things—the beautiful things through which French society defined itself, and the useful ones through which the French state pursued imperial and colonial expansion.[9]

As a historical actor, the *artiste* fits uneasily with the artisan, yet he does not belong with the savant. This was clear to the *philosophes*. Diderot, for example, codified the difference between artisans and *artistes* in the *Encyclopédie*, where readers learned that artisans were "workers who practice the mechanical arts that need the least intelligence," while *artiste* was a "noun that one gives to workmen who excel in the mechanical arts that need intelligence." The difference between artisans and *artistes*, for Diderot, was determined by the art they practiced. This meant that not all artisans could be *artistes*: a good shoemaker could at best be a good artisan, but a skilled clockmaker could be a great *artiste*. Contrary to his straightforward definition of artisans—as workers that carried out mindlessly repetitive manual operations—Diderot's notion of *artiste* was more nuanced. The term could be employed as a mark of appreciation or disapproval. To an artisan who excelled in practical work, "the word *artiste* is always a compliment," he explained. So, "of a chemist who is able to execute skillfully the processes invented by others we say that he is a good *artiste*." If, however, an *artiste* was well versed in the practical part of a science that was also theoretical, the term was "a sort of reprimand, that one masters nothing but the subordinate part of his profession."[10] In the tradition of classical philosophers such as Xenophon, who claimed that workshop techniques spoiled the artisans' bodies and affected their minds, Diderot contended that the art shaped the artisan's intellect, delimiting the horizon of his or her capabilities.[11] Manual work imposed limits on the minds of even the best *artistes*. Diderot conceded that there was intelligence in their activities, but the very practical nature of their work disqualified them in the face of those who dedicated themselves exclusively to *l'esprit*, the savants. By defining and qualifying the difference between artisans and *artistes*, Diderot was in fact reiterating the predominance of theoretical learning over practical activity. This was a *philosophe*'s point of view. How did *artistes* regard themselves? How did they regard artisans and savants?

As a historical actor that is neither properly artisan nor a savant, the *artiste* has been obscured in historical studies by the superimposition of present-day notions of art and science on the past, as well as the conglomeration of various practitioners in the categories of "artisan" and "craftsman." Several scholars have shown that until the late eighteenth century the notion of art differed substantially from that of the "fine arts." Rules and imitation, rather than creative genius, characterized the work of practitioners of the arts. Larry Shiner, in particular, has argued that no artist existed before the late eighteenth century: Leonardo, Michelangelo, and Raphael were, in fact, artisans.[12] This claim is at least partially supported by the absence of an early modern term that gathered together painters, sculptors, and architects, separating them from other practitioners of the arts.[13] Yet other studies have filled this semantic lacuna by showing that the modern idea of the artist, if not the term, has a long history, rooted in poetic texts and historical fictions, such as Homer's *Odyssey*, Ovid's *Metamorphoses*, Dante's *Divine Comedy*, and Vasari's *Lives of the Most Excellent Painters, Sculptors, and Architects*.[14] The efforts to historicize the concept of "art" are attuned with the work that historians of early modern "science" have produced in recent decades. They have meticulously shown that early modern inquiries into the world of nature were entangled in thick networks of trade, exchange, craftsmanship, connoisseurship, and erudition, whose scale ranged from the local microcosm of a London street to the global circuits of colonialism and empire.[15] In this rethinking of the production and circulation of natural knowledge in the early modern period, historians of science have blurred boundaries among sets of dichotomies that have emerged starting in the late eighteenth century: theory and practice, science and industry, science and technology, and science and art. They have regarded these dichotomies as the object of historical inquiry rather than as self-evident explanatory categories. The historical presence of the *artiste*, however, has been overlooked in these studies. Shiner has stated that there were no artists until the eighteenth century. Art historians have limited their analyses to what we now term the fine or visual arts. Historians of science have discussed savants who worked with their hands and artisans who used their minds, but have not considered *artistes*, who worked with hands and mind but did not want to be considered artisans or savants.[16] While Shiner is correct in stating that until the late eighteenth century there was no term that identified the artist in the modern sense of the word, there were certainly *artistes* who defined themselves as such at the time of the Société des Arts.

By retrieving the *artiste*'s voice, this book raises the question, What would the Enlightenment look like from the point of view of artisans who understood themselves as *artistes*? Since the times of Ernst Cassirer's pioneering study, the Enlightenment has been almost invariably considered a philosophical affair. Peter Gay's influential interpretation presented the Enlightenment as a defining moment in Western history, which introduced the values of secularism, reason, freedom, and democracy, and that culminated with the American republic. For Gay, there was only one Enlightenment, and it was brought about by a "family of philosophes."[17] Scholars who agreed with this proposition debated where to locate the intellectual roots of the Enlightenment and how to map its diffusion, considerably expanding Gay's original family of philosophers. Others highlighted instead the heterogeneity of Enlightenment projects, discussing forces of differentiation that originated in national contexts or in political or religious affiliations. The uniformity of the Enlightenment, then, gave way to a multiplicity of Enlightenments, such as the Scottish, the Tuscan, the Neapolitan, the Catholic, the Protestant, and, more recently, the radical, the mainstream, and the conservative.[18] This fragmentation has led scholars to question the utility of the very notion of Enlightenment, or, on the opposite end, to make a case for the Enlightenment based on the common concerns and approaches of great thinkers operating in different political and geographical contexts.[19] Although very different, all these studies have discussed the Enlightenment (or the Enlightenments) from the perspective of the social and intellectual history of ideas, with a sharp analytical focus on the works of philosophers and intellectuals. This book offers a different perspective on the Enlightenment, based on a radical change of historical actors. The artisanal Enlightenment that I discuss, however, should not be understood as yet another definition or qualification of the Enlightenment. With the adjective "artisanal" I do not mean to highlight particular features of the Enlightenment like, most recently, Jonathan Israel's classification of the various projects as "radical" or "mainstream." Nor do I use "Enlightenment" to signal a breakthrough in the artisanal world, in the guise of Joel Mokyr's notion of the "Industrial Enlightenment."[20] Building on the insightful contributions of historians of eighteenth-century science and technology, and of other scholars who have historicized the very notion of Enlightenment, I regard the Enlightenment as a work in progress to which a multiplicity of actors laid claims.[21] These locally inflected and often conflicting claims clustered around questions of knowledge and expertise, and their legitimate sources and methods. They likewise coalesced around the belief that these questions were relevant to the future of humanity. Being enlightened meant to cultivate knowledge as part of one's commitment to moral and social improvement. It also meant to demonstrate such commitment through participation in polite sociability and the ability to discipline one's mind. Much has been written on the *philosophes* and other enlightened men and women of learning. This book explores the *artistes*' participation in, and original contribution to, enlightened projects. By bringing *artistes* to center stage, I argue for the centrality of the mechanical arts and the world of making in the development of key Enlightenment notions, such as improvement, useful knowledge, and progress.

THE *ARTISTE* IN THE *ENCYCLOPÉDIE*

Denis Diderot and Jean Le Rond d'Alembert's *Encyclopédie*, considered a manifesto of the Enlightenment, offers a comprehensive view on how the *philosophes* regarded the mechanical arts and their practitioners. In his *Prospectus* (1750), Diderot set the tone for the consideration of artisans and *artistes*. In a passage that reappeared in the *Preliminary Discourse* (1751), he explained that the *artistes* who had contributed to the *Encyclopédie* were men of learning (*gens de lettres*), but he hastened to underscore that they were exceptional cases. More commonly, artisans could not even describe or explain their own work:

> Most of those who engage in the mechanical arts have embraced them only by necessity and work only by instinct. Hardly a dozen among a thousand can be found who are in a position to express themselves with some clarity upon the instrument they use and the things they manufacture. We have seen some workers who have worked for forty years without knowing anything about their machines. With them, it was necessary to exercise the function in which Socrates gloried, the painful and delicate function of being midwife of the mind, *obstetrix animorum*.[22]

Not all contributors to the *Encyclopédie* believed they were operating as midwives of minds. Because it was compiled by various authors at different times, the *Encyclopédie* constitutes a useful probe into the multiplicity of characterizations of *artistes* throughout the eighteenth century. A basic keyword search through the ARTFL Project's online edition of the *Encyclopédie* confirms that Richelet's definition of the *artiste* as an artisan with *esprit* was widespread in the eighteenth century.[23] In addition to being applied to arts such as painting, sculpture, architecture, dance, and music, the term *artiste* was employed to define artisans working in areas as diverse as chemistry, metallurgy, goldsmithing, enameling, the making of artificial flowers, poetry and fiction (*belles-lettres*), calligraphy, jewelry, anatomy, surgery, pharmacy, engraving, woodcutting, printing, clock-making, coining, drapery, and spectacle- and mathematical instrument making.

The diversity of the arts in which *artistes* operated corresponded to a variety of interpretations of their intellectual abilities and moral qualities. The architect Jacques-Raymond Lucotte, one of the learned artisans whom Diderot regarded as exceptional, embraced Richelet's definition, stating that the ability to think was what distinguished *artistes* from artisans. Quoting Vitruvius, Lucotte pointed out that even though manual work and discernment were equally important in the art of stone-cutting, the former was the province of the artisan, while the latter characterized the *artiste*.[24] *Artistes* like Lucotte strove to differentiate themselves from other artisans, but in the eyes of savants like the military physician Jacques-François de Villiers, the difference was too subtle to grasp. For him, the *artistes* who worked on the science of assaying had neglected to develop new methods and still operated as they had two centuries earlier. Their ignorance hindered the development of metallurgical chemistry. Echoing Diderot's consideration that *artiste* was a pejorative term when it referred to practitioners who neglected theory in favor of practice, Villiers argued that the pursuit of profit and lack of education spoiled *artistes*' minds: "How can one pass the taste for disinterested knowledge [*belles connoissances*] on to people for whom profit is the only motivation and who, besides it, have no other idea, and whose lack of education forbids this acquisition?"[25]

The poet Jean François de Saint-Lambert, famous for his liaison with the marquise Émilie du Châtelet, believed that frequent encounters with educated elites could fulfill the function of uplifting and refining the *artiste*'s soul.[26] Similarly, the chevalier Louis de Jaucourt thought that the manipulation of precious materials and the creation of luxury items distinguished the *artiste* from the rest of the people. For him, the nobility of the materials transferred to the *artiste*'s soul, elevating his or her moral status from the lowly multitude: "Hands that *di-*

vinely paint a carriage couch, that *perfectly* mount a diamond, that *exceptionally* fit a fashionable suit, such hands do not resemble the hands of the *people*."[27]

In drawing the difference between *artiste* and artisan, Diderot emphasized whether the art they practiced required intelligence or not. There were arts that demanded only repetitive actions, while others benefited from "genius," defined by an anonymous contributor to the *Encyclopédie* as "the fire and invention" that *artistes* put in their work.[28] The amateur painter and tax farmer (a private tax collector) Claude-Henri Watelet elaborated on this point, arguing that only intelligent artisans were not slaves to habit and dared to reach beyond the teachings of their masters.[29] The theme of artisans' subservience to rules derived from the classical definition of art, and in particular of the mechanical arts, as *techne*, something that was practiced by following rules. Rules defined the arts, and they were the artisans' damnation. The calligrapher Charles Paillasson wholeheartedly agreed with Voltaire, who complained about the "prodigious number of rules, most of which are useless or false" that overwhelmed the arts. Paillasson added that the abundance of rules in the training process limited the apprentice's intelligence.[30]

Pursuit of profit, lack of education, subservience to rules and habit: this was the damning trinity of the artisan in the eyes of those who lived in the world of words. The *artistes* of the Société des Arts agreed. As the choice of Vulcan as their emblem revealed, they understood themselves as exceptional. They shared the idea—as expressed by Jaucourt in the *Encyclopédie*—that *artistes* who were able to create sophisticated luxury items could not be considered as *peuple*. They extended this concept to sophisticated machines that performed complex, useful operations. Being able to design and make such mechanical marvels work required as much *esprit*—intelligence, knowledge, discernment—as any other products of the mind. In order to advance this image of themselves, they deprecated "mere artisans" who simply executed manual operations.

TRACING THE *ARTISTE*

The combination of manual skill and intelligence was not, of course, an invention of the eighteenth century. Artisans' contributions to the development of empiricism and the experimental method have been widely debated in the historiography of the Scientific Revolution.[31] Neither was it an exclusively French phenomenon, as works by Larry Stewart, Liliane Hilaire-Pérez, Pamela Smith, Celina Fox, Deborah Harkness, and Pamela Long demonstrate.[32] Later versions of Richelet's dictionary acknowledged the non-French origin of the term *artiste*, which derived from the Italian or Spanish *artista*.[33] What is distinctive of

France and of the early eighteenth century is the role played by *esprit* in the characterization of the *artiste*. The *esprit philosophique*, as Daniel Edelstein has recently argued, was a key concept in the early articulation of narratives of the Enlightenment. In his attempt to historicize the notion of "Enlightenment," Edelstein locates the first instances of a discourse on the philosophical spirit that differentiated the modern, enlightened age from classical antiquity in the "Quarrel of the Ancients and the Moderns," a literary debate on the value of Greek poetry and eloquence that divided the Académie Française in the 1680s.[34] I argue that the mechanical arts played an essential role in shaping the discourse on the philosophical spirit of the enlightened age. Not coincidentally, the connection between *esprit* and artisanal work was formalized at that time.

The fact that the formal connection between artisan and *esprit* was a nascent one at the turn of the eighteenth century can be appreciated in the rather different definition of the term *artiste* found in the first edition of the *Dictionary of the French Academy* (1694). There, *artiste* is simply "he who works in the arts," with the specification that the term was especially used to indicate someone who worked with chemical processes.[35] It was only in the fourth edition of the dictionary, published in 1762, that *artiste* was connected to the world of *esprit*. In this new edition *artiste* is defined as "he who works in an art in which genius and hand must collaborate."[36] This delay in changing the definition offered by the dictionary produced by the most illustrious body of Parisian savants points to the lack of urgency on their part to update their own consideration of artisanal work. Artisans, meanwhile, had engaged in various debates aimed at redefining the distinctions among the various arts, and in particular between the liberal and mechanical arts.[37]

Artisans had also circulated self-representations that polemically asserted the collaboration of the mind and the hand in their own work. Late seventeenth-century Parisians strolling on the rue St. Jacques would get the idea of the *artiste*, if not yet the term, in the shop of the engraver Nicolas Guérard. There they would see an image that defied the distinction between the man of the hand and the man of the mind (fig. I.2). Designed and engraved by Guérard himself, the print shows a well-dressed artisan sitting in a chair, with the tools and products of the arts and crafts at his feet. Minerva, the goddess of knowledge, stands on his head, while the personification of industry helps his hand pull a devil by the tail. The engraving is a visual rendering of the French expression *tirer le diable par la queue* ("pulling the devil by his tail"), which meant working hard. The devil has a bag of money in one hand and the flames of honor in the other, with the former pointing downward and the latter pointing upward, to mark a moral hierarchy between them. This was not a stereotypical representation of

Fig. I.2. Nicolas Guérard, *L'artisan tire le diable par la queue*, late seventeenth century. [Bibliothèque Nationale de France, Paris, France; © BnF. Dist. RMN-Grand-Palais/Art Resource, New York]

an artisan. It was perhaps more modest than the Herculean version of Vulcan that the sculptor Coustou submitted for his admission to the Académie, yet it was bolder than Diderot's definition of the artisan as someone practicing the arts that require the least intelligence. With wisdom and knowledge in his head and industry in his hand, Guérard's artisan was definitely not a slave to the rules of the arts, nor an automaton-like workman. Guérard struggled against the commonly held notion, which persisted in the *Encyclopédie*, that in their pursuit of profit, artisans lost their morality and trustworthiness. Indeed, the engraving suggested that the artisan's real hard work lay in his efforts to combine morality and profit, more than in the practice of his art. As the text at the bottom of the image explained, the accumulation of probity was the only worthy outcome of the arts. The top title of the print, "He is not alone," was a moral admonition to the viewer and an invitation to emulate such virtuous efforts to pursue moral rectitude.

At the end of the seventeenth century, rue St. Jacques offered an abundance of visual material for meditating on the arts and those who practiced them. Not too far from Guérard's shop, Nicolas II de Larmessin sold a series of seventy engravings entitled "Fantastic Costumes" representing the arts and crafts.[38] Each plate displayed a human figure dressed in the tools and works of his or her craft, with the title explaining whose costume it was (fig. I.3). With their expressionless faces, all the artisans in Larmessin's gallery were interchangeable mannequins, defined by the dress they were wearing and therefore the art they practiced. Contrary to Guérard's artisan, they were devoid of intelligence, discernment, wit, or—more simply—*esprit*. They offered a face and two hands to their art, no more. Given that Larmessin owed his fame to his portrait engravings, it is all the more relevant that the faces in the plates were not meant to identify any specific individual or to emphasize any specific attitude or mood. With its emphasis on costumes, Larmessin's grotesque gallery was a sarcastic subversion of the common French proverb "l'habit ne fait pas le moine," which in literal translation means "the cloth does not make the monk," or, less literally, "do not judge a book by its cover." Larmessin's plates asserted just the opposite: it was indeed the art (the cloth) that made the artisan (the monk). This, of course, applied to artisans other than Larmessin. Unlike the anonymous practitioners he portrayed, Larmessin distinguished himself in Parisian society. He was admitted to the Académie Royale de Peinture et Sculpture in 1730 and became royal engraver in 1754. With his sarcastic deprecation of other artisans, he elevated himself above them, performing his own *esprit*. Remarkably, there was no costume of engraver in his gallery.[39]

Fig. I.3. Nicolas Larmessin, *Habit de serrurier*, from *Costumes grotesques* (Paris, 1695). [Bibliothèque Nationale de France, Paris]

Diderot's idea that the arts defined the *artiste* rendered the latter an object of study and observation, rather than a potential contributor to the definition or redefinition of the arts themselves. This view similarly underpinned the early encyclopedic projects on the mechanical arts pursued by the Royal Society of London in the 1660s and by the Paris Académie Royale des Sciences in the 1690s.[40] To the savants involved in these encyclopedic projects, artisans were people raised in a culture of secrecy and in subservience to rules. In order to understand the technical processes leading to the production of artifacts, savants believed it necessary to observe how artisans operated and, more than occasionally, to buy information from them. This view was authoritatively voiced by André Félibien, the court historian to the king, a member of the Académie des Inscriptions, and the secretary of the Académie d'Architecture. In his *Principles of Architecture, Sculpture, Painting, and of the Arts That Depend on Them* (1676), he explained that, even though he had to spend a long time reading the numerous texts dealing with relevant topics, the hardest aspect of compiling his book had been the interaction with the men of the arts:

> I admit that when I had to write about them [the arts], entering in the details and explanations of all the terms, and of the different names of several particular things, I was obliged to recur to workmen: I had to enter in their shops, to visit their workshops, to consider their machines and their tools, to consult with them about the various usages, and often enlighten myself with them about the different names they give to the same thing; and it is this that was most distressing.[41]

Félibien did not make his readers guess why these encounters were distressing to him. The artisans' lack of education and their search for profit made the visits quite a different learning experience from the salon conversations or academic meetings where peers instructed one another: "If one thinks of consulting some of them [artisans] to know their tools or to learn something about their art, very often one finds ignorant or bizarre people who, instead of answering the questions one asks and talking honestly about the craft they practice, say things that are opposite to what one wishes to know, and often disguise the truth one searches with mischievousness."[42]

For Félibien, interacting with artisans was the painful but necessary price one paid in order to understand how the arts were practiced. The artisans failed to articulate verbally their embodied skills—whether out of ill intentions or ignorance. Once the savant operated the Socratic action of extracting information from the artisans, acting as an *obstetrix animorum* as Diderot would later write, artisans were no longer necessary to the understanding of the arts.[43] This is clearly illustrated in the plates of Félibien's work. They represent workshops and tools of the arts without human figures. The choice of the Virgilian motto that Félibien placed at the beginning of his text, *mens agitat molem* (the mind moves matter), further suggested that the works of arts were realized by a disembodied, divine spirit. An empty chair in the workshop was an ambivalent allusion to the artisan's presence as well as to his irrelevance (fig. I.4).

The attribution of agency to the arts, and the corresponding irrelevance of artisans' manual work, can be seen in the swarming of putti in visual representations

Fig. I.4. Plate from André Félibien, *Des principes de l'architecture, de la sculpture, de la peinture, et des autres arts qui en dépendent* (Paris: Coignard, 1676). [Yale Center for British Art, Paul Mellon Collection, New Haven, Conn.]

of seventeenth-century artisanal workshops. Art historians have discussed the multiplicity of symbolic meanings of the winged child in early modern artworks, including engravings, painting, sculpture, tapestry, and faience. Associated with both the sacred (as cherubs) and the profane (as cupids), putti commonly carried moralizing messages.[44] As illustrated in *Amorum emblemata* (1609), an influential book of emblems by the Dutch painter Otto van Veen, one of such messages concerned the arts. The book presented several mottos on love illustrated by the actions of putti. One of them, *Amor addocet artes*—"Love is the schoolmaster of the arts"—points to the civilizing effect of love on those who practice the arts:

> All artes almost that bee did first from love beginne
> Love makes the lover apt to everie kind of things.[45]

As the first mover of the arts, love in the shape of a putto is a disembodied creative spirit strikingly similar to the Virgilian *mens* (mind) quoted by Félibien. It ennobles the artisan and distracts the viewer from the artisan's manual work. When depicted operating lathes or drawing tools, putti showed that it was not the embodied skill of the artisan that created the work of art but rather a disembodied *esprit*, or genius, a fine and refined creative intelligence that operated effortlessly on matter. By introducing grace and playfulness in the world of the mechanical arts, putti were visual renderings of *sprezzatura*, the good-mannered nonchalance "which conceals all artistry and makes whatever one says or does seem uncontrived and effortless" (fig. I.5). The presence of putti in

Fig. I.5. Plate from Charles Plumier, *L'art de tourner en perfection* (Lyon: Jombert, 1701). [Bibliothèque Nationale de France, Paris]

mechanical workshops underscored what Baldassarre Castiglione wrote in his *Book of the Courtier*: "True art is what does not seem to be art; and the most important thing is to conceal it, because if it is revealed this discredits a man completely and ruins his reputation."[46]

Félibien's celebrity and the success of his work made his characterization of artisans a commonplace. How did learned artisans, such as Guérard and Larmessin, react to a representation that caricatured and ventriloquized them? It is noteworthy that their engravings of artisans followed the publication of Félibien's text. Their visual materials, which circulated widely across social divides, offered markedly different representations of artisans' *esprit*, and could be interpreted as a response to an early incursion by a savant into the world of the mechanical arts.

This book shows that the emergence of the *artiste* as a self-aware historical actor was connected to the increasing interest on the part of the savants, especially those in the Académie Royale des Sciences, in the mechanical arts at a time when the French state began to regard the mechanical arts as political matters.[47] The *artistes*' response was not a concerted effort to readjust the social perception of artisanal work or the widespread poor consideration of artisans. Rather, it aimed at elevating the artisans with *esprit* from the others, even by mobilizing the discriminating topoi formulated by the savants. As Larmessin's *Costumes* and Guérard's *Artisan* remarkably illustrate, it was by differentiating himself from other artisans that the *artiste* was born.

THE *ARTISTE* AND THE MECHANICAL ARTS

The *artiste* with whom this book is concerned worked in the mechanical arts. But at a time when the classification of knowledge inherited from antiquity and the Middle Ages was undergoing thorough revision, the category was a surprisingly fluid one. As various scholars have shown, until at least the mid-eighteenth century there was no conception of the fine arts in the modern sense of the term.[48] At the same time, there was no widespread agreement as to which individual arts counted as mechanical. The placement of the various arts in the several trees of knowledge that were created in the seventeenth century and onward until the publication of the *Encyclopédie* testifies to a changing cultural landscape for the mechanical arts where their ancient opposition to the liberal arts had been inexorably eroded. Clockmakers and surgeons within the Société des Arts believed that the recent useful discoveries in their profession demonstrated that these arts were gradually becoming "liberal," as they required knowledge of mechanics, mathematics, and anatomy, in addition to manual dexterity. Even the emerging notion of "beaux arts"—which in English indicated the "polite arts" first, the "fine arts" later—did not set itself in contrast to the mechanical arts. On the contrary, when the Société des Arts was elaborating its own program, it thought of itself as a "Société académique des beaux arts," with clockmakers, surgeons, metal workers, engravers, and mathematical instrument makers as well as painters, sculptors, and architects, signing on to the project.[49]

The fluidity of the territory of the mechanical arts points to two related issues. On one hand, the rethinking of a number of arts that were traditionally regarded as mechanical in terms of beaux arts at the turn of the eighteenth century indicates the attempt at concealing the manual, bodily aspect of artisanal work and the emphasis instead on the pleasure, refinement, ingenuity, and luxury materialized in the products of that work. This, as has been argued, was a process of ennoblement of the art and of the *artiste* mediated by the emergence of an art market that privileged luxury over necessity.[50] *Encyclopédistes* such as Jaucourt and Saint-Lambert, as well as Diderot and D'Alembert themselves, made this understanding of the *artiste* their own. On the other hand, the fluidity in the definition of the mechanical arts should be understood in the context of a discourse on useful knowledge that elevated the mechanical over the other arts. The ennoblement of the mechanical arts through the advocacy of their utility was in line with the celebration of Vulcan as the god who produced the weapons of Achilles.

Utility, however, was not set in contrast to luxury, or beauty. Parisians admired sophisticated mechanical devices in salons, coffee-houses, theaters, fairs, gardens, museums, and other venues of social gathering. Over the course of the eighteenth century, these kinds of displays became as popular as contemporaneous public demonstrations of experimental philosophy and natural history collections. Machines instructed and amused, and they inspired new definitions of beauty that emphasized, in the words of Adam Smith, "the fitness of any system or machine to produce the end for which it was intended."[51] The decorated library of Charles Perrault, the cultural consultant of Louis XIV's chief minister Jean-Baptiste Colbert and a member of the Académie Française, offers an effective example of the blurred boundaries between utility, beauty, and luxury at the end of the seventeenth century. Perrault was the initiator of the Quarrel of the Ancients and the Moderns, sparked by his *The Century of Louis the Great*, published in 1687. During the years of the Quarrel, he recruited the best painters in Paris to turn his library into a visual statement of his support for the Moderns.[52] Those who did not visit his library could still get at least an impression of it by looking at the engraved plates of the *Cabinet des beaux arts*, a work he published in 1690 and that, like Guérard's and Larmessin's engravings, was available on the rue St. Jacques.[53] Perrault explained that the notion of beaux arts replaced ancient classifications of human learning which defined the liberal and the mechanical arts as, respectively, the work of the mind and the work of the hand. He rejected the ancient hierarchy that placed the liberal above the mechanical arts because only the former were noble and dignified. The beaux arts, Perrault explained, were those arts that inspired and educated one's taste and genius; they included poetry, eloquence, music, architecture, painting, sculpture, optics, and mechanics.[54] The personifications of Genius and Labor, at the cabinet's entrance, introduced the idea that both manual work and *esprit* were the foundations of the *beaux arts* (fig. I.6).

Perrault's *Cabinet* shows that the recent concern with utility and improvement de facto inverted the hierarchical order among the arts. Mechanics, Perrault maintained, deserved a prominent place among the beaux arts because it made "in reality and in truth all that most of its fellow arts do only by allusion

Fig. I.6. Jean Le Pautre, *Cabinet des beaux-arts*. Plate from Charles Perrault, *Le cabinet des beaux-arts* (Paris: Edelinck, 1690). The personifications of Genius (left) and Labor (right) stand at the entrance. A clock occupies the central position at the bottom, probably to indicate a sense of temporal progress in the arts. [Manuscripts and Archives, Yale University Library, New Haven, Conn.]

and metaphor."⁵⁵ Mechanics was not simply or exclusively the study of the motion of bodies, Perrault reminded his readers. Rather, it was the art that made possible the construction of formidable machines which poets and musicians could only envision in their imagination. These were machines—such as those that built the palace, gardens, and fountains at Versailles—of which the ancient world could not conceive. The Ancients lacked the necessary knowledge that had since accumulated over time. While the Greek poets and musicians could only sing about moving stones and walking trees, the machines employed at Versailles had actually lifted heavy rocks and successfully transplanted and aligned tall firs (fig. I.7).⁵⁶

As Larry Norman's nuanced analysis of the Quarrel of the Ancients and Moderns has shown, the rival parties shared common ground in their consideration of classical antiquity. What really divided them was the concept of human progress, articulated by the Moderns to advance the idea that the present age was superior to any other time in history.⁵⁷ The Moderns harnessed the history of

Fig. I.7. Jean Jouvenet and Louis Simmoneau, *La méchanique*. Plate from Charles Perrault, *Le cabinet des beaux-arts* (Paris: Edelinck, 1690). From left to right, the putti are operating Christiaan Huygens's clock, a stocking machine, and a water-raising pump described in the French translation of Vitruvius. The personification of mechanics is holding a barometer. In the background the scaffoldings and the colonnade underscore the relevance of mechanics to civil and military architecture. [Manuscripts and Archives, Yale University Library, New Haven, Conn.]

the arts to support their vision. Perrault, in particular, did not simply exalt the machines that the Moderns had finally been able to build; he also recruited the art of optics to his program of illustrating the progress of human knowledge. He celebrated optics not so much as the geometry of perspective but rather as the art of making optical instruments such as the telescope and the microscope, which had made the most extraordinary achievements in natural knowledge possible. In order to educate taste and inspire genius, one had to consider carefully not just the arts that were under fire in the Quarrel—poetry and eloquence—but also the ones that made useful contributions to humankind.

The cultural and political significance of the mechanical arts can be better understood if we undo the process of conceptual translation that associates the "mechanical arts" with technology. This anachronistic translation brings along with it the two-cultures divide—science and technology, on the one hand, and the humanities and the arts, on the other—which was not in place in the early modern period. Historians of science have amply demonstrated the intellectual advantages of engaging with actors' categories in the study of the production of natural knowledge. By not projecting present-day notions of scientific knowledge onto the past, they have challenged the heroic narrative of the Scientific Revolution. They have brought to the fore a heterogeneous set of practitioners and savants who approached the world of nature with the intellectual toolkit of the humanist.[58] Here I adopt a similar approach to the mechanical arts. My choice to use actors' categories is not dictated by a desire for philological precision but rather by the ambition to offer a historically accurate and analytically rich interpretation of the *artistes*' political project. It is by historicizing seemingly familiar notions, by making them unfamiliar and therefore deserving of deeper analysis, that I aim to reassess the role of the *artiste* and of the mechanical arts in the French Enlightenment.

As Hélène Vérin has demonstrated, technology as a concept and as an academic discipline first emerged in Germany in the last quarter of the eighteenth century and spread to France in the postrevolutionary period. It was defined as the science that concerned "the transformation and the processing of natural products," and that taught in a systematic and methodical way the principles underpinning the practice of the arts and crafts.[59] I show that already in the early eighteenth century *artistes* were actively involved in "boundary-work" aimed at distinguishing their ways of knowing and operating from what they regarded as the overly theoretical approach of the savants and the rote practice of other artisans. As Eric Ash has argued, this kind of boundary-work characterized early modern claims to expertise. Putative experts in search of legitimation by the state emphasized the practical nature of their knowledge, while also digressing on the theoretical aspects of their work. The relationship between the state and the expert was one of mutual benefit: as experts contributed to the state's imperial and commercial expansion, they obtained legitimacy and social status. The explanation of how something worked and why was at once a public statement of the experts' distinctive abilities and a prescriptive systematization of their body of knowledge, aimed to exclude competing practitioners. I argue that the articulation of what kind of knowledge counted as useful was a crucial element of the *artistes*' boundary-work. *Artistes* presented their expertise as essentially different from those of philosophers and other theoreticians, insisting on its practical nature. Their claim to make machines that worked in practice and not only in principle demonstrated to the French state and, more generally to the French public, their commitment to the common good. Useful knowledge belonged in the realm of the mechanical arts, not in the leisure world of natural philosophy.[60]

I regard the *artistes*' claims to expertise as essentially political. By presenting themselves as the most reliable experts on the mechanical arts, *artistes* challenged the authority of the savants, particularly the members of the Académie des Sciences, and presented themselves as the only experts who could advise the state on how to pursue its colonial and commercial visions. The attempt to establish the Société des Arts as a state institution dedicated to the advancement of the mechanical arts was the most visible expression of the *artistes*' ambition to play an official role in the decision-making processes of the French state. Even if their project did not directly challenge Old Regime institutions, the *artistes*' aspiration to participate in the administration of the French state implied a reshuffling of France's social fabric. *Artistes* were artisans, even if learned and polite, and artisans were traditionally excluded from the institutions of the Old Regime, including the Académie des Sciences.

The methodological choice to historicize the notion of useful knowledge and the mechanical arts distances this book from other discussions of useful knowledge and its role in what has been defined the "Industrial Enlightenment." In the attempt to find an answer to the question of why the Industrial Revolution occurred in Britain in the nineteenth century, Joel Mokyr has introduced the notion of Industrial Enlightenment as the causal connection between the Scientific Revolution of the seventeenth century and the technological and economic developments of the nineteenth century. In his analysis, the notion of useful knowledge is an ad hoc category that essentially indicates any kind of natural knowledge that has the potential of being applied to the manipulation of materials, artifacts, energy, or living beings. The Industrial Enlightenment indicates the social changes that such useful knowledge underwent as a result of the diffusion of the scientific method, scientific mentality, and scientific culture that spurred from the Scientific Revolution.[61] Although I share Mokyr's understanding of technology as knowledge, and his recent emphasis on improvement as a distinctive feature of the Enlightenment, my approach diverges from his retrospective analysis of the Industrial Enlightenment as an intellectual process that originated in the world of learning and that effected changes in the artisanal world. As Liliane Hilaire-Pérez has noted, Mokyr attributes causal agency only to philosophical and scientific theories, while he regards the artisanal world as a homogeneous context where such theories were put to work for practical purposes.[62] I show that *artistes* opposed the idea that improvement derived from the application of theoretical knowledge. They countered that the only useful knowledge was that which made machines work. It was not Newtonian. It was not Cartesian. It was the knowledge that they produced and that only they could assess. This knowledge was useful because, they claimed, it could effectively contribute to the economic and imperial advancement of the French state.

ARTISANAL ENLIGHTENMENT

This book focuses on the *artiste* to explore the relationship among science, the mechanical arts, and the state in Old Regime France. Its main chronological arc is between the foundation of the Académie Royale des Sciences in 1666 and the publication of the first volumes of the *Encyclopédie* in 1751. During this era, the mechanical arts became objects of political interest and learned scrutiny. Several encyclopedic projects on the mechanical arts were carried out by ad hoc associations working within or around the Académie des Sciences. The book connects the social emergence of the *artiste* to such activities, framing them within the political economy of the French state. I discuss the Société des Arts, founded in Paris in 1728 and active during the first years of Louis XV's reign, as the most visible attempt to turn the expertise of the *artiste* into an instrument of the state. The Société obtained the patronage of the Count of Clermont, a prince of the blood, and within a few years it attracted about two hundred members from France and other European countries. The society's ambitious program, together with its distinguished membership and the until now obscure reasons for its quick decline, have attracted the attention of several Enlightenment scholars, who have pointed to the intellectual connections between the program of the Société des Arts and Diderot and D'Alembert's *Encyclopédie*.[63] The scope of these early studies was limited by the very few documents on the Société des Arts that were then available. Recently, a large body of primary sources has surfaced, which consists of more than five hundred pages of minutes of the Société's meetings, lists of members, letters, and memoirs.[64] Thanks to the dedication of Oliver Courcelle, the vast majority of these documents have been transcribed and made publicly available.[65] Although these new sources are still fragmentary, they offer enough documentary evidence not only to demonstrate the relationship between the Société des Arts and the *Encyclopédie* but more generally to place the Société des Arts in the cultural, political, and economic history of Old Regime France. *Artisanal Enlightenment* brings these new documents into dialogue with a range of contemporaneous works on the mechanical arts, published or unpublished, written, visual, and material. By retrieving the continuity between the *Encyclopédie* and the long tradition of writings on the mechanical arts that Diderot effaced in his programmatic article "Art," it contributes also to the history of encyclopedism.

Artisanal Enlightenment consists of three parts. The first, "The Natural History of the Arts," examines early encyclopedic works carried out by savants between 1664 and 1727. It shows that these works were conceived as a Baconian "history of the arts" that would offer descriptions of machines, tools, and artisanal practices. These descriptions, or "histories," were to be published together as an interconnected, multivolume work. The notion of natural history that underpinned such projects shaped the ways in which information was gathered and artisanal work understood. Similar to natural specimens in works on natural history, savants regarded artisans as objects of observation: they were replaceable, mindless pieces of machinery. Chapter 1, "Lost Knowledge and the History of the Arts," places the emergence of the projects on the "history of the arts" in the political and cultural contexts of the 1660s, particularly with respect to the memorandum on trade that Colbert prepared for Louis XIV in 1664. It addresses the ongoing discourse on lost knowledge, revived in the late 1680s by the Quarrel of the Ancients and the Moderns, which divided the Académie Française and intrigued the reading public. The mechanical arts offered an abundance of evidence to support the new idea that human knowledge was cumulative, which was the preliminary step for the elaboration of the notion of progress. In their attempts to make the mechanical arts an attractive subject for the reading elites, savants emphasized the "immaterial ingenuity" of mechanical inventions, bypassing the people who performed manual work.

These early encyclopedic projects on the arts differed from technical dictionaries, or even from Ephraim Chambers's *Cyclopedia*, in two main respects. First, they relied on visits to artisanal workshops and direct observation of artisans at work, similar to the way in which naturalists studied the kingdoms of nature by observing animals, plants, and minerals. Second, these works were characterized by the search for an organizing principle that would show the interconnectedness of all the arts. The search for this principle, I show, was informed by the belief that the mechanical arts were part of the world of nature. Just as the notion of the "chain of being" organized the natural world hierarchically, so too the *enchaînement* (interconnection or mutual dependence) of the arts would reveal a "natural" hierarchy among the arts. This hierarchy would guide savants in articulating rules for the improvement of the arts.[66]

Chapter 2, "Réaumur and the Science of the Arts," discusses the connections between natural history and the mechanical arts in the work of René Réaumur, a prominent member of the Académie des Sciences. It argues that his understanding of natural materials created connections among the natural world, the mechanical arts, and the world of industry and trade. For Réaumur, the history of the arts was strategic to the economic advancement of France because it would stimulate artisans to improve their techniques and promote technical literacy among entrepreneurs who owned manufactures. He conceived of the encyclopedia of the arts as an instrument that would enable the learned public to understand the true value of labor and to appreciate the actual cost of useful items. By being educated about the best processes for making things, readers would also be aware of common frauds and forgeries. For all these reasons, Réaumur claimed that the history of the arts was a project in useful knowledge that should be supported by the state. Réaumur's vision became reality in 1723, when the Royal Manufacture of Malleable Cast Iron and Steel was established at Cosne, near Orléans, on the principles that he had outlined in his *The Art of Converting Wrought Iron into Steel* (1722). Réaumur moved to Cosne in order to help entrepreneurs implement his written directions, and helped them instruct workers to operate according to his theoretical principles. The failure of the industry in 1727 halted the encyclopedic work on the arts and crafts that Réaumur had directed at the Académie des Sciences.

Réaumur's work addressed mainly entrepreneurs and administrators. Artisans, in his vision, were to be directed by savants, just as artisans themselves directed tools and machines. Artisans were also excluded from the planning and organization of the encyclopedic works on the arts discussed in the book's first section. They were involved only as auxiliaries executing specific tasks, or as the objects of learned observation and scrutiny. The second section of the book, "The Société des Arts," shifts the focus from work *on* artisans to work *by* artisans. It shows that the savants' involvement in the mechanical arts prompted the public emergence of the *artiste*. Yet the *artistes*' response to the savants' understanding of artisanal work was not a concerted effort to readjust the social perception or widespread poor consideration of artisans. Rather, it aimed at elevating the artisans with *esprit* from the others, even by deploying the discriminating motifs formulated by the savants. The Société des Arts articulated a political project that claimed a central role for the mechanical arts in the administration of the state and that positioned the *artiste*—instead of the savant—as the expert who could evaluate technical innovations. Writing about the arts was a significant component of the Société des Arts's mission. For its members, the compilation of volumes with descriptions of best practices and tools for each art was an important step toward the articulation of formal regulations for the practice of the arts and the manufacture of goods. The debacle of the Royal Manufacture of Malleable Cast Iron and Steel in 1727 demonstrated the inability of savants to articulate such rules. It was now time for *artistes* to propose new ways of writing about the arts: the Société des Arts was created one year later, in 1728.

Chapter 3, "Theory, Practice, Improvement," unearths early attempts to constitute associations dedicated to the improvement of the mechanical arts, casting light in particular on the earlier, little-known Société des Arts founded in Paris during the Regency (1718). It frames such associations within the broader economic history of France, with special attention to the history of artisans' migrations between France and England, and the demands deriving from France's colonial expansion. *Artistes* involved in these associations advocated a clear distinction between theoretical and practical knowledge, which aimed to discredit the expertise of academic institutions in technical matters. Appropriating Francis Bacon's distinction between "operative" and "speculative" knowledge, *artistes* contrasted the philosophers' search for truth to their expertise on practical matters and their concern with public good. Addressing many themes that would characterize the debate on the *noblesse commerçante* later in the century, they advanced the idea that the merging of collective and individual interest was necessary to the pursuit of useful knowledge.[67] The establishment of the Royal Manufacture of English Watches at Versailles, entirely managed by *artistes* and supported by John Law (France's general controller of finances), was presented as a model of the interactions among artisans, *artistes*, entrepreneurs, and state authorities that the Société des Arts wanted to promote. The choice of working on watches was particularly strategic. The technology of watchmaking seemed to hold the secret for finding longitude at sea, a crucial problem at a time of imperial competition and colonial expansion.

Much like a business partnership, the Société des Arts united individuals who worked together toward a common goal: *la perfection des arts*, the French term for technical improvement. Chapter 4, "Society of Arts," discusses the Société des Arts as a microcosm of overlapping networks modeled on the working practices of the *artiste*, particularly clock- and watchmakers. It compares the

Société with other institutions of artisanal and intellectual sociability—such as guilds, royal academies, commercial societies, and the Republic of Letters—and reveals the role that the group played in the constitution of the Académie de Chirurgie in 1731. It also offers a nuanced discussion of the relationship between the Société and the Académie des Sciences, and of the reasons that led to the early dissolution of the Société des Arts.

The third part of the book, "Writing and Making," examines the legacy of the Société des Arts in the works that former members produced after the Société's dissolution in the early 1740s. Chapter 5, "The Politics of Writing About Making," argues that writing about their own art was a political act through which *artistes* presented themselves and their work as central to France's projects of improvement. They did so by discussing the features that distinguished them from other artisans and by articulating a theory of cognition based on sensorial intelligence. This was an open-ended approach to problem-solving in which the body and the mind operated in synergy. At stake, I argue, was the refutation of Bacon's argument that improvement in the mechanical arts occurred slowly and only by chance. Based on this theory, savants thought that in order to promote advancement in the mechanical arts, it was their duty to search for new rules that would then be imposed on artisans. *Artistes* reacted to this conception of a slowly moving artisanal world, countering that in artisanal workshops, opportunities for improvement occurred frequently though they often went unnoticed because of the artisans' ignorance and attachment to routine. *Artistes*, by contrast, were able to improvise and be in the moment when serendipitous opportunities presented themselves. In their writings, *artistes* articulated rules for the practice of the arts and the manufacture of goods, continuing a long tradition of technical writing that has been elegantly discussed above all by Pamela Long and Hélène Vérin.[68] But *artistes* nourished also the political ambition to become major actors in the implementation and enforcement of such rules. In the pursuit of this goal, they employed the denigrating representations of artisans and workmen that savants had articulated in the previous decades. At the same time, they eagerly emphasized the practical nature of their knowledge, which distinguished them from theory-inclined savants. They wanted their readers to understand that *artistes*, and not artisans or savants, were reliable experts for the assessment of technical knowledge.

Chapter 6, "*L'Esprit* in the Machine," examines the role of the mechanical arts in the articulation of educational programs and of projects of reform of the manufacturing system. It shows that around the 1730s, mechanical devices such as clocks and watches were considered as material metaphors of the rational mind. A positive conception of the repetitive nature of artisanal gestures underpinned the introduction of machines in popular educational programs, such as Louis Dumas's method for teaching children how to read, or the abbé Nollet's course of experimental physics. Later in the century, gloomier visions of the analogies between humans and machines became more widespread.[69] This was particularly evident in the case of Jacques Vaucanson's inventions for the silk industry. I argue that Vaucanson's machines relocated skill from the body of the artisan into the machine, and so introduced a new system of production that mechanized human labor and silenced potentially riotous workers. These machines embodied a political vision that was consistent with the *artistes'* understanding of other artisans as mindless automata.

Taken together, the chapters in this and in the previous sections present the *artiste* as an educator who employed texts, images, and artifacts to enlighten the public—especially French administrators—about the particular nature and usefulness of his knowledge. A ruling class able to see the *esprit* at work in machines and other practical inventions would readily understand the role that *artistes* could play in guiding France toward a golden age of imperial and commercial expansion. The self-portrayal of the *artiste* that emerged from these works stood in stark contrast to the views of savants and *philosophes*. If they saw the artisan as someone to enlighten, this book shows that *artistes* enlightened too. By placing *artistes* center stage, *Artisanal Enlightenment* challenges the conception of the Enlightenment as an exclusively intellectual movement. It shows that, from the *artiste*'s point of view, the Enlightenment was less about ideas and more about political economy, projects of improvement, and policies of inclusion or exclusion in the decision-making practices of the state.

Part 1

The Natural History of the Arts

Chapter 1

LOST KNOWLEDGE AND THE HISTORY OF THE ARTS

In his *Encyclopédie* article "Art" published in 1751, Denis Diderot argued forcefully for the importance of a general treatise on the *history* of the mechanical arts.[1] He believed that such a treatise would accomplish two related goals. It would compensate for widespread ignorance about these human activities, which were often neglected or even despised by the learned, and it would prevent the loss of knowledge that humanity had experienced as a result of the lack of written descriptions of ancient techniques. The term "history" that Diderot employed did not mean a chronological account. On the contrary, he believed that the actual history of the individual arts was pointless because it would only highlight long, uneventful intervals between momentous discoveries that clustered closely together in time. Furthermore, a chronological account of the development of the arts was almost always impracticable because of the lack of written information. Diderot's "history of the arts"—like his general conception of the mechanical arts—was indebted to Francis Bacon's understanding of natural history. In *The Advancement of Learning* (1605), the English philosopher defined natural history as the study "of nature in course, of nature erring or varying, and of nature altered or wrought; that is history of creatures, history of marvels, and history of the arts."[2] Bacon believed that the history of the arts should be regarded as part of natural history because the raw materials upon which the arts worked were natural objects. Since the arts transformed those raw materials, the systematic observation of artisanal procedures would yield information on natural properties that would not otherwise be understood. This information could not be obtained by studying nature in its ordinary course, or even in its preternatural manifestations. Bacon argued, and Diderot reiterated, that no work on natural history could be considered complete unless it included the history, or description, of the mechanical arts.

In later works such as the *Novum organum* and the *Parasceve ad historiam naturalem et experimentalem* (*Preparative Toward a Natural and Experimental History*), Bacon elaborated on the notion of the "history of the arts," which he also called "experimental history." As Diderot would later underscore, the term "history" did not indicate a chronological account of the development of the arts but the description of artisanal practices and workshop tools. Since all that took place in artisanal workshops could be regarded as a series of experiments carried out on the natural materials that artisans worked upon, Bacon defined this history as "experimental" in order to distinguish it from other branches of natural history, and underscored its relevance for the advancement of natural knowledge. He believed that the collection of mechanical experiments emerging from the history of the arts would contribute to natural philosophy "like streams flowing from all sides into the sea." He even envisioned a state-funded, collaborative enterprise that would take on the task of surveying all the arts and trades.[3]

The intellectual lineage that Diderot traced with his numerous references to Bacon effaced the long history of written descriptions of the mechanical arts that even predated the invention of the printing press. Literate artisans such as Filippo Brunelleschi, Antonio di Pietro Averlino (called Filarete), Francesco di Giorgio, Leonardo da Vinci, and Bernard Palissy—to mention a few—had produced influential treatises that modeled how to write about the arts.[4] By closely associating his encyclopedic project with Bacon's "experimental history," Diderot offered a rationale for writing the history of the arts. Not only would a general treatise on the mechanical arts contribute to "true philosophy," as Bacon had stated; it would also, and above all, illuminate the connections between the various arts and thus contribute to the improvement of the arts themselves. By adding "order" and "method" to what chance alone had produced, Diderot believed that the history of the arts would lead to otherwise unimaginable innovations.[5]

The Baconian ideal of "experimental history" had informed several projects, begun in Paris and London in the mid-seventeenth century, on the history of the arts and trades. The French projects culminated with the *Descriptions of the Arts and Crafts*, a multivolume work initiated at the time of the foundation in Paris of the Académie des Sciences (1666), which began publication only in 1761, prompted by the success of the *Encyclopédie*.[6] At the beginning of its history, when it was a project in the making, the *Descriptions of the Arts and Crafts* was referred to as the "History" or "Description" (note the singular) of the arts. I refer to these early, unpublished projects as *History of the Arts*, in order to highlight their encyclopedic nature and, at the same time, to distinguish them from the concept of the "history of the arts." These early encyclopedic projects have been the topic of two areas of overlapping research: the prehistory of the *Encyclopédie* and the early history of the Académie des Sciences.[7] Scholars who worked on the genesis of the *Descriptions of the Arts and Crafts* have shown that a substantial amount of work had been produced before the publication of the first volumes of the *Encyclopédie*, and they have pointed to connections and overlaps between the encyclopedic projects of Diderot and of the Académie. The analysis of the plates in the *Descriptions of the Arts and Crafts* and in the *Encyclopédie*, in particular, has demonstrated that Diderot and his collaborators in some cases pirated the visual material produced by the Académie. Scholars of Diderot and the *Encyclopédie*, on the other front, have highlighted the intellectual and material debts that Diderot owed to the lesser-known actors who had worked for several decades before him in an attempt to create an encyclopedia of the arts.[8]

This chapter and the next discuss writing projects on the "history of the arts" asking different questions and adopting a different methodology. I argue that writing about the arts was not merely an academic exercise. It was a political act that impinged on visions of how improvement could, and should, be achieved. The encyclopedic projects articulated within or around the Académie des Sciences excluded practitioners of the arts from the thinking processes leading to the organization of the work, relegating artisans to the role of objects of observation and description—not too different from workshop tools and machines. As news of the savants' interest in the mechanical arts spread through Paris and the rest of France, however, learned artisans reacted to the savants' indiscriminate representation of workshop workers as mindless pieces of machinery. They consistently presented themselves as *artistes*, sharply different from other workers in virtue of their own knowledge, or *esprit*. This chapter and the next provide the foundation for this argument by analyzing the savants' early work on the history of the arts. They ground such work in its political context and in the world of erudition and learning that the early authors of the history of the arts inhabited. I show that the savants' attempts to make the mechanical arts a subject worthy of the reading elites compelled them to bypass the people who performed manual work. They thus constructed an image of the artisan as a mindless provider of mechanical power, and of the mechanical arts as a world where improvement occurred only slowly and merely by chance.

By employing the notion of "savant" I do not mean to reassert the untenable dichotomy between people of the mind and people of the hand. The savant in this chapter and in the rest of this book may well have had some practical knowledge and expertise in the mechanical arts, just as the artisan could have been knowledgeable in various domains of human learning.[9] However, unlike savants, artisans made their living by the work of their hands. "Savant" and "artisan" were categories that my historical actors employed, often to articulate epistemological claims about the superior nature of their own working practice. In other words, "savant" in this book is an ethnographic category, not an epistemic one. It identifies the social status that some people, such as the academician, enjoyed and that others, such as the artisans who understood themselves as *artistes*, aspired to obtain.

LOST INVENTIONS, TRAVEL, AND THE HISTORY OF THE ARTS

A century before the publication of Diderot's "Art," the Baconian notion of "experimental history" was common currency in learned Parisian circles. Through correspondence networks and travel, savants in the French capital were well connected with their equivalents in Britain. As news of the posthumous publication of Bacon's works reached France, the Parisians hastened to request copies from their London correspondents.[10] Thanks to these networks of exchange, they were *au courant* with the "History of Trades," a project that the Royal Society had begun shortly after it was established in 1660 and that was modeled on Bacon's "history of the arts." Several eminent fellows of the Royal Society (FRSs), such as Sir William Petty, John Evelyn, Robert Boyle, and Robert Hooke, contributed to the project. When Boyle's query list for the compilation of mining histories was published in the *Philosophical Transactions* in 1666, Henry Oldenburg, the Royal Society secretary, endorsed this publication, hoping that "several foreigners of his acquaintance" would participate in the project.[11]

Although Bacon's concept of a "history of the arts" circulated between London and Paris, the encyclopedic project elaborated in France should not be seen as a French appropriation of an English idea. In 1659, before he became the Royal Society secretary, Oldenburg visited Paris, where he participated in the activities of several learned circles. He interacted, in particular, with the astronomers Pierre Petit and Adrien Auzout, the Dutch mathematician Christiaan Huygens, and the traveler Melchisédech Thévenot, who were main actors in the articulation of a project on the history of the arts to be carried out in Paris, with the support of the state. The French project was characterized by the idea of an interconnectedness of the arts, something that did not concern the FRSs, who published their descriptions independently of one another.

Specific political and cultural circumstances informed the project on the history of the arts that emerged in France in the 1660s. In 1664, Louis XIV's chief minister, Jean-Baptiste Colbert, prepared a memorandum on trade that called the king's attention to the fact that the "manufacture of cloths, serges, and other textiles of this kind, paper goods, ironware, silks, linens, soaps, and generally all other manufactures were and are almost entirely ruined." The poor condition of the arts called for swift action on the part of the king toward reestablishing the primacy of French internal and international trade. In his thorough analysis of the contemporary global economy, Colbert highlighted the lack of a powerful French fleet that might compete with the flourishing Dutch and offered quantitative data on the French imports of finished goods that called for urgent, corrective interventions. Embellishing his dispassionate analysis with the language of the courtier, Colbert gently but firmly urged the king to devote his attention to a subject—trade—that was "disagreeable to hear" about.[12] In what was, de facto, a mercantilist plan for establishing governmental control over trade and manufactures, Colbert compiled a list of measures to reinvigorate trade and devised strategies aimed at allying the court with the social actors—merchants and inventors—through which such control could be exerted. He believed that these alliances would bolster the glory of Louis XIV by offering to the public an image of a benevolent king, invested in the public good. Colbert advised, first and foremost, the dissemination of the notion of the magnanimous Louis XIV. The king should publicize his support for innovations in the arts and manufactures, as well as his intention to reestablish internal and international trade, through royal decrees. He should take steps to demonstrate his involvement with trade and manufactures, and should display his good disposition toward merchants and inventors seeking patronage. He should welcome merchants at court, and his retinue should include merchants' representatives, whom the king's council should at least occasionally consult. In exchange, those who had "the honor of serving him . . . will talk and publish about the advantages that the king's subjects will receive."[13]

Colbert's support of the useful arts and his plan to revive French trade spread quickly across the communities of savants in Paris and, through them, reached Oldenburg in London. In September 1665, Oldenburg wrote to Robert Boyle:

> [The French] seem to be set upon it [trade] more, yn ever, to advance ye same in their Contry: for which purpose orders are already issued, as I find by letters, not only to promote the Manufactures of Cloath and Silk, but also to sett up severall others, of which they specifie Glasmaking, in so much yt no Venice-Glas is any more to be imported into France, as also Lace-making, of wch shall be made their owne people, and so of other manufactures, yt soe they may keep their moneys at home, and make conquests wth it abroad.[14]

Oldenburg had witnessed firsthand the excitement that Colbert's memorandum stirred among the group of savants who met in the house of Thévenot. They welcomed Colbert's interest in trade and manufactures as an opportunity to institutionalize their knowledge and expertise. The compilation of a *History of the Arts,* an encyclopedic project inspired by Bacon's principles, was one of the key activities discussed in the "Projet d'une Compagnie des Sciences et Arts" (Project of a Company of Sciences and Arts) they presented to Colbert. They aspired to obtain royal support for the establishment of an academy mainly dedicated to the improvement of the mechanical arts.[15]

The Thévenot group was one of several informal assemblies that animated seventeenth-century cultural life in Paris. In addition to the aforementioned Huygens, Petit, and Auzout, its members included the Danish naturalist Nicolas Steno and the Dutch microscopist Jan Swammerdam. Their work focused on experimentation, travel writing, and the mechanical arts (in particular navigation and cartography), yet none of the members was a practicing artisan.[16] Thévenot was a wealthy royal officer whose cultural interests ranged from cartography and hydrostatics to the voyages of exploration.[17] He enjoyed an international reputation as a collector and travel writer, and had personal connections with some of the most distinguished members of the Republic of Letters, including John Locke, Gottfried Wilhelm Leibniz, Vincenzo Viviani, and Lorenzo Magalotti, as well as Oldenburg and Huygens. Thévenot's collections of books, instruments, and machines were all available for the group's use.[18]

Auzout, Huygens, and Thévenot authored the "Project of a Company of Sciences and Arts" when Colbert's vision for French trade began to spread. The influence of Colbert's memorandum on the "Project" cannot be overstated. In line with Colbert's expectation that the people who received royal support would "talk and publish about the advantages that the king's subjects will receive," the "Project" detailed how the Compagnie would contribute to the public good.[19] Its members would work to create new roads, bridges, ships, geographical maps, and tools for navigation; they would gather information, in France and in foreign countries, about medical remedies, mines, agriculture, and methods to drain marshes and make rivers navigable. They would conduct experiments and observations on astronomy, geography, chemistry, anatomy, and medicine, and they would invent new machines and new technical processes.[20] Their statement of intentions addressed key issues discussed in Colbert's memorandum by harnessing Bacon's idea of a state-funded academy dedicated to the sciences that advanced the public good. Partly reflecting the activities of the Thévenot group and partly presenting new ones, the "Project" explained that the Compagnie's mission was to "work towards the improvement [*perfection*] of the sciences and the arts, and in general [toward] all that can be of use and of convenience to mankind and particularly to France." Its goal was to become "a Council able to give [the king] sincere and truthful advice" on the improvement of the arts. This, they believed, was the preliminary step to the advancement of internal and international trade.[21]

The *History of the Arts* was a strategic component in the association of the "useful" (i.e., mechanical) arts with the public good that characterized the "Project." Its authors explained that a *History of the Arts* would contribute to the public good in two key ways, which would later be also underscored by Diderot in his *Encyclopédie* article "Art." First, by making public the best productive practices and the common frauds, it would compel laborers to work more honestly, and it would allow readers to guard themselves against deceptions. Hence, it was of central importance to Colbert's program for French trade. Second, a *History of the Arts* would protect against the loss of technical knowledge.

The theme of lost knowledge resonated across the Republic of Letters. If Pliny's *Natural History* bore testimony to the many ancient inventions that the moderns were not able to retrieve, the success of Guido Panciroli's *Two Books on Memorable Things Lost and Found* was a more recent reminder of the ingenious accomplishments of the past.[22] Reissued several times since its posthumous Latin publication in 1599, and translated into English, Italian, and French, Panciroli's work was a monument to the achievements of the classical period. It offered a long list of inventions and techniques that had been lost since antiquity, ranging from natural substances such as purple dye and asbestos to inventions such as malleable glass and Archimedes's burning mirror. It listed also *nova reperta*, or recent inventions, that were overwhelmingly outnumbered by the lost things. Thus, the confrontation with the splendor of antiquity made the sense of loss only more humbling.

By articulating their project of a *History of the Arts*, the authors of the "Project" participated in contemporary debates on lost knowledge. Their proposed work encompassed a revised list of lost inventions, based on their careful scrutiny of Panciroli's list, from which they would expunge the lost things that had never actually existed. The authors of the "Project" thereby showed that they were attentive readers of Bacon who, fascinated with the lost inventions of antiquity, had nonetheless cautioned against Panciroli's credulity and errors.[23] Following the English philosopher's recommendations, the Compagnie also would create a list of desiderata, a wish list of all the innovations that had been tried unsuccessfully since the beginning of human history.[24]

The arts and crafts were salient features in the emerging culture of travel, which informed the proposal for a *History of the Arts*. Fashionable Parisians were avid consumers of travel literature. They were enamored with tales of loquacious voyageurs, returned to Paris after Grand Tours or journeys through exotic lands, and their "universal libraries" reserved special rooms for imaginary "Orients" such as India, China, and the Ottoman world.[25] Together with natural wonders and rarities of various kinds, the ways in which foreign human hands created artifacts and manipulated natural substances excited readers' curiosity. Thévenot was a prominent contributor to this emerging culture of travel. His own role in the group that gathered in his house was to compile the first collection of travel accounts in French, which he published in 1664 by drawing directly from the minutes of the group's meetings.[26] The gathering of information on arts and crafts that the Compagnie planned required firsthand observations of workshop practices and tools, a practice that depended on what I have elsewhere defined as "intelligent travel": journeys aimed at intelligence-gathering, sometimes but not always secret, carried out by people who were able to understand the processes they observed and to record them.[27] Intelligence was an essential prerequisite for this kind of travelers because in the world of the arts and crafts, they would face forgeries and trickeries, which they were expected to identify and expose. The Compagnie envisioned that "intelligent persons" would join overseas expeditions, whereas the most expert among its members would travel through France and the rest of Europe to document where and how the arts and crafts were best practiced. During their inspections of workshops and manufactures, they would interview artisans and workmen about their practices and operations, and they would produce drawings of all the machines and tools in use.[28]

The project of a Compagnie des Sciences et Arts did not materialize in the Académie des Sciences that Colbert and the king founded in 1666, but the project on the history of the arts was not altogether lost. Colbert and Louis XIV preferred an alternative project that centered on chemistry, astronomy, anatomy, and physics, rather than on arts and manufactures. Some members of the Thévenot group were offered membership in the new institution, but no practicing artisan was admitted to the Académie.[29] Colbert probably believed that the compilation of a *History of the Arts* could become one of the tasks that the newly founded institution would adopt. In 1675, he asked the Académie to produce a treatise on the arts and crafts that were practiced in France, complete with visual and textual descriptions of workshops, machines, and tools. The project however did not take off. After Colbert's death in 1683, his successor, the marquis de Louvois, requested in vain that the academicians devote themselves to "works of discernible and prompt utility."[30] Historians of the Académie have remarked that the early members despised the idea of working on a treatise on the mechanical arts because of the low status of manual work.[31] It was probably because of their reluctance to engage in this kind of work that the new president of the Académie, the abbé Jean-Paul Bignon, formed an external committee to which he entrusted the task that Colbert had envisioned for the Académie: the compilation of an encyclopedic work on the arts.[32]

STYLE AND MEMORY

Bignon addressed several concerns when he assembled the committee that would work on the project of a *History of the Arts*.[33] The work had to be complete yet accessible, informative yet entertaining. It had to appeal to artisans who were willing to improve their art but also to administrators and entrepreneurs who wanted to police how laborers worked within their manufacture. Bignon was aware that the inhabitants of the world of trade, manufactures, and the arts—whether human or material—were not attractive subjects for the French reading elite. His fellow members at the Académie des Sciences regarded the history of the arts as a "dry, thorny, and by no means brilliant" subject.[34] Even Robert Boyle, who regarded the history of trade and manufactures as "a very great design," nonetheless admitted that it was a "very hard, difficult, and tedious task."[35] Unlearned artisans and workmen who repeated the same operations over and over again did not seem to offer much in the way of amusement to a society that valued *esprit* above all virtues. Self-interest, fraud, and deception were attitudes that the educated French had learned to despise (at least in public), yet they seemed to be ever-present in the world of the mechanical arts.[36]

Writings about arts and crafts, however, did not necessarily need to be tedious. As a frequenter of Parisian salons, Bignon knew that an appropriate literary style could turn even the mechanical arts into a fashionable topic. Parisian elites raved over Bernard Le Bovier de Fontenelle's *Conversations on the Plurality of Worlds* (1686), a witty work on Cartesian philosophy in the form of dialogues between a gentleman and a marquise that took place in the lady's villa.[37] They also enjoyed Wednesday afternoons at Jacques Rohault's, where they learned Cartesian physics by experimental demonstrations staged as theatrical performances. The sociability of the Parisian elites included educational games and conversations with travelers, as well as anything else with *esprit*. There was no room for pedantic disquisitions on machinery or philosophical systems, unless presented in the form of amusing entertainment.[38]

With the intention to captivate the attention of the reading elite, Bignon drew the three initial members of his Commission des Arts from the fashionable Parisian circles he attended, demonstrating a preference for people who had some involvement with the mechanical arts and a propensity for rational recreations. Gilles Filleau des Billettes, Jacques Jaugeon, and Père Sébastien Truchet epitomized this combination. Des Billettes, whose brother had translated *Don Quixote* into French, belonged to a noble family. He attended the circles of the Duchess of Longueville and the Duke of Roannez, and he cultivated interests in geography and in the genealogy of European households.[39] In 1690, he invented a hydraulic machine that he presented to the Académie des Sciences.[40] Jacques Jaugeon held the title of *écuyer* (which indicated an ennobled family) and was the author of *The Game of the World*, a book published in 1684 that described a table game for children and adults. Dedicated to the Sun King, *The Game of the World* was a playful map of knowledge that included a long section on inventions and the mechanical arts. As the game progressed, reader-players toured the world like travelers, learning various legends, myths, arts, and histories of European and non-European countries.[41] Around 1690, Jaugeon created and dedicated to the young Duke of Bourgogne a perpetual calendar that showcased his ability to combine utility and playfulness.[42] The third member of the group, the Carmelite friar Père Sébastien Truchet, was a skilled mechanician. Originally from Lyon, Truchet learned practical mechanics in the cabinet of Grollier de Servière, a Wunderkammer of superb mechanical devices that was widely known in France and the rest of Europe.[43] Once in Paris, his manual skills became known to Colbert, who introduced him at court. Subsequently, Truchet was involved in constructing the aqueducts of Versailles and later the canal d'Orléans.[44] The Commission des Arts met near the Louvre in the hôtel of the abbé Bignon, who attended most meetings. They each received a salary of one thousand livres from the king, which was a respectable amount, considering that full-fledged members of the Académie des Sciences received, on average, six hundred livres.[45]

The Commission des Arts commenced its work early in 1693, when the Quarrel of the Ancients and the Moderns was heating up.[46] The party of the Moderns had strong connections with the Académie des Sciences. Charles Perrault, the initiator of the debate, had played a crucial role in the design of the Académie des Sciences. Fontenelle, a strenuous advocate for the Moderns, was the perpetual secretary of the Académie des Sciences.[47] Even though the Quarrel had started as a literary debate within the Académie Française, the mechanical arts soon came into the crossfire in this learned battleground. If the Moderns emphasized the holy trinity of inventions that were unknown to the classical world—gunpowder, the compass, and the press—supporters of the Ancients commented on the innumerable inventions that humanity had lost since antiquity. They harnessed the rhetoric of lost knowledge, elaborated during the Renaissance, as a mantra. Supporters of the Moderns engaged in the critical examination of ancient inventions that the Compagnie des Sciences et Arts had planned to carry out. Citing Panciroli, Perrault explained that there was no need for triremes, catapults, or battering rams in the age of gunpowder. The modern discovery of cochineal supplanted the lost ancient secret for purple dye. The Moderns could easily rebuild ancient architectural masterpieces such as amphitheaters, obelisks, and triumphal arches, as Perrault's brother had demonstrated with his design for a grand triumphal arch.[48] As Chandra Mukerji has argued, the France of Louis XIV came to think of itself as the new Rome, in its literary as well as in its built landscapes. The Moderns did not dismiss the achievements of the Ancients. They believed that the technical knowledge that was available at the time of Louis XIV allowed them to remake—and improve upon—anything that made the Ancients great.[49] Knowledge of the mechanical arts was crucial for a better framing of the debate.

Like the Académie des Sciences, the Commission des Arts sided squarely with the Moderns. In the confrontation between classical antiquity and the France of Louis XIV, the history of the arts offered new weapons for the arsenal of the Moderns. It could dispel the myth of extraordinary inventions that had been lost and simultaneously instill the idea of a cumulative advancement of learning manifested in modern machines and instruments that performed operations unthinkable in the past. As Des Billettes noted, the very idea of a published description of the arts underscored the Moderns' superiority: unlike the Ancients, they were working toward the preservation of their technical achievements for posterity. Des Billettes, who directed the works of the commission, studied Panciroli's text carefully and concluded that the lost inventions of the past did not constitute a threat to the Moderns. The fact that the Ancients were not able to preserve their knowledge supported the idea of progressive, cumulative knowledge that culminated with the age of the Sun King. Thanks to the work of the Moderns on the history of the arts, the notion of lost knowledge would finally disappear. As Des Billettes wrote, "The great result of this plan could be that of transmitting to posterity the best uses and practices of the arts and of not leaving any matter for new books *de rebus perditis* [on lost things]."[50]

Des Billettes believed that the preservation of technical memory would be the most valuable and noble result of a *History of the Arts*. He did not think that the commission's work necessarily had to result in the improvement of the arts. Its focus should be on the "pure practice of the arts" in order to "explain everything so well as to make it palpable even to those who have no smattering of it."[51] There may well have been practical advantages from this kind of work. The few artisans with the ability to understand the larger scope of their practice—whom Des Billettes defined as "workmen with *esprit*"—would improve their own craft, creating products of better quality. Also, those who employed workmen and artisans would be able to direct and supervise their work more effectively, discerning good and bad practices. This might result in the overall improvement of the arts, or not. Improvement was not Des Billettes's main preoccupation. The most significant contribution of the commission was that the best artisanal practices would be handed down to posterity. With the publication of the *History of the Arts*, there would be no more nostalgia for lost inventions.[52]

The motif of the preservation of technical knowledge for posterity was central to the rhetoric of the commission, as it was their response to Colbert's request that they highlight how the king's support of specific activities would benefit the public. The commission underscored the fact that the history of the arts so conceived was a testament to the greatness of the French king. Unlike other monarchs who merely protected the arts, the patronage of the Sun King did not limit itself to supporting the current arts and sciences. It did not live just in the moment. The Sun King's sponsorship of the arts was forward-looking and unprecedented. By making the history of the arts publicly available, his magnanimity extended not only through space but also through time, reaching to posterity.[53]

IMMATERIAL INGENUITY

The commission's coupling of the history of the arts with the preservation of knowledge spoke directly to the themes debated by both parties involved in the Quarrel of the Ancients and the Moderns. However, the commission could not ignore the contempt for the mechanical arts that was so widespread among polite society. Des Billettes's solution was to borrow the philological tools of the party of the Ancients to offer an etymological analysis of the term "mechanical arts," which emphasized the ingenuity of the inventive process at the expense of practical execution and manual operations. In his "Preface to the History of the Arts," he explained that "mechanical arts" derived from the Greek *mēkhanē* for "machine." As his readers knew well, a machine was the outcome of inventive design, an intellectual process that presupposed ingenuity and creativity, just like any other work of *esprit*. The manual work that was necessary to execute machines was less relevant than the mental plan that they actualized: "In all machines we are less concerned with materials than with the artifice and the intelligence [*esprit*] of the author."[54]

This immaterial ingenuity that Des Billettes emphasized had its roots in the successful literary genre of the "theaters of machines," dating to the later sixteenth century. The "theaters" were books that contained several illustrations of machines of various kinds, each accompanied by a short descriptive text. Page after page, the books presented the same pattern: a full-page image of a machine in perspective accompanied by a short description that explained its function. Although the authors of the "theaters" were often practitioners in the mechanical arts with experience in the building of actual machines, the devices that were presented in the books were meant to highlight the ingenuity of conception—the mechanism—rather than represent an actual working contrivance. The immaterial ingenuity of the machines was intended to puzzle and surprise the reader's mind by the sheer possibility of each machine's function. Whether it was raising or channeling water, moving heavy objects, grinding, polishing, sawing, or otherwise transforming materials, the wonder that the machine elicited derived from what it purported to do, not from whether it could actually do it (fig. 1.1). As with treatises on theoretical mathematics, readers were expected to admire the originality of the project and its internal logic, not to assess its feasibility. Indeed, normally these "theaters of machines" did not describe existing devices. The machines they presented were *jeux d'esprit*, playful manifestations of the author's ability to devise purposive function. In contrast to the mechanical projects submitted to obtain *privilèges* (early modern patents), which had to demonstrate that they could successfully be put to work, reduction to practice was irrelevant to the ingenuity of the design.[55]

This did not mean that the machines described in the "theaters" were never built. In 1683, the court officer (*maréchal-des-logis*) Jean-Baptiste Picot organized in rue de la Harpe in Paris an exhibition of small-scale, working models of machines. Most of them were drawn from the theaters of machines by Jacques Besson, Agostino Ramelli, and Salomon de Caus. There is frustratingly little material on this interesting exhibition, maybe the first of its kind. We do know that the organizers were motivated by Colbert's support of the arts and manufactures, and that they believed that the display of models would contribute to educating the public on the importance of mechanics. They intended to fight the prejudice that machines were a subject more appropriate to "artisans than to a gentleman or a bourgeois." So, they displayed mills, pumps, sawing devices, and other potentially useful contrivances, and added models of recent machines in use at the Paris arsenal. All models were made of "wood, iron, and copper," without further specification about the materials that constituted them or how the choice of different materials might affect the machine's function. Just like the "theaters," the display emphasized function and design, erasing the manual work and material knowledge necessary to build working machines.[56]

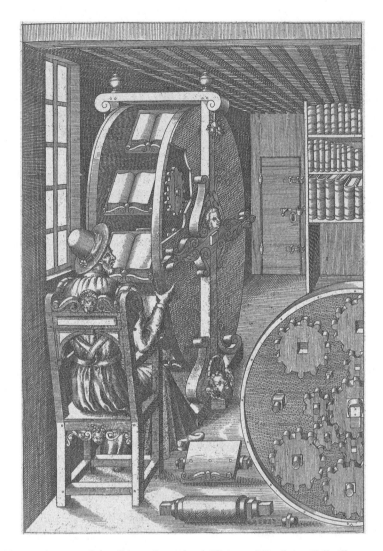

Fig. 1.1. Agostino Ramelli's reading wheel. This machine beautifully illustrates the erudite aspirations of the authors of the theaters of machines. It also provides an effective example of immaterial ingenuity. As the Lyon mechanical inventor Grollier de Servière would remark decades later, Ramelli's design failed to perform its function when built. Yet Ramelli's idea inspired several mechanicians, who made reading wheels through the eighteenth century. There are several existing models in various libraries worldwide, which are all modified versions of Ramelli's original design. They typically present four shelves, a lot fewer than in Ramelli's ideal machine. Agostino Ramelli, *Le diverse et artificiose machine* (Paris, 1588), plate 188. [Beinecke Library, Yale University, New Haven, Conn.]

Even before the exhibition on rue de la Harpe, the utility of models had been criticized by various mechanicians, who emphasized that large-scale machines built from models often failed to perform as expected. Picot forestalled this "vulgar objection," explaining that mechanical principles were "the soul of all machines."[57] Even though some models may fail when built in larger scale, they still had something to teach by inspiring further inquiry. The display of models was intended to benefit many different kinds of minds (*esprits*): artisans and people of the arts would draw inspiration to improve and devise new machines; the curious would familiarize themselves with mills and other machines that could be seen around Europe; land owners would acquire ideas on how to better exploit their properties.[58] By calling attention to the immaterial ingenuity inscribed in machines, then, the Commission des Arts reiterated a familiar refrain. Although this exhibition was short-lived, its success indicated a widespread taste for the display of immaterial ingenuity; the exhibition catalogue was present in the libraries of the upper classes in France and abroad.[59]

The immaterial ingenuity that Des Billettes emphasized infused Diderot's well-known *Encyclopédie* article on the stocking machine, "Bas." Struggling with the limitation of words and images to explain the functioning of a machine that consisted of some 2,500 parts, Diderot wanted to convey to his readers his fascination for a device that he had to assemble and disassemble numerous times in order to understand its hidden workings. Diderot's stocking machine was highly idealized. As "the brilliant creation of a single mind," it manifested "a single and unique chain of reasoning, of which the fabrication of the machine is the conclusion."[60] By learning about the stocking machine, Diderot's readers were invited to admire the faculties of the human mind. They would not pay any attention to the "messy processes of experimentation" associated with artisanal work on materials.[61]

THE *ENCHAÎNEMENT* OF THE ARTS

The search for an ordering principle that would link the arts to one another was an essential component of Des Billettes's project. He conceived of the *History of the Arts* as "a kind of encyclopedia" that would be published as an interconnected, multivolume work, not as a series of individual articles on the various arts.[62] As Bacon had stated in *The Wisdom of the Ancients*, the search for an ordering principle entailed entering the labyrinth of mechanical ingenuity: "All ingenious and accurate mechanical inventions, may be conceived as a labyrinth, which, by reason of their subtilty, intricacy, crossing, and interfering with one another, and the apparent resemblances they have among themselves, scarce any power of the judgment can unravel and distinguish; so that they are only to be understood and traced by the clue of experience."[63] The "clue of experience" was to be found in artisans' workshops. The members of the commission, like André Félibien and Antoine Furetière decades earlier and Diderot and D'Alembert decades later, visited workshops and interviewed artisans. In preparation for these visits, Des Billettes compiled query lists to submit to the artisans.[64] The commission anticipated that this information would have to be purchased from "the people of the art"; he envisioned a special fund for "drawing from artisans all knowledge necessary to an exact description."[65] The members of the commission did not regard the artisans as peer contributors; rather, the members thought that they had to ply the artisans with money in exchange for information.

The labyrinth of mechanical ingenuity presented pitfalls. As Bacon indicated, the "intricacy, crossing, and interfering" among the arts challenged the intellect of the learned. It was therefore essential to find an ordering principle or, in Des Billettes's words, "an *enchaînement* or dependence among the arts" that would lead "naturally from one art to the other." This search was a speculative part of the project that could not be avoided.[66] During a time of economic hardship for the book market precipitated by years of war, the commission was eager to make the mechanical arts a subject of learned interest, something to delight the reading public.[67] Des Billettes thought that the *enchaînement* of the arts would allow the *esprit* to move effortlessly from one art to the next: the description of one art would stimulate "the appetite [*goust*]," then provide "the introduction [*entrée*], and the knowledge [*lumières*]" for the art that would follow in the encyclopedic work.[68] The *enchaînement* of the arts would make reading and talking about these topics a worthy activity for the elites concerned with the competition between the Ancients and the Moderns. It was the notion of the *enchaînement* of the arts that distinguished the commission's work from any previous written work on the arts.

How to establish the principle that would bind the arts together? Des Billettes was aware of the arbitrariness of the choice. He considered a number of possibilities: one could start with the art that had developed first (agriculture) and proceed chronologically; start with the art that worked on the noblest materials (goldsmithing) and proceed to those that operated on less noble matter; or start with the art that depended the least on the other arts but offered the most to them (metallurgy) and proceed in increasing order of complexity. Des Billettes favored the last option as the one that most effectively grounded the mechanical arts within the natural world. As the notion of the great chain of being underpinned conceptions of the order of nature, so too the *enchaînement* of the arts constructed the relationship of mutual dependence among the arts as a natural, timeless hierarchy. Des Billettes explained that the art of metallurgy, in particular the forging of iron and steel, was necessary for making instruments and tools employed in other arts. Hence, the *History of the Arts* should begin with it and immediately proceed to blacksmithing and locksmithing. Then, it would branch out in three directions, following the order of the three kingdoms of nature: the history of the arts that operate on animal parts, those that operate on minerals, and those that operate on vegetables. Within each branch, the commission would start with the simplest art and move progressively to the most complex ones. To support his vision, Des Billettes created a "tree of the arts" and demonstrated that it would include between 100 and 120 crafts.[69] By presenting the history of the arts as a branch of natural history, Des Billettes hoped to make the mechanical arts a topic of erudite learning.[70]

Pivotal to Des Billettes's vision for the *History of the Arts* was the notion that mechanical arts changed only slowly. The idea of the *enchaînement* of the arts was grounded in an understanding of the arts as a timeless world of repetitive actions, where change occurred only as a result of chance. According to him, knowledge was not created in the world of the arts, and ingenuity expressed itself only in the minds of the learned. In the world of the mechanical arts, where no one cared for recording information, knowledge could only get lost.

THE COMMISSION DES ARTS AND THE ACADÉMIE DES SCIENCES

When the Commission des Arts started its work, Bignon was planning a reorganization of the Académie des Sciences. He asked Des Billettes's advice on the role that the *History of the Arts* should play in the renovated institution, and they discussed three possibilities. Des Billettes proposed that the commission could be incorporated within the Académie des Sciences, it could become an academy of its own, or it could be an ad hoc committee working under Bignon's orders. Des Billettes believed that the first scenario was the most "natural." The members of the reorganized Académie des Sciences who worked on the *History of the Arts* could constitute a new department. This would be nothing new, he remarked, as in the existing Académie, "the chemists have nothing to do with the geometers or the herbalists with the astronomers," yet this distinction of roles did not prevent "the confraternity, the liberty of thinking and of suffrage."[71]

The creation of a new Académie des Arts, instead, would entail expanding the commission to more members because the *History of the Arts* was a collaborative project that benefited from a variety of expertise. Furthermore, its dependence on the notion of the *enchaînement* of the arts distinguished it from the work that was carried out in the unreformed Académie des Sciences, "where everybody gets away with some work of the mind [*production d'esprit*] or experiment of no consequence or connection with the others." A dedicated Académie des Arts would produce coordinated, collective work, "which must be well digested, precise and complete."[72]

If Des Billettes believed that the plan of a special department within the reformed Académie des Sciences was the most "natural" possibility, the establishment of an Académie des Arts especially devoted to the production of the *History of the Arts* was the option that he preferred. In detailing how the Académie des Arts should organize its activities, Des Billettes repeatedly emphasized the collaborative nature of the project and presented the work of the commission as exemplary of how this could be done.[73] The *History of the Arts* was more than a collection of individually authored essay, so it required a uniformity of style that could only be achieved with the intervention of an editor. This figure was crucial to the success of the project; he had to be knowledgeable in all the subject matters and not miss anything essential. The contributing authors, on their part, had to understand and accept the role of the editor. They should not feel that the editor's interventions diminished their work, and, conversely, the editor should not regard them as mere compilers. Each author would circulate his work among the others, who would offer comments and suggestions for revision. A draftsman would also be involved in the project because visual descriptions of workshops, machines, and tools were essential. In budgeting costs for this work, the state should separate authors' compensations from the amounts required for buying information from artisans. When René Réaumur eventually assumed the direction of the project, as we see in the next chapter, he maintained the organization that Des Billettes outlined. Diderot, of course, also shared many of these views, but his "société de gens de lettres" would not depend on the state or on one individual editor.

The third possibility that Des Billettes presented to Bignon was the least appealing to him. Bignon could have his own group of people who would work on priorities decided by Bignon himself. In this case, there would be no need for regular meetings as the group would not need to devise strategies for coordinating the work. The abbé would decide the roles, and he would arbitrarily assign pensions. In effect, this was how the commission was already working. A few months after the constitution of the commission, Bignon decided to direct the group's attention to the revitalization of the typesetting of the Royal Press, a project that undermined Des Billettes's vision for the *History of the Arts*. This was one of the manufactures that Colbert had singled out for urgent corrective intervention and that the new general controller of finances, Louis Phélypeaux de Pontchartrain (Bignon's relative), wanted to revive.[74] Bignon realized that in order for the commission to propose viable solutions for the improvement of the art of printing, it had to include people who were directly involved with the art, such as the director of the Royal Press, Jean Anisson, and the type-founder, Philippe Grandjean. Bignon also included the engraver Louis Simonneau, who eventually would produce several illustrations for the *Descriptions of the Arts and Crafts*. Unlike the members of the commission, these artisans were remunerated on the basis of the work they produced. During the next six years the group worked intensely and produced a large amount of material, introducing the visual language for the representation of labor in the workshop that would ultimately characterize the *Encyclopédie*.[75] Des Billettes drafted a new preface that offered the rationale for the work. The emphasis was now on the art of printing as the activity that allowed for the preservation of historical memory of all the other arts.[76] The *enchaînement* of the arts had to be reconfigured according to the patron's demands.

With the reform of the Académie des Sciences of 1699, Bignon tried to implement the possibility that seemed "most natural" to Des Billettes. He offered membership to Des Billettes, Jaugeon, and Truchet, but he did not create a special class for the compilation of the *History of the Arts*.[77] The trio continued to work on the project within the Académie. For a decade, they read to the Académie numerous articles on the mechanical arts, ranging from bookbinding and decoration to papermaking, from typesetting and pin-making to the manufacture of silk.[78] Yet these works did not elicit any enthusiasm among the other members of the Académie. Its official publication, the *Histoire et memoires de l'Académie Royale des Sciences*, devoted only a few lines to report their activities. Jaugeon, in particular, was embittered by the hostility of his fellow academicians toward his work.[79]

France's economic situation did not help the completion of the *History of the Arts*. Years of war—the Nine Years' War (1688–97) followed by the War of the Spanish Succession (1701–14)—strained France's finances, putting the country on a path of steady economic decline.[80] The compilation of the *History of the Arts*, with its reliance on detailed visual material and information bought from artisans and laborers, was costly. In 1709, Jaugeon was frustrated to learn that he had been refunded only 360 livres of the 900 he had advanced for several illustrative plates for the *History of the Arts*.[81] Bignon knew that the slow pace with which the project proceeded was due to "the state of affairs [that] does not allow at this time to make the expenditures that are necessary for this kind of drawings and engravings." These explicit references to economic hardship cast light on why the project languished at the beginning of the eighteenth century.[82]

Yet the fact that years later Bignon decided to place the leadership of the project in the hands of his protégé René Antoine Ferchault de Réaumur suggests that he was still hopeful that the *History of the Arts* could be completed within the reformed Académie des Sciences. As we see in the next chapter, Réaumur's "natural history of the arts" revived the utilitarian project of the Compagnie des Sciences et Arts. For Réaumur, the natural history of the mechanical arts was not a strategy to win the interest of the reading elite; it was functional to France's economic advancement. Réaumur's failure to implement this vision paved the way for the constitution in 1718 of the first Société des Arts, which realized Des Billettes's idea of a separate academy dedicated to the mechanical arts.

Chapter 2

Réaumur and the Science of the Arts

The Quarrel of the Ancients and the Moderns, along with the discussion about Guido Panciroli's *Two Books on Memorable Things Lost and Found*, highlighted the fact that lack of documentation about past techniques hindered technical advancement. Information about processes, materials, machines, or tools in use anywhere in the world and anytime in human history was necessary to prevent the loss of knowledge so common in oral traditions. In many respects, the gathering of technical information needed for the compilation of the *History of the Arts* was similar to the collecting practices of natural history. In its efforts to represent the entire world of the arts and crafts, the encyclopedic project directed by Gilles des Billettes shared the universalistic ambitions of early museums of natural history. As in naturalistic collections, the *History of the Arts* relied on "intelligent persons" who, being able to identify and gather relevant information, acted as connoisseurs and collectors involved in the global trade of natural specimens. The *History of the Arts* would thus offer a timeless archive of technical knowledge, a sort of Wunderkammer where all that there was to know about the arts would be made easily accessible to the reader. As the concept of the order of nature underpinned the organization of early "theaters of nature," so the notion of the *enchaînement* of the arts—the idea that the arts were all connected together—provided an ordering principle for the *History of the Arts* that responded to the demands of managing and making available collected information.[1] The *enchaînement* of the arts created a stable architecture for the world of arts and crafts that would allow readers to find their way through the various volumes in order to access specific bits of information. The underlying assumption that made the *enchaînement* of the arts so relevant, and the encyclopedic project appear feasible, was that in the world of the mechanical arts, changes occurred only very slowly, and almost always as the result of chance. The compilation of the *History of the Arts* might well take some time, but never so long as to render it worthless or outdated.

The arrival of René Antoine Ferchault de Réaumur marked a shift in the approach to the *History of the Arts* within the Académie des Sciences.[2] For him, the encyclopedic project was an essential contribution to the pursuit of improvement. Born in 1683 into a noble family at La Rochelle, Réaumur initially studied philosophy at the Jesuit College of Poitiers. When he moved to Paris at the turn of the century, he dedicated himself to mathematics and physics. He joined the Académie des Sciences in 1708 as the protégé of the geometer Pierre Varignon and soon afterward started to contribute to the encyclopedic project. By 1718, his leadership of the Académie's *History of the Arts* was well established.[3] Bernard de Fontenelle's records indicate that for a few years Réaumur collaborated with Jacques Jaugeon and Des Billettes, presenting his unpublished reports on various arts, such as making mirrors, gilding leather, drawing gold wires, and making artificial pearls.[4] It is likely that he worked initially under Des Billettes's supervision and that he assumed the directorship of the project during the Regency (1715–23), when Des Billettes began to withdraw from the Académie's activities.[5] Réaumur was a generation younger than his predecessors. Unlike them, he was neither an eclectic inventor nor a fashionable author with some technical knowledge. He was a practicing mechanical philosopher with a background in mathematics and active interests in natural history. In his approach to the history of the arts he put all these literacies to work.

Although Réaumur valued what the Commission des Arts had done until he took over the editorship, his vision for the *History of the Arts* was more ambitious than that of Des Billettes. Whereas the latter prioritized the preservation of technical knowledge and was skeptical about the functional relationship between the history of the arts and improvement, Réaumur firmly believed that the encyclopedic work would play an important role in fostering improvement. Réaumur shared with his predecessors the idea that the *History of the Arts* would prevent the loss of technical knowledge. But he also believed that the encyclopedic work should offer guidelines for introducing innovations that would optimize production and quality. Such guidelines constituted what he regarded as a "science of the arts," a new kind of certain knowledge that would promote technical and economic advancement. The *History of the Arts*, in Réaumur's hands, was one of the ways through which the Académie des Sciences would create useful knowledge for the French state.

MATERIALS AND *LA PERFECTION DES ARTS*

The key difference in Réaumur's approach to the *History of the Arts* was the importance he gave to the materiality of artisanal practice. Des Billettes and his collaborators were interested in the thought process behind the invention of machines. For them, the ingenuity of invention resided in the machine's function and design, not in its materiality. The materials with which artisans worked provided the ordering principles for the *enchaînement* of the arts, yet they were not regarded as sources of new knowledge. For Réaumur, instead, the observation of artisanal practice was not an end in itself. Unlike Des Billettes, he did not intend to produce descriptions of "pure practice." The study of how artisans worked with specific materials was to be combined with systematic research on the natural properties of the same materials through laboratory work. Visits to workshops offered only preliminary information for *la perfection des arts*, the French phrase for improvement that hinted at the existence of an achievable status of maximum efficiency and best quality for the arts. With his experimental work, Réaumur wanted to learn how to achieve such perfection. Laboratory research on materials was central to the history of the arts.

So conceived, the Académie's work on the arts could only be a natural history of the arts: "The history of the arts that we have undertaken to write is not that of their progress or decadence: it is that of the practices that are currently in use. We have called this kind of description history, as we call natural history the description of the works of nature, such as of animals, of minerals, etc."[6] As Mary Terrall has shown, Réaumur's notion of natural history merged observation and experiment. In his work on insects, Réaumur did more than simply observe insects' body parts and behaviors in their natural settings. He created artificial environments that modified the ordinary course of nature, thereby imposing constraints on what insects would naturally do in order to learn more about them. He operated on insects as artisans operated on natural materials. Intervening by artifice on nature's ordinary course forced the subjects of his observations to reveal nature's hidden ways.[7] Similarly, his work on the arts relied not only on visits to artisanal workshops to inspect artisanal practices but also on laboratory work designed for the exploration of the natural properties of materials.

In the world of the mechanical arts, Réaumur was especially interested in the manufacture of the seemingly simplest things, just as in the realm of natural history he focused on insects, the smallest and apparently most irrelevant inhabitants of the animal world. If the discussion of the marvels of the microcosm was an argument for God's wisdom and intelligent design, as Lorraine Daston has argued, Réaumur's emphasis on the complex processes for making ordinary things was a celebration of human ingenuity and of the possibility of mastering nature.[8] In Réaumur's words, "Art, like Nature, has wonders that often we do not perceive because they are constantly under our eyes. The satisfaction of our needs and luxury makes us content, so we hardly recommend the investigation of the ingenious practices to which we are indebted for the things we ordinarily use."[9] The ingenious practices to which Réaumur referred consisted of the artisans' manipulation of natural materials. Where Des Billettes saw the ingenuity of design and function, Réaumur saw the power of art to transform matter.

Pamela Smith has argued that early modern artisans produced a "vernacular science" of matter based on their bodily engagement with materials. Crucial to the articulation of this science was the notion that matter was alive. The manipulation of materials and their transformation into works of art resulted from the artisan's interactions with life forces. The very ability to produce artifacts was the expression of the artisan's inquiry into the relationship between matter and spirit. The vernacular science of matter was an unsystematic "lived theory" that blurred boundaries between knowing and doing.[10] Réaumur's work on materials offers an example of what changed when new conceptions of matter as inert—stemming from René Descartes's philosophy—started to prevail. The artisanal epistemology discussed by Smith had its roots in the alchemical tradition. Réaumur's natural history of the arts was grounded instead in the Académie des Sciences's public rejection of alchemy and in a simultaneous interest in *la perfection des arts*, an ethos of improvement that called for relentless experimental work.[11] In his admiration of ingenious practices, as well as in his formulation of a natural history of the arts, Réaumur was not simply appropriating Francis Bacon's idea of "wrought nature." He was reviving some ideas of Bernard Palissy, a Huguenot craftsman and natural historian who worked in Paris in the sixteenth century. Bacon himself, who visited Palissy's collections while in Paris, may have been influenced by his work. A skilled potter whose extraordinary artworks earned him the patronage of the French queen Catherine de Medici, Palissy was also a successful public lecturer, a collector of natural and mineral specimens, and the author of three books. His *Admirable Discourses on the Nature of Waters and Fountains*, published in 1580, discussed various subjects in natural history and the arts, such as springs and fountains, tides, metals and alchemy, drinkable gold, theriac, ice and salts, enamels, and fire. The text advanced the idea that natural knowledge emerged from practical work on materials, the bodily engagement with matter exemplified by artisanal practice. In order to acquire knowledge of nature it was not enough to read and write; it was essential to engage in practical work on materials.[12] After more than a century of oblivion, French savants rediscovered Palissy's works in the early eighteenth century. Although they did not generally share his conception of living matter, they valued his thinking on natural history to the point of defining him "as great a physicist as nature can form one." Réaumur read Palissy's works and acknowledged his theory on the formation of fossils in an essay he read to the Académie in 1720.[13]

Réaumur appropriated Palissy's idea that the practical investigation of materials was key to advances in natural knowledge. Yet for Réaumur, this work took place in the savant's laboratory, not in the artisan's workshop. In his approach to materials and workshop practices, he disregarded the body of the artisan as a source of knowledge. His work on metals, in particular, rejected the alchemical understanding of matter even when retaining alchemical processes or instruments. Réaumur was eager to demonstrate that one could work on metals "without searching the Philosopher's Stone or breaking the bank."[14] His dealchemization of the study of metals made the body of the artisan a subject of observation and experimentation, just like materials and insects. This conception of the artisan was explicitly laid out by Fontenelle in his presentation of Réaumur's work on metals:

> Workmen do not invent anything, unless they are rare geniuses. They are kinds of automata designed for a certain sequence of movements. A skilled physicist who observes them at work, and who knows how they should be observed, cannot help inventing, especially if he can carry out all the experiments that his reflections demand, and if he has the soundness of judgment [*sagacité d'esprit*] and the ability of execution that his experiments require.[15]

This conception of artisans went hand in hand with Réaumur's great goal: the transformation of the history and practice of the arts into a "science," something that the Académie had done in the case of geography and navigation but, as he noted, not with the mechanical arts.[16] The term "science" here is to be understood in the early modern sense of "certain knowledge," something that once achieved would inform the creation of guidelines for practicing the arts. Indeed, the "science of the arts" was how Johann Beckmann, commonly regarded as the founder of technology as an academic discipline, defined the neologism "technology" in 1770.[17] For both Palissy and Réaumur, certain knowledge was not exclusively theoretical. On the contrary, it was firmly grounded in experimental work on materials. This knowledge yielded new ways of manipulating materials to produce artifacts, and it would ultimately be tested in manufactures.[18] Réaumur, however, did not believe that artisans could articulate the science of the arts. He shared the widespread idea, also voiced by Diderot in his *Encyclopédie* article "Art," that innovations in the arts occurred sparsely and mainly by chance. The slow progress of the arts was a motivating factor for the compilation of the *History of the Arts*. Without external guidance the arts developed imperceptibly, so the *History of the Arts* served the dual function of preventing the loss of knowledge and of offering directions for improvement.[19]

Réaumur's early work at the Académie offered an example of what the science of the arts might look like. Just like the insects in his later work, Réaumur systematically observed and experimented on natural materials. In his study of ductility, published in 1713, Réaumur investigated various substances such as gold threads used for embroidery, glass, and spider silk. He was interested in finding out the possibilities that nature offered—yet concealed—for the artificial manipulation of these materials. His experimental work subjected ductile materials to specific constraints that revealed unexpected opportunities for exploitation. In the case of gold, for example, he observed how craftsmen stretched gold to make threads for embroidery. He then devised new techniques and tools for producing the thinnest gold threads that were resistant enough to embroider on silk. He measured the corresponding amount of gold and concluded that artisans' imperfect tools and practices caused them to employ much more gold than needed. Through his experiments he obtained knowledge on the properties of materials and gained useful insights on the economic viability of technological projects. In his view, this was the goal of a natural history of the arts. Under his leadership, the *History of the Arts* would contribute to improving the arts and therefore to the economic and technical advancement of the French state.[20]

Réaumur's practical research on materials pointed to the inadequacy of current instruments and craft practices, and at the same time unveiled a realm of possibilities for the improvement of the arts. His work on the ductility of glass, for example, showed that the material became more flexible as it became thinner. So, he remarked, with the proper tools one could produce glass threads that could be woven into wearable fabric. Réaumur did not detail whether glass fabrics could be economically advantageous, but he offered this kind of analysis in his work on François-Xavier Bon de Saint-Hilaire's idea of producing fabrics with spider silk. Saint-Hilaire was a physician from Montpellier who had even sent samples to the Paris Académie to demonstrate the possibility of using the fiber that enveloped spider eggs for knitting stockings and mittens. In order to assess the viability of the idea, Réaumur compared the strength of spider silk with that spun by silkworms and studied the amount of labor needed to feed large quantities of spiders. He concluded that Saint-Hilaire's idea was doomed to be a catastrophic investment.[21]

In an undated, unpublished draft of a *Preface to the History of the Arts*, Réaumur emphasized the economic relevance of the encyclopedic project and underscored the fact that Jean-Baptiste Colbert had included the compilation of an encyclopedic work among the most important task of the nascent Académie des Sciences. Réaumur's argument unfolded as a response to a recurrent objection against the *History of the Arts*. Since the work would make public detailed descriptions of best practices and tools, it could backfire if it fell into the hands of foreign countries, which could rapidly catch up with French technologies. The more accurate the descriptions, the more concrete the risk. Réaumur countered that in order for new manufactures to flourish, it was not enough to learn about the best techniques and tools. It was not even enough to have skilled workers perform all the processes, as demonstrated by the migrations of skilled workers from Lyon to Tours and vice-versa, which had not erased the different quality of the damask produced in the two cities. An "infinite number of circumstances" had to concur in order for new manufactures to flourish in any country: Réaumur highlighted royal interest, the specificities of place and time, and the fortunes of entrepreneurs. The real value of the *History of the Arts*, then, rested in the opportunity it offered savants (*gens d'esprit*) to contribute to the progress of the arts. Thanks to the new knowledge they would acquire, they could offer precious directions for revitalizing the stagnant world of the mechanical arts. In conjunction with his disregard for embodied knowledge, Réaumur also believed that in the hands of clever entrepreneurs, the *History of the Arts* would enable any worker to develop skills appropriate to any art. This would make a skilled workforce virtually always available to new or existing manufactures. Hence, the countries that would benefit the most from this diffusion of technical knowledge would be the most densely populated, the most industrious, and those "most dominated by luxury." This country was France.[22] The (unrelated) annotations on the verso of the pages suggest that Réaumur wrote the draft around 1721. By that time, he had completed his work for the Regent's Survey and was ideally situated to propose a comprehensive plan for the Académie's natural history of the arts.[23]

THE REGENT'S SURVEY

Between 1716 and 1718, Réaumur and Jean-Paul Bignon, on behalf of the Académie des Sciences, supervised the so-called Enquête du Régent, a survey of France's natural resources and industries ordered by Philippe II, duc d'Orléans, the regent of France. It is difficult to establish whether the regent himself conceived of the plan or if Réaumur—through Bignon—suggested it.

The latter option seems plausible. In the few years before the survey, while he was working on the *History of the Arts* under Des Billettes, Réaumur realized that direct examination of specimens from the French provinces could greatly contribute to the encyclopedic project. He related that while he was writing on the arts that work with precious stones, he felt that he "had to investigate the Kingdom's best products in this field." So he convinced Bignon to intercede with the regent, and obtained several samples from provincial intendants. The mobilization of the intendants would be the most salient feature of the survey.[24]

The regent had risen to power in 1715, at the end of an era of wars that had drained the state treasury. He was eager to reestablish French internal and foreign trade, to rebuild the essentially nonexistent Royal Navy, and to experiment with new methods. A survey that mapped France's natural resources and industries was a strategic move in the longer term plan of encouraging investments in the arts and manufactures. The survey would also identify industries that could strengthen the Royal Navy. The regent was deeply interested in the experimental sciences, in particular transmutational alchemy (fig. 2.1). In his youth he studied with the (al)chemist Guillaume Homberg, a member of the Académie des Sciences who became his personal physician and with whom he performed experiments in a dedicated laboratory at the Palais Royal. With a glass mask on, the duke spent hours experimenting on metals and other substances and boasted of having succeeded in making the Philosopher's Stone. This was a claim that various contemporaries confirmed.[25] Proficient in the art of distillation, the duke delighted in making perfumes, fake gems, and copies of medals and cameos. He created also a mercury amalgam with which he could open letters without breaking the seal. His large burning mirror, which he made available to other experimenters and with which he performed experiments on metal alloys, earned him an international reputation as a patron of the arts and sciences.[26]

At the death of Louis XIV in 1715, Philippe's involvement in alchemy induced many at court to suspect him of having poisoned his direct rivals to the Regency, all of whom died unexpectedly within a small space of time. In response, the duke stopped his involvement in chrysopoeia. Meanwhile, Homberg, who continued to exert a strong influence on him, advocated for the role that the Académie des Sciences could play for the improvement of the arts and crafts. Just a few weeks after the duke assumed the regency, Homberg died, and soon afterward the regent issued a royal decree that placed the Académie des Sciences under his own direct patronage, unlike the many other state institutions that he entrusted to the general director of the Bâtiments du Roy.[27] Réaumur's

Fig. 2.1. Workshop furnace from the Château de Villers-Cotterêts (Aisne), residence of the regent, dating from 1715. Decorated with fleur-de-lys and three escutcheons holding the letters "P.P.," the coat of arms of Philippe II, duc d'Orléans, regent of France. [Cité de la Céramique, Sèvres, France. © RMN-Grand-Palais/Art Resource, New York]

interest in the arts and manufactures, and his nonalchemical approach to metallurgy, matched perfectly well the new priorities of the duke.

The Regent's Survey realized one of the projects laid out by the Compagnie des Sciences et Arts in the 1660s: the gathering of information on the arts and crafts through networks of correspondence and intelligent travel. The Compagnie's plan had been a response to Colbert's analysis of France's financial decline. Similarly, the survey stood as one of the regent's several measures to revive French trade and upgrade the Royal Navy. The survey relied on the administrative structure of the French state. France's network of provincial intendants, who mediated between the central administration and the local social fabric, constituted the perfect architecture for it.[28] As in the Compagnie's plan, savants and merchants working abroad, as well as "intelligent persons" who traveled to specific sites, were also recruited to the operation.[29]

In Paris, Réaumur (with Bignon's approval) prepared a circular, signed by the regent, that instructed informants about the survey. The letter was sent to all intendants (the administrative officials who represented the king in the French provinces) and to any Frenchmen abroad who could serve as informants. The letter introduced the regent's plan to improve the sciences and the arts, and requested the addressees "to consider with attention all that nature and art produce in the region where they are; to draft exact reports on these matters, and to send them to him [the regent] together with samples."[30] A list of topics of interest followed. The regent was interested in "natural things," including precious and semiprecious stones, minerals, fossils, medicinal herbs and other plants, and animals, as well as in "works of art." Regarding this last, informants were asked to describe the practice of the arts with "the most scrupulous exactitude, without fear of entering into the minutest detail"; to include samples; and, if possible, to append drawings of tools, machines, and people at work.[31] The regent forwarded the materials received from the intendants to Réaumur, who examined reports and samples and followed up with more specific questions and requests. Réaumur supervised the experiments carried out by his assistant, the otherwise unidentified Monsieur Fousjean, in the laboratories of the Académie, the mint, and his own.[32] He then reported his conclusions to the regent during weekly meetings. A contemporary magazine reported on the survey's logistics: "Every week [Réaumur] gave to this enlightened prince reports concerning the various provinces of the kingdom, to demand information on what they produced of earths, stones, mines, and any mineral ore. These reports were sent to the Intendants, their responses to the reports were accompanied by samples of all the requested matters, and these samples now form M. de Reaumur's equally curious and useful cabinet."[33]

The regent listened carefully to Réaumur's reports, many of which addressed metallurgy, a subject of great interest for the French state. The manufacture of heavy objects of forged iron, such as cannons and anchors, was crucial to the navy. France was a great importer of expensive steel and steel items from the northern European countries, especially Germany. The opportunity to produce domestic steel (at lower cost) was an attractive possibility for the regent. Metallurgy was, not coincidentally, a topic that Réaumur was developing for the *History of the Arts*. It is unclear if his focus on metals derived from Des Billettes's original idea of organizing the encyclopedic work around natural materials, starting with metallurgy. In any case, in the three years following the completion of the survey, Réaumur worked extensively on mineral ores and other samples sent by the intendants. Based on his experiments and observations, he produced a history of cast iron and steel, followed by a study on anchor production. Both were intended as contributions to the Académie's *History of the Arts*.³⁴

The Regent's Survey and Réaumur's work on steel and cast iron unfolded at a time of financial optimism, fueled by John Law's experiments on public finances. With the support of the regent, Law launched a large-scale financial operation aimed at reducing France's national debt by the introduction of paper money and the encouragement of trade and investments. Between 1716 and 1720, he created a national bank backed by the king, established the Company of the West—a trading company that operated a monopoly on foreign and colonial trade—and nationalized the collection of taxes. These steps pumped bank notes into the economy, encouraging wild speculation first and creating great inflation soon afterward. The so-called Mississippi bubble peaked in 1720, when the regent appointed Law as the controller general of finances, and the bubble burst later that year.³⁵ As we see in detail in the next chapter, Law also advanced a plan for the development of French manufactures. The Regent's Survey offered quantitative data for pursuing that goal.

THE POLITICAL ECONOMY OF METALLURGY

Réaumur intended his experimental work on steel and cast iron for implementation at the production level. In line with his vision of a science of the arts, the essays he produced on the topic did not consist of mere descriptions of artisans' practices and techniques; they offered concrete proposals for setting up a manufacture of steel and cast iron, based on new principles:

> I admit, and how could I help admitting it, that most of the observations and reflections in these essays cannot benefit the ordinary artisan. They are almost exclusively for those who are capable of managing a business enterprise, who can put artisans to work, just as artisans put their tools to work. . . . It is necessary to enable those who are capable of seeing it through to build the establishments in which this [making steel and soft cast iron] is done, and I dare say that my essays are sufficient for that.³⁶

Réaumur collected his studies in *The Art of Converting Wrought Iron into Steel*, published in 1722. As he explained, there were several reasons that propelled him to publish the work as a freestanding volume, before the completion of the *History of the Arts*. He had presented his findings at various public meetings of the Académie des Sciences and, since then, "a number of workshops have already been started in the Kingdom. . . . There has been a clamoring for a selection of fine domestic steels; all the people would have liked to see the shops all filled with the most perfect cast iron products."[37] Hence, Réaumur hastened the preparation of the volume, postponing his updates and corrections until the publication of a *New Art of Making Steel and Cast Iron*.[38]

Despite its title, *The Art of Converting Wrought Iron into Steel* did not describe any existing art. It was the description of an art that could be. It contained discussions of several experiments that Réaumur had carried out in order to devise new, effective methods for converting iron into steel and for softening cast iron. And, most importantly, it was an example of how the *History of the Arts* could foster technical and economic advancement. Réaumur presented the descriptions of artisanal practices, required for all projects in the history of the arts, dispassionately, as necessary fieldwork that informed and supported more sophisticated laboratory analyses. In the laboratory, he tested new techniques that improved on artisans' routines, based on his investigations into the natural properties of metals. The novelty of his work consisted in offering methods for making steel and malleable cast iron that were economically profitable to entrepreneurs in France. In calculating the economic advantages, mathematical computations and natural history demonstrated that some mercantilist dogmas required revision. For example, imports should not always be seen as a loss: "Should it be necessary to resort to the irons of Sweden [for making steel], the inconvenience would not be great, as it can be acquired in our ports at approximately the same price as our own."[39] Yet the work Réaumur carried out on the samples sent from the provinces demonstrated that the French did not need Swedish iron at all; they had plenty of their own. His metallurgical account of France emphasized the prospects of producing entirely domestic steel.

> I have tried the irons of several forges in Berry and have been very successful with them. I am experimenting with good iron from Nivernais . . . which I was able to convert into steel. There is no better steel than the kind I have made from the irons sent me from Vienne in Dauphiné, which came from the raw iron of Bourgogne. The Lavard irons, also from Dauphiné, were a complete success. There is a forge in Painpont in Brittany whose iron, which was sent to me, has made good steel. Of all the irons of Angoumois, I have tried only that from the forge of Rancogne . . . , an iron which can be safely converted into steel. . . . I have experimented with iron from the Roc forge in Périgord, which has proved to be good. I know of no iron more suitable for being converted into steel than that produced from the ore of Biriatou, from around Labour, near Bayonne. . . . I could continue for much longer with this enumeration; but it is enough to convey the idea that in most provinces of the Kingdom good steels can be made.[40]

Réaumur advanced a proposal for making domestic steel by enumerating the economic advantages of establishing a manufacture for converting iron into steel based on his principles. He factored in the cost of materials, labor, heating, transportation, and building furnaces. The regent deeply appreciated Réaumur's work, and in 1721 endowed him with the annual amount of twelve thousand livres in recognition of the merit of his research. Réaumur accepted, under the condition that this sum would be transferred to the Académie upon his death. This move, which he emphasized in his published works, manifested his strong commitment to the *History of the Arts* as a project in political economy.[41]

Because *The Art of Converting Wrought Iron into Steel* had been published as a freestanding volume, Réaumur warned that "some parts may seem insufficiently explicit or even obscure." He then explained how the work would have figured within a multivolume *History of the Arts* published by the Académie. His views on the *enchaînement* of the arts were similar to Des Billettes's: The arts would be grouped according to the materials they worked, and the progression from one art to the next would be organized by degrees of complexity. So, the work on iron and steel would have been sandwiched between the description of iron ores and the methods for smelting them and the description of the arts that utilized these materials, such as the arts of making anchors and cannons.[42] Regardless of any shortcomings, however, *The Art of Converting Wrought Iron into Steel* made clear how Réaumur's conception of the natural history of the arts would benefit the French state.

Réaumur's vision for the *History of the Arts* was intertwined with his ideas on the role that the Académie should play in the administration of the state. Most likely when the Regent's Survey neared completion, Réaumur penned a document tellingly entitled *Reflections on the Usefulness That the Académie des Sciences Could Have for the Kingdom, If the Kingdom Gives It the Support It Needs*.[43] The document analyzed the causes of the Académie's decline since its establishment. Chief among them was the lack of state funding, a concern that Bignon had expressed earlier.[44] Invoking the unwavering support that the Académie had received from Colbert, Réaumur advanced the idea that all academicians should be salaried and that the state should give them "tasks that oblige them to instruct themselves in what is practical."[45] Thus, savants would work on the specification of rules for technical processes that would replace inefficient hearsay in artisanal practice. Conversely, their visits to artisans' workshops would make them "think better" and direct their reasoning to practical matters. But the most relevant result was the new role of the Académie as a reserve of strategic personnel for the state.

One should make a rule . . . of always giving to academicians the position of director of the Mint, like Mr. Newton, who has it in England; of entrusting them with the positions of inspectors of the various manufactures, with the general inspection of roads, bridges, and carriageways; would it be too much if one gave occasional admittance to the Bureau of Commerce . . . to the savants . . . that occupy themselves with learning thoroughly about the manufactures of the kingdom.[46]

The functional connection between the state and the Académie would also facilitate the production of the *History of the Arts*. As inspectors of manufactures, the authors would have free access to workshops to learn all they needed.[47] As directors of the mint, or as expert consultants for the Bureau of Commerce, on the other hand, they would embody the connections between the world of learning and the world of trade that were necessary for the progress of the French state. Improvement was precisely what Réaumur's "science of the arts" intended to achieve.

Réaumur's vision for the relationship between the Académie des Sciences and the French state started to become a reality in the years following the Regent's Survey. Members of the Académie were regularly consulted by various administrative offices of the French state and offered official positions as inspectors of manufactures. The mint, instead, constituted an interesting exception, where conflicting approaches to metallurgical knowledge coalesced in the name of the pursuit of profit. Independently of one's belief in chrysopoeia, the mint was the place where metals and profit coexisted in a very material alliance.

The Paris Mint was a citadel of metallurgy. Located in the Hôtel des Monnoyes, this state manufacture included various laboratories for smelting, casting, punching, assaying, measuring, and all the other activities related to minting coins. The Hôtel des Monnoyes was the property of the king, who granted lodging there to the mint director, his family, and all workers under the director's supervision. Among them, the engravers enjoyed special privileges; they could pass their positions on to their children for six hundred years. The hôtel housed the standard weights and measures, and the office of the receveur au change, who exchanged foreign currency (which was then recast as French coin). It was the official address of the Cour des Monnoyes, a court of magistrates that adjudicated all cases of counterfeit money and oversaw all the guilds of arts and crafts that worked precious metals, from goldsmiths and silversmiths to clockmakers and mathematical instrument makers, locksmiths, and apothecaries.[48]

At the mint, metallurgical expertise yielded profit. Seigniorage, the main source of revenue for the mint and its personnel, depended on the difference between the face value of coins and the cost of minting (which included the value of the precious metals). The director, who participated in decisions about minting processes and metal alloys, had an obvious interest in being up to date with metallurgical techniques. Knowledge of metallurgy, however, contributed in several other ways to increase his personal gains. Gold and silver coins were supposed to have standard weights and fineness, yet the king allowed for some deviations from their nominal values. The director could mint gold or silver coins that weighed slightly less than expected. He could—and was expected to—mint gold and silver coins whose fineness was inferior (within established limits) to the standard set by the king. These deviations generated revenues for the king, who then gave the director a percentage. These practices established a strong connection between the director of the mint and the king; however, the director could easily exploit his expertise at the expense of the metallurgically novice king. It was the responsibility of the general controller of the mint to prevent any wrongdoing on the part of the director. However, the fine measurements involved in these processes were often imprecise, and the director could discreetly take more than his due. This was so common that in 1719, the regent signed an edict that exonerated the directors of all French mints from the charge of illegally appropriating more than their due share.[49]

The regent had strong interests in the mint, and the fact that he did not choose an academician as the new director, as instead suggested by Réaumur, speaks of the ambiguous relationship of the Académie des Sciences to alchemy. Larry Principe has shown that, in spite of the Académie des Sciences's public rejection of alchemy, several academicians engaged in transmutational work during the eighteenth century and, in particular, in the 1720s.[50] This was the time when, according to contemporary gossip, the regent reverted to his interest in alchemical transmutation. Now that Law's attempt to turn gold into paper had failed, it made sense to foreign observers that the ancient desire of creating gold from base metals should haunt the regent's mind. In 1722, the German doctor Johann Thomas Hensing related that "the current Regent has himself requested the members of the Académie Royale who apply themselves to chemistry to seek out the Philosopher's Stone."[51] Around the same time, even the English ambassador in France was eager to record the regent's interests in alchemy. The regent's revived interest in alchemy in the aftermath of the collapse of Law's scheme seems to find confirmation in the appointment of Mathieu Renard du Tasta as the new director of the mint. In Paris, there was no better place than the mint to experiment on metals.

Du Tasta's interests in alchemy and metallurgy are well documented. His library included key titles of the alchemical corpus, from Paracelsus to the Petit

Fig. 2.2. *Le chimiste peint par David Teniers et terminé par Lorieux*, seventeenth century. [U.S. National Library of Medicine, New York]

Albert.[52] His alchemical work also informed his collecting tastes: in his apartments at the Hôtel des Monnoyes he displayed *The Chemist*, by the Dutch painter David Teniers, a painting the regent acquired following Du Tasta's death—underscoring their common interest in alchemy (fig. 2.2).[53]

During the years of Law's system and the fluctuating fortunes of metal coins, the Paris Mint was under extraordinary pressure. In his attempt to concentrate and centralize all the revenues of the French state, Law had incorporated the mint within the Mississippi Company. The scheme included a plan for the establishment of a larger building for the mint at the Pépinière du Roule, a large estate owned by the king (fig. 2.3). In its new location, which would more comfortably accommodate large machines and the animals that operated them, the mint would additionally carry out the minting of medals (traditionally carried out at the Louvre). Although the collapse of the system halted the plan, Du Tasta obtained permission to use the area where some of the equipment had already been moved. The Pépinière du Roule was conveniently distant from the center of Paris and would have allowed for alchemical experiments

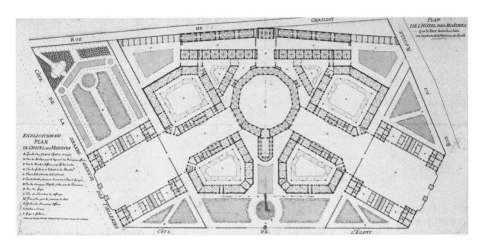

Fig. 2.3. Plan of the new mint at the Pépinière du Roule.
[Bibliothèque Nationale de France, Paris]

to be carried out far from indiscreet eyes.[54] Transmutational alchemy was by no means the only kind of metallurgical work that took place at the mint. Du Tasta experimented with Réaumur on methods to extract gold from the wastes of goldsmiths' and mint workers' workshops. But he also provided access to the mint to alchemically minded academicians such as Charles de Cisternay Dufay, who experimented there on the purification of gold.[55]

Although the relationship between the Paris Mint and the Académie des Sciences differed from Réaumur's recommendations, the mint was crucial to his attempt to demonstrate that the *History of the Arts* was a project in political economy. The strong links between the regent and the mint officers created a favorable context for the establishment of a manufacture of cast iron and steel that operated according to Réaumur's principles. Du Tasta traded in diamonds and had invested in the Company of the West in its early years, when it was still the Company of West Indies. His wealth during the Mississippi bubble was considerable. He bought positions in state administration for all his younger brothers (whom he had initially employed at the mint) and was generous with the branch of his family that resided in Bordeaux.[56] It is very likely that in 1720, he received timely information about the upcoming financial collapse, as had many other investors who were close to the regent.[57] Du Tasta was certainly not ruined after the collapse. On the contrary, he was ready to seize opportunities for profitable investments. Du Tasta was one of thirteen investors who in 1723 obtained a privilege for establishing at Cosne, in the Nivernais, the Royal Manufacture of Orléans for Making Steel and Malleable Cast Iron. The investors intended to produce ironworks by implementing the principles that Réaumur had laid out in *The Art of Converting Wrought Iron into Steel*. At least eight other investors were connected with the mint, including Du Tasta's brother. The other four investors included state administrators and naval officers close to the regent and to Réaumur. One of them, the chevalier de Béthune, had witnessed the regent's success with the Philosopher's Stone. In addition to the privilege, the Parlement granted them an initial sum of fifteen thousand livres.[58]

FROM WRITING TO MAKING: THE ROYAL MANUFACTURE OF ORLÉANS

The establishment of the Royal Manufacture, with the various difficulties it experienced, offers a particularly rich case for discussing Réaumur's understanding of the relationship between writing and making, or, more specifically, between the *History of the Arts* and the pursuit of improvement. It shows, in particular, that his envisioned science of the arts provided practical directions for the onsite management of new enterprises.

Réaumur's work on steel and cast iron initially addressed the manufacture of small luxury items, such as locks, door knockers, keys, balconies, chandeliers, hammers, garden vases, and other household items usually made of wrought iron or copper. His business idea was to replace these materials with cast iron that had been softened in order to smooth its rough appearance. The softening would allow final polishing. Hence, the artifacts of soft cast iron could be made as aesthetically pleasing as those made of wrought iron or copper, but they would be produced with a fraction of the labor involved in making wrought iron, or at a fraction of the cost of copper. In *The Art of Converting Wrought Iron into Steel*, Réaumur produced an arresting example of the enormous cost savings of malleable cast iron compared with wrought iron. He chose the extraordinarily expensive door knocker on the carriage door of the Hôtel de la Ferté, well known to wrought iron experts as a piece of excellent craftsmanship. Réaumur had it made in malleable cast iron and explained that its total cost was thirty-five livres versus the seven hundred livres of the original (fig. 2.4).[59] The small luxury items business at the manufacture at Cosne was a good opportunity for the investors to conduct metallurgical experiments leading to a more ambitious goal. In Réaumur's vision, the work on these smaller items was preliminary to the development of another manufacture that would produce anchors and cannons.

Anchors and cannons historically had been made of cast iron, a heavy and breakable material. The French had wanted to develop new methods for pro-

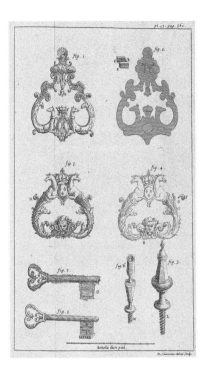

Fig. 2.4. Copy of the door knocker from the carriage door of the Hôtel de la Ferté in cast iron. From René Réaumur, *L'art de convertir le fer forgé* (Paris: Brunet, 1722). [Beinecke Library, Yale University, New Haven, Conn.]

ducing lighter and more durable cannons and anchors since the 1660s. Cosne was home to a manufacture of anchors that the royal investors sought to take over. Indeed, the privilege they obtained gave them monopoly rights over a small area around Cosne where, as Réaumur knew from the Regent's Survey, there were excellent iron mines and several operating forges. Investors such as the chevalier de Béthune and Nicolas Thibault had the military and naval expertise to supervise the developments in this area.[60]

The translation from Réaumur's written instructions to the actual production of malleable cast iron, however, was much more complicated than the investors expected. Pierre Jarosson, one of the investors who was also Réaumur's close friend, moved to Cosne to follow the initial phases of the operation and updated Réaumur with all failed attempts. After several months, he could not hide his frustration, remarking that "of cast iron works, we only have two rosettes as big as a hand."[61] Réaumur, however, attributed these early failures to the amount of time needed to properly train workmen, "who do not educate themselves by books." So he moved to Cosne where he spent several months in 1724 and 1725.[62] He carried out experiments in situ and was eventually successful in obtaining malleable cast-iron samples that he brought back to Paris as tangible instances of how the history and the improvement of arts were interrelated. He eagerly sent samples to his learned correspondents abroad, as material demonstrations of his commitment to useful knowledge.[63] When the malleable cast-iron manufacture was finally in its productive phase, the chevalier de Béthune moved to Cosne to supervise the establishment of the manufacture of cannons and anchors.[64]

The personal connections among the regent, Réaumur, and the investors blurred any possible distinction between public and private interests. In 1725, the Royal Manufacture started to produce the first items of steel and malleable cast iron. The investors then petitioned the Parlement de Paris to grant them an extension of their privilege, from the initial radius of a few miles around Cosne to the entire French territory, though for the duration of ten years instead of twenty. The Parlement, in turn, asked the Académie des Sciences to assess the utility of the investors' request for "the public good."[65] The state institutions that adjudicated the effects of the manufacture on the public good could not be neutral with respect to the private interests of the investors. The investors were too close to the regent for the Parlement or the Académie des Sciences to express an impartial opinion. Jarosson was a member of Parlement and Réaumur's closest friend; Cyprien Bénezet was a member of the Royal Council; the chevalier de Béthune, who managed their interactions with the Parlement, was the regent's *valet de chambre*. The other investors were connected to the mint. Not surprisingly, the Académie's positive evaluation of the Royal Manufacture of Orléans acknowledged the contributions of Réaumur's "excellent treatise" to the state and to the public good. The privilege was registered by the Parlement in 1726.[66]

Réaumur believed that the blurring of public and individual interests was necessary for the successful implementation of the guidelines he articulated in his *The Art of Converting Wrought Iron into Steel*. He articulated this point clearly during a public meeting of the Académie des Sciences in 1726: "The attention and the intelligence of those who undertook it [Manufacture d'Orléans], the attitude I know they have of regarding their own interests as inseparable from those of the public and, something that well deserves to be taken into account, the large amounts that they want to invest, seem to guarantee the good state of our art [of making steel and malleable cast iron]."[67] Improvement was useful knowledge turned productive. The transition could only happen when private and public interests merged, and artisans were disciplined according to the savants' directions.

The Hôtel d'Uzès, the Paris storefront of the Royal Manufacture, showcased the manufacture's contributions to society. Located in rue Saint-Thomas du Louvre, it was a theater of metallurgic virtuosity. There, Réaumur's principles materialized in works of steel and cast iron designed "by the greatest masters."[68] Contemporary magazines reported that the store had been visited by "the most distinguished at court and in Paris," and further that "the beauty and style of the works found there" struck everyone.[69] Like seventeenth-century Wunderkammern with an alchemical laboratory, the cellars of the Hôtel d'Uzès housed a large furnace and several workshops swarming with some forty workmen.[70] These cellar-dwellers gave the final touches to the cast-iron and steel works from Cosne. They filed, chiseled, and turned at the lathe, working under the supervision of Pierre de la Tour, one of the investors and the general controller of the mint.[71] The final products for sale in the store were as "sophisticated and polished as items made by goldsmiths."[72] There were one-piece balconies "enriched by all that Sculpture can make in wood," hammers, door knockers, locks, and chandeliers with decorations that "until now nobody was able, or dared, to try." Objects such as paper weights and vases, now made of cast iron instead of copper, enabled customers to avoid the "hideous smell that [copper] leaves on fingers."[73] The price list of the items for sale at the Hôtel d'Uzès emphasized the difference between the old and the new materials: "To make the public aware of the beauty and inexpensiveness of these works, we will remark that the first of the [listed] hammers that represents two dog's heads [at 35 livres] is the same as the one at the Hôtel de la Ferté, which cost 750 livres."[74]

In assuming that their private interests coincided with the public good, however, Réaumur and the investors overlooked the power of the Paris guilds. The presence of a forge at the Hôtel d'Uzès violated the regulations of the locksmiths, who in 1726 petitioned the palace for a formal inspection. Official police reports confirmed the violation (only locksmiths could work on iron) and formally acknowledged the economic damage that the workshops at the Hôtel d'Uzès could do to the locksmiths. The investors called on their connections in Parlement to outweigh the influence of the guild, but a couple of years later the storefront at Hôtel d'Uzès was shuttered.

The locksmiths were not the only cause of the failure of the Royal Manufacture. The investments that it required had not yielded the economic advantages the investors expected. At the end of 1727, the abbé Bignon sent an alarmed note to Réaumur, who was at Cosne to monitor the establishment of the manufacture of anchors. Bignon warned, "I do not like the news from Cosne; everything there goes too slow and it will not give you the honor you deserve."[75] A year later, Réaumur moved into the Hôtel d'Uzès.[76] It is unclear if some of the equipment employed for finishing the items from Cosne had remained there. What is certain is that, from then onward, he dedicated his attention to natural history. In 1728, one year later, the newly established Société des Arts was ready to resume the writing projects on the mechanical arts, building them on new foundations.

RÉAUMUR'S *HISTORY OF THE ARTS* AND THE ARTISAN

Three decades after the constitution of Bignon's commission, Réaumur still echoed its members' sour remarks over the contemporary disregard for the mechanical arts, within and beyond the Académie des Sciences. In his draft preface to the Académie's *History of the Arts*, Réaumur noted that he could not understand why the Académie had not "made a science of the arts, as we have made a science of geography, history, etc."[77] The public did not demonstrate any interest in the manufacture of objects for everyday use, even though Des Billettes had illustrated that something as trivial as the pin passed "through more than twenty hands" before taking its final shape. The pin was in fact a wonderful example of ingenuity and efficiency. For Réaumur, writing the natural history of the pin was a self-evident argument for the ingenuity of the arts, just as the description of the magnified flea offered a self-evident argument for intelligent design.[78] He noted with disdain that the learned raved over Benoît Baudouin's *On Ancient Footwear*, a treatise on ancient Roman footwear, but did not think much of contemporary shoe making.[79] Réaumur concluded ironically that "1500 or 2000 years give some nobility to things." Nobody objected to the relevance of history to the study of ancient artifacts, but the history of the arts did not elicit the same sort of interest.[80]

For Réaumur, disregard for the mechanical arts was the result of ignorance. In drafting his 1721 (unpublished) preface, Réaumur insisted on the educational mission of the Académie's *History of the Arts*. By illustrating the various stages of manufacturing and by offering a vocabulary with the terms of the arts, the *History of the Arts* would promote technical literacy and emulation. Technical literacy was relevant to entrepreneurs who would learn how to estimate the value of labor.[81] Emulation would result in better artisanal practices and therefore in products of higher quality. Since the *History of the Arts* would divulge best methods and practices, "the good workmen" would gain prestige and income, and the bad ones would be compelled to improve.[82]

The Art of Converting Wrought Iron into Steel was the only portion of the *History of the Arts* to be published in Réaumur's lifetime. Réaumur continued to oversee the Académie's project on the arts until his death in 1757. In 1733, the Prussian traveler Charles-Étienne Jordan was impressed by the large number of "magnificent plates" that Réaumur had amassed. He recorded that the materials, which represented in detail the artisans' tools and their gestures, were part of Réaumur's *History of the Arts and Crafts*, a project that had been underway for some time.⁸³ Réaumur's commitments to other projects are commonly cited as the reason for the delayed publication of the *Descriptions of the Arts and Crafts* (as the *History of the Arts* was known when it was finally published), which began in 1761 in the aftermath of the *Encyclopédie*'s phenomenal success. Scholarship on this subject has assumed that the published *Descriptions* corresponded to Réaumur's plan for the *History of the Arts*. However, Réaumur was as committed as his predecessors to the idea of presenting the multivolume work as an interconnected encyclopedia. It was not because he turned his attentions to other topics in natural history that the *History of the Arts* was not published during his lifetime.⁸⁴

Réaumur's plan all along was to accumulate all the descriptions of the individual arts first, and then to organize the publication according to an ordering principle that was based—as for Des Billettes and Jaugeon—on the natural material each art worked upon. As he explained to the audience that attended the public meeting of the Académie des Sciences in April 1727, this plan entailed a long-term commitment.⁸⁵ The plan took even longer than Réaumur anticipated. When he died, he left behind numerous manuscripts on the mechanical arts, some of which dated from the times of Des Billettes, Jaugeon, and Sébastien Truchet. When the Académie finally began to publish the *Descriptions of the Arts and Crafts* under the direction of Henri-Louis Duhamel du Monceau, the introduction of the plural in the title marked the end of the *enchaînement* of the arts as a precondition for publication. Duhamel still felt the need to explain the reasons for publishing individual, disconnected descriptions of the arts in an essay he read to the Académie:

> Had it been possible to produce simultaneously all the descriptions of the arts, the natural order of their publication would have been to organize them, so to speak, by subject; that is to say, to place in sequence those that could have an essential relationship with each other; but it would have taken too long to publish them this way, and the Académie believed it ought to publish each art separately, as early as it was ready to be published, and without making the page numbers follow from one [volume] to the next.⁸⁶

In 1761, the notion of a "natural order" that connected the arts together was still a valid principle. Duhamel did not hide his preference for an interconnected work, which would present all the arts with the necessary *enchaînement*, yet he also admitted that the endeavor would take too long.[87] It was only the pressure of the moment that induced the Académie to approve the publication of individual volumes on the arts and crafts. A large quantity of the work that had been carried out since the 1690s was now published, often by new authors who simply revised old materials.[88]

Since the times of the Compagnie des Sciences et Arts, the attempts to produce a natural history of the mechanical arts had not contemplated the possibility that artisans could be enrolled as authors. Réaumur reiterated the theme of workmen as obstacles to the encyclopedic endeavor: in a vain attempt to gain importance, they refused to explain their operations and retarded the project.[89] In general, Réaumur regarded artisans as automata who carried out mechanical operations to convert natural materials into artifacts. There was ingenuity neither in their hands nor in their minds. They were working materials to be observed, corrected, and directed—just as natural materials were to be experimented on in order to determine how they could be most effectively exploited. At best, artisans could execute rules conceived elsewhere. If "good willed," they could improve their techniques by following the savant's directions. Yet there was nothing unique in their work. Artisans could be easily replaced. As Christiane Demeulenaere-Douyère and David Sturdy have noted, there is no mention of the dangers of mining work in the Regent's Survey. Although all the operations occurring in the mines were described in great detail, the fact that many miners were injured or even died while performing their work was not relevant enough to be mentioned.[90]

The comparison between Réaumur's consideration of two illustrators he worked closely with illustrates even more clearly his conception of artisans' work. Hélène Dumoustier, the woman who contributed most drawings to his *History of Insects*, was Réaumur's intimate friend and produced this work, in Mary Terrall's words, not as an artisan but as a "lady of leisure." Having spent many hours observing and experimenting with Réaumur, Dumoustier knew, by his own admission, "how to enter into my eyes, she knew and knows how to predict them, since she knows how to recognize what is most remarkable about an insect, and what position it should be represented."[91] Réaumur acknowledged Dumoustier in several circumstances and demonstrated his deep gratitude in his will, which named her his heir and required that she receive "all that laws and customs permit me to give her."[92] Réaumur's interactions with the engraver Louis Simmoneau, who produced drawings and engravings for the Académie's

History of the Arts, were markedly different. In a letter that explained why he would not pay Simmoneau the amount the engraver requested, Réaumur declared that he had asked the engraver to simply deliver the drawing just as "we request a workman whom we ordered some works to deliver them when he has completed them." In his reply, Simmoneau explained that his request concerned drawings that he gave Réaumur in good faith when Réaumur expressed his desire to see them. Réaumur, however, had the drawings engraved without Simmoneau knowing and even made a profit from their sale. Simmoneau remarked that it was not honorable "for Mr. Réaumur to erect himself as superior on this occasion, and to use the comparison with works ordered by a master to a workman, because Mr. Simmoneau is not inferior to him in any respect, being in his own genre an academician like him [Réaumur]."[93] Simmoneau's reply highlights the difference between Réaumur's conception of the engraver and Simmoneau's own. Simmoneau's understanding of himself and his work was similar to that of a number of *artistes* who, at first during the Regency, and then again soon after the failure of the Manufacture at Cosne, established the Société des Arts. As we see in the next section, the Société believed that writing on the mechanical arts was a necessary step for the project of improvement. Unlike Réaumur, the members of the Société were convinced that *artistes*, not academicians, could effectively lead the French state toward technical and economic advancement.

Part 2

The Société des Arts

Chapter 3

Theory, Practice, Improvement

In 1718, the British clockmaker Henry Sully obtained the French regent's permission to establish a Société des Arts in Paris. Gathering at the Louvre, like the most important academies, and working under the leadership of the abbé Jean-Paul Bignon, the Société des Arts assembled a small group of *artistes* who worked toward the improvement of the mechanical arts. The Société realized one of the options that Gilles Filleau des Billettes had discussed with Bignon at the turn of the century, when they reflected on the role of the encyclopedic *History of the Arts* within the renovated Académie des Sciences: the constitution of an independent association that would become an academy of the mechanical arts. There are very few sources that document the activities of this Société des Arts. The most relevant is an anonymous booklet printed in London in 1722, only three years after its establishment: *Three Letters Concerning the Forming of a Society, to Be Call'd the Chamber of Arts*. The booklet was not intended for the broad public; it was printed only in order to circulate rapidly among interested parties and to generate support for the constitution, in London, of an association similar to the Société des Arts. The author of the *Three Letters* admitted that his proposal was modeled on the Société des Arts, whose regulations he used as a blueprint for those of the Chamber of Arts (which remained only a proposal until the establishment of the Royal Society of Arts in 1754, thirty-two years later).[1]

The London booklet therefore offers precious insights into the goals of the elusive Société des Arts that was constituted during the Regency and lasted only a few years. In his outline of the program of the Chamber of Arts, the compiler of the *Three Letters* consistently translated the French *artiste* as "artist." The English language accommodated the French distinction between *artiste* and *artisan*, compensating for the lack of an English term for *esprit* by referring to "ingenuity" as the element that distinguished "artists" from other practitioners of the arts and crafts. French-English dictionaries translated the French *artiste* as "artist," adding that the term meant "an ingenious workman." By contrast, they offered a plethora of possibilities for the French *artisan*, ranging from craftsman to tradesman, as well as workman, artificer, author, contriver, and finally artisan.[2] The word choices of the author of the *Three Letters*, then, demonstrate his desire to differentiate those artisans endowed with intelligence, wit, politesse, and discernment from all other figures that inhabited the composite world of the arts and crafts.[3]

The *Three Letters* characterized the Société des Arts and the Chamber of Arts as associations that contributed to improvement, or in the words of the anonymous author, the "bettering of posterity." Improvement, however, was not conflated with innovation. The strategy through which these associations purported to improve the human condition was not only by the encouragement of new inventions but also via the preservation of technical knowledge. The members' collective work would focus especially on the publication of "registers" that recorded and made available the best processes and machines used in all arts, trades, and manufactures. These registers would also serve as valuable instruments for stimulating innovation by offering to future inventors a "continual fund" of information, directions, and inspirations.[4] The *Three Letters*' insistence on the compilation of "registers" appropriated the rhetoric of the savants who had worked on the *History of the Arts* in the previous decades. By capturing in texts and images the best artisanal practices, the double goal was that of preventing the loss of technical knowledge and of inspiring artisans to imitate best practices. The broader diffusion of state-of-the-art knowledge would result in better average quality of artisanal products, a desirable achievement in the mercantilist framework within which these associations operated.[5]

At the same time, and more importantly, publications of this kind defined the state of the art. The registers that the authors of the *Three Letters* envisioned, and therefore the writing projects of the Société des Arts, were not conceived as a representation of the myriad techniques that artisans employed to make specific artifacts. They were intended as a normative selection of what the editors considered the best practices in use. By describing methods, tools, and procedures, they enunciated the rules that defined each art. If these publications could boast of a royal endorsement, as in the case of the works published under the banner of the Académie des Sciences, they would enjoy unparalleled reputation and prestige. In addition to compelling artisans and workmen to emulate best practices, they would play an important role in directing the taste of elite consumers, thus exerting an enormous influence over the broader public. Réaumur's attempt to establish new rules for the manufacture of goods exemplifies the normative character of publications on the arts and their tremendous effect on the entrepreneurial elite. The fact that the author of the *Three Letters* regarded the compilation of the registers as a defining feature of the Chambers of Arts, as it was of the Société des Arts, points to the *artistes*' ambition to take the definition of the state of the art into their own hands.

This ambition, the author of the *Three Letters* emphasized, did not overlap with the activities and goals of existing institutions devoted to the advancement of natural knowledge, such as the Royal Society of London and the Académie des Sciences in Paris. The Chamber of Arts was conceived as a society "for the preserving and improvement of operative knowledge, the mechanical arts, inventions, and manufactures." As such, it was predicated on the Baconian difference between "speculative" and "operative" knowledge, which, the *Three Letters* explained, were defined, respectively, as "the inquisition of causes" and "the production of effects."[6] Natural philosophy inhabited the realm of theory, the search for causes, and the pursuit of philosophical truth. The mechanical arts inhabited instead a world of improvement and productivity. This was a separate, if neighboring, sphere of operation, characterized by a different moral system where the mantra of a disinterested pursuit of truth lost its edifying value.[7] Monetary profit—taboo in the world of the savants—was one of the benefits that members of the Chamber of Arts would enjoy: "If [members] produce any thing useful and beneficial to mankind, either by their ingenuity, labour, or expense, they shall be entitled to a proportionable share of the advantage arising thereby." The project, in other words, emphasized the collective nature of inventions and supported a reward system that distributed revenues among inventors and investors.[8] This idea stemmed from the consideration that without prospects of "private advantage, some recompense for their labours, and return for their expences," members would not be motivated to share their expertise.[9] The economic rewards proposed for the members of the Chamber of Arts were not to be conflated with the selfish pursuit of profit that had haunted the morality of artisans in the public sphere, as highlighted by Nicolas Guérard's print (fig. I.2). The project of an association dedicated to the mechanical arts pivoted around the idea that individual and collective profits should be one and the same thing. The difference between speculative and operative knowledge, then, came down to the fact that only the latter offered concrete contributions to the common good. Improvement and useful knowledge fell in the field of *artistes*, not in that of savants.

The *Three Letters* specified that Francis Bacon himself judged "it most requisite that these two parts [of knowledge], *Speculative* and *Operative*, be separate." As Bacon had indicated, the Chamber of Arts would take up what the Royal Society had left out: while the Fellows of the Royal Society were "enquirers into causes," the members of the Chamber of Arts were "producers of effects." If the former were "recommenders of curious speculations, or censors of philosophy," the latter were "the preservers of useful arts and inventions."[10] Therefore, the two associations would complement each other. It is likely, then, that the relationship between the Société des Arts and the Académie des Sciences, both of which operated under the directorship of the abbé Bignon, was intended to be one of complementarity, as proposed in the *Three Letters*.

The relationship between theory and practice, the mind and the hand, and science and technology has been the focus of scholarship that has forcefully argued for historicizing such dichotomies. Blurring the boundaries between such opposites, historians of science have challenged the association of savants with the world of learning and artisans with the world of making.[11] They have shown that the artisanal world was a site of knowledge production where the ability to manipulate materials and to make things provided an intellectual foundation for understanding nature and its laws. In the making of natural knowledge, artisans and artisanal expertise played a significant role.[12] The focus of these historical studies has been the epistemological foundations of "natural philosophy," or the investigation into the causes of natural phenomena. The establishment of the Société des Arts and the project of the Chamber of Arts offer the opportunity to shift perspective and examine an emerging discourse on technology, intended as the systematic study of manufacturing processes. Hélène Vérin has dated the emergence of technology as a separate field of systematic inquiry into the world of the mechanical arts to the 1780s, yet a discourse on the distinctive features of this kind of inquiry was already taking place in the 1720s.[13] By focusing on the ways in which *artistes* understood and articulated the difference between theory and practice, we can detect a discourse that defined the mechanical arts as a domain of expertise poised to contribute to the public good through the pursuit of commercial and imperial expansion. Although neither the Chamber of Arts nor the Société des Arts intended to challenge the authority of existing institutions dedicated to the advancement of natural knowledge, they characterized the expertise of the Royal Society or the Académie des Sciences as too theoretical to effectively serve the imperial and colonial interests of the state. Conversely, they presented their own practical knowledge as a separate sphere of action where public and private interests coincided.

By examining the work of the *artistes* involved in the Société des Arts, then, the relationship among *artistes*, the mechanical arts, and the state comes sharply into focus.[14] The Société des Arts expressed the *artistes*' ambition to participate in state administration at a time when the reorganized Council of Commerce drew expert opinion on technical matters from the Académie des Sciences. Although the reorganization of the council demonstrated a renewed interest in commerce in the years of the Regency, its preference for the savant academicians excluded *artistes* from formally contributing to the political economy of the state.[15] Through the constitution of associations dedicated to the mechanical arts, *artistes* aimed to attract the attention of state administrators and the general public alike to the public potential of their expertise. In so doing, they constructed a difference between theory and practice that highlighted their knowledge and problem-solving abilities as essentially different from that of academic savants or other members of the artisanal world. *Artistes* highlighted the boundaries of natural philosophy—which philosophers like Bacon had delineated—in order to claim for themselves the task of guiding humankind toward technical and economic progress.

THE PARIS-LONDON CONNECTION: HENRY SULLY AND JOHN LAW

The Chamber of Arts and the Société des Arts may both have been Henry Sully's projects.[16] Sully's trajectory exemplifies the connections among the mechanical arts, manufactures, trade, and the state that formed the foundation of the *artistes*' political and cultural project. After his apprenticeship in London, Sully traveled throughout Europe and settled in Vienna as the personal watchmaker of the Duke of Arenberg, a French aristocrat and military officer. In Vienna, Sully published *The Artificial Regulation of Time* (1714), a work that highlighted the importance of horological knowledge for the educated. The book anticipated several motifs of the *Three Letters*. It presented the learned watchmaker as an *artiste* who possessed the intellectual and manual skills to translate the abstract notions of astronomy, mathematics, and physics into a working device. The learned watchmaker did not search for the causes of phenomena (which is what a philosopher would do) but offered solutions to practical problems. A good watch resulted from mastery of both practice and theory because the learned watchmaker was as skillful with his hands as he was quick with his mind.[17]

Thanks to the patronage of the Duke of Arenberg, Sully established connections with learned circles in Vienna and met Gottfried Wilhelm Leibniz, who acted as his sponsor and patron. *The Artificial Regulation of Time* included Leibniz's favorable review of Sully's work. When Sully moved to Paris with Arenberg, Leibniz recommended Sully to his friend and correspondent Nicolas-François Rémond, the first counselor to Philippe d'Orléans, the regent of France. Leibniz suggested that Sully should be made a member of the Académie des Sciences, where he could complete the encyclopedic work on watchmaking he had already planned. Leibniz firmly believed that Sully's inventiveness would greatly contribute to horology.[18] Not surprisingly, given his status as a craftsman, Sully was not admitted to the Académie. Nonetheless, in 1716 he presented to the Académie des Sciences a watch that he had modified so as to reduce friction among its parts. Because friction was the most significant source of watches' irregularity, Sully's was an important improvement, and the Académie praised it as a demonstration of his "intelligent discernment" (*sagacité d'esprit*).[19] Sully's watch attracted the attention of the regent, who introduced him to John Law, the minister of finances and, like Sully, a British émigré.[20]

That a watch could so rapidly elevate Sully's career should not be a surprise. Watches and clocks proved crucial to several branches of natural knowledge as well as in human affairs. Pendulum clocks had changed the practice of astronomy and, by extension, cartography and geography, while portable watches and clocks offered new possibilities for the military arts. Lack of synchronization, when no standard time was available, proved notoriously disastrous in war. The watchmaker at the service of an army officer was a strategic figure who repaired and adjusted watches to ensure that coordinated actions would happen at the planned time.[21] Watches also seemed to promise a solution for the problem of finding longitude at sea, a pressing concern for colonial and commercial powers seeking to expand their domains through navigation. In the late sixteenth century, naval powers such as Spain and the Netherlands had offered prizes to stimulate human inventiveness in this field. The prizes, however, had never been granted for lack of viable proposals. So elusive was the solution that it prompted satirists to liken finding longitude at sea to chimerical investigations: squaring the circle, finding the Philosopher's Stone, producing perpetual motion, or predicting the future.[22] Meanwhile, deadly shipwrecks continued to be attributed to the lack of reliable methods for finding longitude at sea. Most notably, in 1707 four warships of the British Royal Navy went missing off the Isles of Scilly near Cornwall; thousands died, and Britain endured significant economic losses. In 1714, the same year in which Sully published his book, the British Parliament issued an act that established that a prize of up to twenty thousand pounds (a sensational amount at the time) would be awarded to the inventor of a reliable system for finding longitude at sea. A special committee, the Board of Longitude, was assigned to arbitrate the contest. Isaac Newton, a member of the board, believed that the most promising methods were astronomical and horological.[23]

The horological method relied on the fact that the time difference of one hour between any two locations on the globe corresponded to a difference of fifteen degrees of longitude. This was due to the rotation of the earth around its axis. In order to find longitude, all that was needed was a timekeeper that would always tell the time of a reference location, together with astronomical instruments to tell local time. The resulting time difference would then be converted to degrees of longitude. The main problem that this method presented was the unavailability of timekeepers that continued to perform as desired in spite of irregular movements on board the ship and changes of temperature and humidity across the globe.

Sully had been convinced that clock- and watchmaking offered the solution to the problem of longitude since the end of his apprenticeship, at the beginning of the eighteenth century. While in London, he had worked on the arrangements of watch components so as to improve watch regularity and had even met Newton to discuss the various techniques.[24] In France, he collaborated with the watchmaker William Blakey, another British émigré who specialized in making steel springs for watches. Blakey's deep knowledge of steelwork was very important for Sully's commitment to improving watch technology. The irregular movements of a ship at sea affected pendulums far more than spiral springs, even though the spring's oscillations changed with temperature and were not as regular as those of pendulums.[25]

The frontispiece of Sully's *Artificial Regulation of Time* was a bold statement of his belief in watch technology (fig. 3.1). The image portrays Sully comfortably seated on a pendulum clock, handing the project of a watch over to the personification of Time while pointing to a group of putti engaged in telling time by means of astronomical instruments. This was entirely consistent with the contents of the book, which explained the relationship between time measured by astronomical observations (solar time) and time measured by clocks and watches (mean time). There was probably another, subtler, message in the frontispiece. The project that Sully was offering Time likely represented the marine watch he hoped to perfect in order to solve the problem of longitude. The context of navigation is evoked in the illustration through a telescope, compass, and what appear to be ship's log books on the ground. In 1726, when Sully became convinced that he was very close to making timepieces that could be used at sea to calculate longitude, he announced his plan to publish an entire volume with descriptions of such machines.[26]

Fig. 3.1. Frontispiece from Henry Sully, *Règle artificielle du temps*, 1714. [Österreichische Nationalbibliothek, Vienna, Austria; © ÖNB Vienna: 51.X.16 Alt, Frontispiz]

Sully's commitment to watchmaking and his connections with London artisans were a perfect fit for John Law, who was eager to implement his plan to establish new manufactures in France. Law's financial scheme has been widely analyzed, but his interest in the world of manufactures has received scant attention.[27] Law's financial reflections relied on a comparative and historical analysis of trade and manufactures in various European countries that he had presented to the French general controller of finances in 1715. Law's analysis focused in particular on Holland and Britain, and it argued that their commercial superiority was due to their strong fleets and successful naval trade. Referring explicitly to Jean-Baptiste Colbert's plan for the encouragement of French trade, Law proposed a program for the development of new manufactures and the improvement of navigation that would complement the establishment of a national bank. Unlike Holland, Law remarked, France abounded with raw materials and workers. It was therefore imperative, according to the basic principles of mercantilism, that France should minimize exports of raw materials and maximize exports of finished goods. To this end, Law believed it crucial to establish in France industries styled on successful foreign models.

In Law's analysis, skilled workers were essential resources for advancing commerce. The wars of religion of the sixteenth century had had a catastrophic impact on French trade, not only because of their cost. One of the consequences of those wars, he explained, had been the revocation of the edict of Nantes in 1685, which caused Huguenot artisans to leave France, bringing their expertise to foreign countries. It was now essential, according to Law, to start a new policy aimed at importing into France skilled foreign workers, regardless of their religion. Law acknowledged that Colbert had already initiated this plan, which was interrupted by the war of 1672. He intended to resume it, and Sully was the perfect go-between for coordinating operations on the British front. He was familiar with watchmakers in London and was eager to demonstrate his worth to the French state.[28]

Law planned to cultivate in France the areas in which the British trade was particularly robust: steel making, wool textiles, flint glass, iron works, watchmaking, and naval design. The latter two activities were particularly relevant to his vision of promoting French colonial trade through the establishment of a large holding company. The Mississippi Company would merge the three French trading companies operating in Asia, Africa, and America.[29] Watches promised a solution to the problem of finding longitude at sea, and the new technique of bending timber by steam heat offered new possibilities for naval design. In collaboration with Law's brother William, who was in London at the time, and with William Blakey, Sully traveled to Britain where he selected two hundred workers to bring to France.[30] With the support of the regent, Law offered the communities of immigrant British workers special allowances: they were granted religious freedom, a minister of the Church of England, and the suspension of the *droit d'aubaine*, according to which the estates of non-naturalized foreigners reverted upon death to the French Crown. Within a few months, Law employed the British workers in manufactures that he set up throughout France: a manufacture of wool at Charleval (Provence) and another on his estate at Tancarville (Normandy); one of steel- and ironworks at Harfleur (Normandy), directed by Blakey; and a foundry at Chaillot (near Paris).[31] He also established a Manufacture of English Watches at Versailles and made Sully its director.[32] The operation coordinated by Sully was so vast that it alarmed the British government; in response, in 1719 the British government issued its first legislation prohibiting the emigration of skilled workers.[33]

The Société des Arts was created in 1718, a few months after Sully successfully concluded his mission. Just as the Académie des Sciences had emerged from Colbert's "Memorandum on Trade," the Société des Arts seems to have arisen from Law's plan for the development of French trade. The few sources that document its existence describe a group that worked on the areas at the core of Law's plan for reinvigorating France's international and colonial trade: geography, cartography, and navigation. A tentative list of members that we can extrapolate from later sources shows that, in addition to Sully, the Société likely included the clockmakers Julien Le Roy and Henri Enderlin; the maker of steel springs for watches William Blakey; the map engraver, lecturer, and royal geographer Henri Liébaux; the mathematician Jean-Baptiste Clairaut (a lecturer of mathematics at the Académie du Roy); and possibly the royal goldsmith Thomas Germain.[34] The areas of expertise of these members suggest that horology and its applications were of primary importance for the Société des Arts. Horological advancements depended on geometrical models, to which Clairaut could have contributed, and on the use of frictionless materials, a subject on which Germain could have advised. Diamonds and rubies, as Sully learned from Newton, were particularly useful in reducing friction in watch mechanisms. Sully was also convinced that gold chains could be usefully employed to increase the regularity of watches.[35] Mapmaking and geography were intertwined with navigation, areas in which Liébaux would have offered expertise. The attempt to find a solution to the problem of longitude at sea through horology also seems to link the Société des Arts with the Manufacture of English Watches at Versailles, directed by Sully and supported by Law.

THE MANUFACTURE OF IMPROVEMENT

In 1719, at the apex of his popularity, Law commissioned a plafond (a decorated ceiling) for the Royal Bank he directed. Now the general controller of finances—the position formerly held by Colbert—he had guided France through a few years of unprecedented financial excitement. In the style of contemporary allegories glorifying monarchs, the now-lost ceiling celebrated the bank's prosperity and associated it with the wise governance of the regent and the infant king. Amid several groups of human figures personifying the virtues of trade and the manufactures that flourished under an enlightened monarch, a putto representing Genius guided Commerce, supported by Invention, Arithmetic, and Industry, and followed by Wealth, Security, and Credit.[36] With this stunning visual statement, Law emphasized the significance of his Royal Bank to the history of France. He acknowledged the crucial role of invention and industry in supporting the mind (genius) as it guided commerce. Wealth, security, and credit—the most immediate outcomes of thriving commerce—owed as much to the mindful management of manufactures and the arts as to the arithmetical mastery of state finances. Law's role in the establishment of French manufactures modeled on successful British ones was a case in point.

The majority of British workers (about seventy) who settled in France under Sully's auspices specialized in watchmaking. Most of them were employed in the Manufacture of English Watches at Versailles, which relied on the steel springs produced at Harfleur by the manufacture directed by Blakey.[37] The Versailles manufacture epitomized an alliance among people possessing different types of capital—financial, political, artisanal, and managerial—and it had been fostered by the relationship between an entrepreneurial artisan and a statesman investor. This was precisely the model of operative knowledge that the anonymous advocate for the Chamber of Arts wished to implement in London. Watchmaking required proficiency in mathematics and physics, knowledge of materials, mastery of cutting machines, and skill in working with tools. But in addition to the combination of theoretical and practical knowledge, it also required, on the part of the director, the ability to supervise all the processes of production. Contrary to the views of savants like Réaumur, Sully believed that improvement could occur in any and all the phases leading to a watch. It was therefore essential that the director of the manufacture should be able to identify machines or processes to be improved.

Sully conceived of the Manufacture of English Watches not just as a production site but above all as a research and learning center. As outlined in the anonymous *Three Letters*, improvement was a project that combined learning, education, and experimentation. Similarly, the manufacture was a place for improving the art of watchmaking in France. The British artisans at Versailles were employed with the understanding that they would teach their methods and practical skill to their French counterparts by offering demonstrations and lectures. The courses offered at Versailles were open to any French watch- or clockmaker. Thanks to this policy of openness, the manufacture became an axis of socialization—not of competition—between the watchmakers of Paris and those at Versailles.[38] We know, for example, that Julien Le Roy, a highly regarded Paris watchmaker, attended these gatherings.[39]

The Manufacture of Watches at Versailles merged collective and individual interests, as did the Royal Manufacture of Orléans for Making Steel and Malleable Cast Iron established at Cosne on Réaumur's principles a few years later. It proved profitable for Law and Sully but, more substantially, it set new, higher standards for watchmaking. It soon became an exemplary model that stimulated the ingenuity and industry of local watchmakers. French watchmakers, both masters and workmen, learned the most efficient techniques and applied themselves to improving the art. An adulatory article in the 1719 issue of the *Mercure de France* remarked that the "benefit of the state" was the main objective of the manufacture.[40] Several decades later, Parisian watchmakers acknowledged the manufacture's role in advancing the quality of French watchmaking by encouraging best practices. Emulation, they admitted, had been a significant engine of innovation.[41] Although it is difficult to assess the actual impact of the manufacture at Versailles beyond the acknowledgments of later Parisian clockmakers, it is a matter of fact that France became a leading power in the manufacture of watches and clocks (especially ornamental clocks) during the eighteenth century.[42]

There was a profound epistemological difference between Réaumur's vision for the manufacture of steel and malleable cast iron at Cosne and Sully's vision for the manufacture of watches at Versailles. It was a difference that was rooted in their contrasting understanding of artisanal knowledge and the role it should play in the project of improvement. Réaumur experimented on materials in his laboratory, and it was there that he articulated the directions that artisans and workers in the manufacture should follow to produce malleable cast iron. For him, any worker should be instructed in the processes of production. When confronted with the difficulties of converting laboratory practices into large-scale production processes, Réaumur moved to Cosne to offer his expert supervision. In other words, the manufacture at Cosne was conceived as a savant-directed enterprise. The Manufacture of Watches at Versailles, in contrast, was a place where all workers were encouraged to perfect their art. It was from within the manufacture, in the very process of making, that improvement would emerge.

A research and learning center as well as a profitable business, the Manufacture of Watches demonstrated that *artistes* such as Sully could effectively contribute to the economic advancement of the state. Whereas Réaumur's vision implied a hierarchical relationship between savants and artisans, Sully's project was built on the autonomy of *artistes* from savants and the *artistes*' close collaboration with state administrators. Sully had selected the workmen who operated the manufacture and brought them to France. Workers, for him, were not replaceable automata. On the contrary, artisanal knowledge resided in their bodies and could only be passed on through a learning-by-doing approach, typical of artisanal apprenticeships. Embodied skill was so precious as to be worth the cost and energy of the operation that Sully had coordinated.[43] Like the loyal "mystery men" described in Bacon's *New Atlantis,* who searched for secrets of the arts in foreign countries, Sully also acted as an intelligencer, a manager, and an entrepreneur. In other words, he was an expert in operative knowledge—the very knowledge the state needed in order to implement Law's plan for reviving French manufactures.

The watches made at Versailles were material representations of the social, political, and epistemological views of Law and Sully. They demonstrated that *artistes* had much to offer to the state. They spoke of the *artistes*' ability not only to design and execute machines and devices but above all to conceive, plan, and manage manufactures that contributed to the wealth of the nation. In 1719, when the Manufacture of Watches at Versailles peaked in popularity, Law, in his role as the general controller of finances, implemented drastic measures to discourage the accumulation of precious metals and to simultaneously promote the circulation of paper money.[44] The manufacture was integral to the plan. At the time, gold and silver specie that no longer circulated as currency could be sold to the Paris Mint. However, the price offered in Holland for the same amount of precious metal was higher. Those who could export out-of-circulation specie did so, with the undesirable result that large quantities of precious metals left the country. The manufacture at Versailles constituted a viable alternative to the Dutch market. It accepted old coins that it melted and converted into the fine components of luxury watches, thus keeping the precious metals within France. With the escalating stylishness of these watches, this option seemed an easier investment and became popular.[45]

Law and Sully presented the first watch produced at Versailles to Louis XV, who received them both at court and praised Sully for all he had done to make the manufacture a reality.[46] The gift of the watch to the king was more than routine homage to the monarch. As material evidence of Sully's ability to set up and run a manufacture dedicated to the perfection of the art, it was an implicit request to acknowledge the role that *artistes* like him could play for the state. The establishment of the Société des Arts in the same year as the Manufacture of Watches suggests that Sully's proposal for an association dedicated to the mechanical arts was endorsed by the regent. However, the fact that its leadership was entrusted to Bignon indicates that Sully's program was accepted with some reservations. Unlike the Académie des Sciences, the Société des Arts maintained a low public profile, operating in a regime of quasi-secrecy.[47] Indeed, the Société seems to have acted as the Académie des Sciences's shadow institution—a role that their common location at the Louvre highlighted.[48] If Sully the *artiste* was rewarded with the establishment of the Société des Arts (whose leadership went to Bignon), Law's contributions to the improvement of the arts were acknowledged by the regent with his election as an honorary member of the Académie des Sciences.[49] The different rewards underscored the superior institutional role of the Académie des Sciences as well as its unmatched social prestige.

The Manufacture of English Watches was as short lived as it was successful. After some disagreements with Law, possibly due to his frustrated ambition to participate in state administration, Sully left the manufacture in 1719 and started a similar enterprise at Saint-Germain, under the protection of the Duke of Noailles. Neither manufacture survived the collapse of Law's system in 1720. The changed financial circumstances made it difficult to support the families of the foreign workers, and the British government quickly seized the opportunity to reverse the flow of worker immigration. With the collaboration of the British ambassador in Paris, Sully reverse-engineered the operation that he had supervised a few years earlier, returning to Britain the majority of the people he had previously enticed to France.[50] It is therefore likely that it was Sully who, after working on the establishment of the Société des Arts in Paris, attempted to establish a similar association in London.

It is significant that the author of the *Three Letters* criticized Bignon's Société for limiting membership to "artists and mechanicks" and proposed that the Chamber of Arts open its doors to the aristocracy, the gentry, and landowners—all people who had vested interests in its project of improvement aimed at trades and manufactures. They would readily recognize the need for a place where those with common interests and different expertise could work together: "They [aristocracy, gentry, and land owners] will be at once informed of the best manner of improving their lands, of employing their Poor, of draining and working their mines, etc. where they may find models of every engine, draughts and descriptions of all manufactures, and accounts of the produce of every singular part of their own and foreign countries."[51] This statement addressed a longstanding French debate on the *dérogeance*, the debasement of the nobility caused by their participation in commercial activities. During the seventeenth century, several authors had attempted to reverse the social stigma against commerce by arguing that it was legitimate and useful for the aristocracy to engage in trade. Colbert was particularly active in the promotion of commerce among the aristocracy, and it was during his ministry that some exceptions to the *dérogeance* were granted: in addition to mine exploitation and glassmaking (which were traditionally accepted), the aristocracy was allowed to engage in maritime and wholesale commerce. Law continued Colbert's promotion of commerce among the aristocracy. Sully, for his part, praised his patron, the Duke of Arenberg, for not being one of "those Great men, who believe that knowledge of the arts is derogatory [*déroge*] of their status."[52] Nonetheless, resistance to the idea of a *noblesse commerçante* was rampant, and the debate would erupt again in later decades, famously involving the abbé Coyer and the Marquis of Lassay, as well as *philosophes* such as Voltaire.[53] Several scholars of this debate have highlighted the Anglophilia of the supporters of the *noblesse commerçante*. What has not been noted until now is that this Anglophilia had precedents in the history of Law's plan for French manufactures, his relationship with Henry Sully, and the establishment of the Société des Arts.

COLLECTIVE INVENTION AND THE STATE

Sully returned to Paris after realizing that he could not thrive in London. When his patrons in Britain were replaced with less sympathetic officers, France once again became his land of opportunity. It is unclear if the Société des Arts in Paris continued to operate after Sully's departure for England, but it would be difficult to believe that it survived the collapse of Law's system. In any case, when he was back in Paris, Sully resumed his horological work for navigation, collaborating with Bignon and the Académie des Sciences. In 1723, he presented to the Académie a lever pendulum clock for use at sea, which was tested at the Royal Observatory and received positive evaluation by the institution (fig. 3.2).[54] Encouraged by the result, Sully continued to work on improving the clock mechanism. In 1724, he presented another invention to the Académie: a maritime watch to be used aboard together with the lever pendulum (fig. 3.3).[55] Although watches did not keep time as regularly as clocks, Sully believed that during extreme weather, when the lever pendulum was likely to lose its regularity, his watch would be a reliable substitute. After the storm, it could be used to reset the pendulum. Sully was convinced that he was very close to the solution of the problem of finding longitude at sea and obtained Bignon's support to test his machines at sea, under the supervision of the Académie des Sciences at Bordeaux.[56] In line with the early program of the Société des Arts, Sully presented his horological work as intertwined with navigation, cartography, and geography. His inventions, both accurate and portable, were meant to facilitate the production of more detailed nautical maps, useful for calculating distances among ports. As he explained, the importance of these charts would become greater with increased accuracy in the calculation of longitude at sea.[57]

Sully's research on the problem of longitude was discussed in French journals as well as in the periodical publication of the Académie des Sciences. His lever pendulum earned him a pension from the king and became a sought-after diplomatic gift that several ambassadors tried to secure by subscription.[58] The

Fig. 3.2. Sully's lever pendulum clock; from Henry Sully, *Description abregée d'une horloge d'une nouvelle invention* (Paris: Briasson, 1726). [Bibliothèque Nationale de France, Paris]

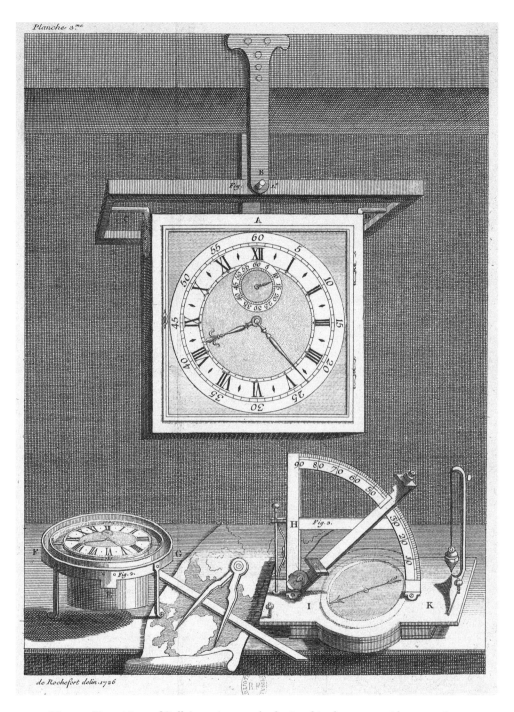

Fig. 3.3. Two visions of Sully's marine watch, depicted in the center with suspension apparatus, and at lower left; from Henry Sully, *Description abregée d'une horloge d'une nouvelle invention* (Paris: Briasson, 1726). [Bibliothèque Nationale de France, Paris]

geographer Joseph-Nicolas Delisle admired Sully's work on watchmaking and tried, unsuccessfully, to secure a position for him at the Académie des Sciences in St. Petersburg.[59] In 1726, after the tests at Bordeaux seemed to demonstrate the reliability of his inventions, Sully published *Brief Description of a Clock of New Invention,* a collection of materials related to his contributions to solving the problem of longitude. The book offered French audiences the opportunity to appreciate the *artistes'* approach to knowledge-making and improvement. In it, Sully published the essays he presented to the Académie des Sciences, together with his correspondence with *artistes,* savants, and statesmen who discussed his works. The materials portrayed Sully as an *artiste* who mastered notions of physics and mechanics, who read widely in mathematics and natural philosophy, and who understood the problem of longitude in the broader context of French colonial and commercial expansion. He was familiar with the publications of the Académie des Sciences and the Royal Society of London, as well as those of the most celebrated savants of the time, including Isaac Newton, Christiaan Huygens, Philippe de la Hire, Gottfried Leibniz, and René Descartes. The Swiss mathematician Johann Bernoulli expressed great admiration for Sully's knowledge of the sciences, adding that it was "something rare among people of your profession."[60]

Sully's discussion of theoretical principles not only showcased his learning; it also emphasized the limitations that theoretical knowledge imposed on the process of improving the mechanical arts. Sully articulated this point by discussing Christiaan Huygens's approach to horological improvements. Although he conceded that Huygens's introduction of the pendulum clock had contributed enormously to the art, he disagreed with Huygens's insistence on theory-driven solutions to the problem of maintaining the clock's regularity. Huygens had demonstrated that the oscillations of a pendulum were isochronal (of equal duration independent of their length) if the pendulum oscillated along a curve that he called cycloid and which he defined geometrically. Isochronal oscillations were crucial for the use of pendulum clocks at sea, where the ship movements caused timekeepers to lose their regularity. However, trials made at sea with clocks built on his principles had all failed, with considerable waste of time, money, and energy. Sully, in contrast, did not spend any time studying the geometrical properties of the curve described by his pendulum; he manipulated the metallic plates until he was satisfied that the clock worked as desired. When he presented his lever pendulum to the Académie des Sciences, however, Sully was asked to provide the curve's mathematical description. Sully's response underscored that the search for the curve's mathematical properties was the savants' task, whereas his goal was to make things work as desired: "I leave to geometers the more detailed analysis of this composite curve. . . . For the use I intend to make of it at present, it is enough to be able to build it mechanically, with all the required exactness, and to demonstrate by experiment that it produces all the effects I claim, and that we wish for."[61] Only experiments at sea could demonstrate whether his lever pendulum preserved the uniformity of oscillations. By carrying out proper trials, it would be possible to draw conclusions about the clock's properties more efficiently than just formulating "highly sophisticated conjectures, founded upon the most learned theories."[62]

During their practical work on materials, *artistes* often encountered "unexpected accidents" that proved long-held theoretical beliefs wrong.[63] Hence, *artistes* could help savants make sense of the natural world thanks to their experiential approach to knowledge-making. This approach ran counter to Réaumur's understanding of the role of the artisan as discussed in the previous chapter. Sully's correspondence with the London clockmaker and fellow of the Royal Society George Graham demonstrated that *artistes* agreed on the epistemological superiority of practice over theory. Graham praised Sully's improvements to pendulum clocks and expressed his desire to be informed about the outcomes of trials at sea. "These experiments," he added, "will bend all conjectures of mere theory. . . . They will give better ideas of the excellence of the machine to those who know nothing about our art."[64]

The *Brief Description* anticipated Sully's vision of a state-funded institution for the improvement of the mechanical arts, which would result in a new Société des Arts in 1728, founded by Sully ten years after his first attempt. The *Brief Description* included his translation of the Longitude Act into French and a dedication to the French king in which he expressed his hopes that Louis XV would support research on the arts as his illustrious predecessor, the Sun King, had done. Sully did not expect Louis XV to establish the equivalent of the prize to be awarded by the Board of Longitude. He believed that similar awards excited the "industry of skilled and ingenious people" but failed in supporting the necessary collective work leading to important inventions.[65] Whereas the Board of Longitude would award the prize following tests that proved the reliability of the new method, Sully wanted to attract his readers' attention to the collective work needed to perfect existing machines: "Who is the man, if he is aware of the difficulties to overcome in such an enterprise, who can hope for a full and perfect success? And even if he could hope, until what age would he need to live in order to have enough time to make all the necessary experiments by himself, and to give sufficient evidence to convince everybody of his success?"[66] In contrast to the "heroization of inventions" that underpinned the establishment of the prize as a reward to a successful individual, Sully advocated for a different system that would encourage collaboration among *artistes*, and among *artistes* and savants who were pursuing inventions as contributions to the public good. The solution to the problem of longitude, Sully believed, would not emerge from the work of an isolated man.[67]

The collaboration Sully advocated for included the participation of the state. A "wise government" could not expect great successes from "*one man only . . . without somehow collaborating with him, by extending to him patronage and support.*"[68] The history of the mechanical arts showed that improvement occurred relatively rapidly when people of "different talents" worked together. This was the case of instruments and machines such as clocks, watches, telescopes, and microscopes, which had required the work of *artistes* and savants from various nations over the course of several years. But the history of the arts showed also that improvement occurred slowly because "there are only a few people who get involved in improving the arts, while all others believe themselves exempt from thinking about these subjects accordingly."[69] The best thing the state could do to stimulate improvement, then, was to encourage the inventiveness of *artistes* by supporting their mutual collaboration and their interactions with savants: a new Société des Arts.

THE NEW SOCIÉTÉ DES ARTS

Sully's understanding of improvement as the result of collaborative research sponsored by the state was at the heart of the new Société des Arts he founded in 1728. This happened shortly after he realized that his lever pendulum did not perform as expected over an extended period of time.[70] Having failed to obtain a privilege from the Count of Maurepas, in 1728 Sully approached Jean-Joseph Languet de Gergy, the parish priest of Saint-Sulpice, with the idea of building a gnomon and meridian for the church.[71] Languet de Gergy had been Sully's confessor during the process leading to his conversion to the Catholic faith, a necessary step to benefit from the king's pension.[72] The partnership between the two men proved foundational for the establishment of the new Société des Arts. Languet de Gergy shared with Sully an interest in the world of industry and manufacturing. When Sully worked as the director of the Manufacture of Watches at Versailles, Languet de Gergy obtained from the regent a thirty-year privilege for the establishment of a Manufacture of Muslin in Paris that would employ destitute girls from the city and countryside. As in the case of Law's and Sully's manufacture, Languet de Gergy's project relied on the association among economic returns, morality, and the public good. By employing the poor, the manufacture claimed to contribute to collective welfare.

Languet de Gergy conceived of the new church of Saint-Sulpice as a central pole of Parisian religious life. He funded the ambitious project by a lottery and a remarkably well-publicized subscription plan. The meridian that he commissioned from Sully served his centralizing plans. The instrument would grant the church a prominent role in Paris by enabling the precise determination of the day of Easter as well as the time of local noon. Since watches and clocks were not easily synchronized, the bells of the church of Saint-Sulpice would establish the time of noon for the entire city of Paris. The precise measure of time provided by the meridian would also serve the organization of labor in the Manufacture of Muslin.[73]

The work on the construction of the Saint-Sulpice meridian offered Sully the opportunity to reconnect with his fellow clockmakers and reconstitute the Société des Arts on new foundations. With the approval and the support of Languet de Gergy, who initially headed the group, Sully sought the collaboration of the clockmakers Pierre and Julien Le Roy, as well as of his fellow members of the earlier Société des Arts: the geographer and now royal censor of books, Henri Liébaux, and the geometer Jean-Baptiste Clairaut, who brought in his talented son Alexis. The group met regularly every Sunday at the Liébaux home in the Saint-Germain area and was active in reaching out to other Parisian *artistes*, savants, and state officers, as outlined in the project of the Chamber of Arts.[74]

By November 1728, the new Société des Arts had recruited a total of twenty-five members, including surgeons, geometers, engineers, clockmakers, mechanicians, and state officers. Among them were the would-be first editor of the *Encyclopédie*, the abbé Jean Paul de Gua de Malves; the future perpetual secretary of the Académie des Sciences, Jean Paul Grandjean de Fouchy; and a group of mint officers and *artistes*: the director, Mathieu Renard Du Tasta; his brother; the general treasurer, Le Normand; the engraver Roettiers; and the assayer Jacques Louis Pelays. Liébaux acted as the Société's secretary.[75] It is likely that Du Tasta became interested in the Société des Arts after the debacle of the manufacture at Cosne, in which he was deeply involved, and that he gathered other members from the mint. He would act as the director of the Société des Arts in 1731 and most likely shared his alchemical interests with other members.[76]

All members of the new Société des Arts subscribed to Sully's project of improvement based on the collaboration among *artistes* and savants. The Société's wide-ranging membership indicated the broader scope of the group as well as its ambition to move away from the invisibility of its 1718 ancestor. The presence of state administrators and the ambition to attract aristocrats underscored the Société's intention to become a noticeable state institution—an Académie that would pursue the public good through the improvement of the mechanical arts. As in the 1664 project of the Compagnie des Arts, membership would be extended to distinguished foreigners who would report on the state of the arts in their countries.[77]

Sully died unexpectedly in October 1728, but his fellow members of the Société des Arts continued to pursue his goal of a state-funded academy for the advancement of the mechanical arts.[78] They sought the support of a powerful patron who, by connecting them to the aristocracy and the administration, would help realize their objective. They proposed this position to Bignon, a choice that they presented as "natural" since they understood the new Société des Arts as the "renaissance" of the association Bignon had directed during the Regency (which they described as a mere "draft" of the present project).[79] The fact that they addressed Bignon also indicates that they did not believe that their work would overlap with or challenge that of the Académie des Sciences. Rather, it suggests that they presented themselves as an alternate group of experts who would complement the Académie's work. Bignon, however, declined the proposal. Years later, he also declined the invitation to join as an honorary member. Although he regarded the Société's "object important and its establishment useful," he believed that its goals should be pursued—as had happened during the Regency—without "so much fracas."[80] For him, an association composed mainly of artisans should have continued to work in the shadows.

Languet de Gergy's connections to the Parisian elites offered an alternative patron for the new Société. His fundraising activities for the church of Saint-Sulpice had put him in close contact with the former prime minister, the Duke of Bourbon, who, in the years of the Law System, had gained considerable wealth and been Law's great supporter.[81] The duke's younger brother, the nineteen-year-old Louis de Bourbon, Count of Clermont, seemed a perfect match for the Société des Arts. A prince of the blood of considerable wealth, the count is mostly remembered as a military officer who led a dissolute life and as the fifth grand maître of the Grand Loge de France (a position he first occupied in 1743).[82] In his youth, however, he cultivated strong interests in the sciences and the arts. He assembled a cabinet of natural curiosities,[83] and owned the Forges de Claviére, a large complex of foundries in Ardentes (central France) that had a monopoly supplying the French arsenals.[84] He was interested in metallurgical chemistry and, more broadly, in the world of manufactures.[85] His inventory at death indicates that he continued to cultivate an interest in the mechanical arts and, possibly, alchemy over the course of his life. He owned a "cabinet de méchanique," which included models of machines and scientific instruments, and maintained a fully equipped chemical laboratory in his palace at La Roquette.[86]

The Count of Clermont was eager to sponsor the group of "savants in all the arts" that Languet de Gergy presented to him on 7 December 1728.[87] At least since the time of Leopoldo de Medici's patronage of the Accademia del Cimento, the classical model of an individual aristocrat bestowing his protection on an assembly of men pursuing knowledge had been variously replicated in Europe.[88] As Languet de Gergy reminded the count, the patronage of this association would bring him immortal glory: "The Ceasars, the Mecenas, the Bourbons owe to the love and gratitude [of their protégées] this kind of immortality that they still enjoy today."[89] This form of patron-client relationship brought to the patron visibility and prestige among his peers. It was a form of moral ennoblement for aristocrats who wanted to demonstrate that their nobility was not just a matter of blood but also of virtue. The young Count of Clermont was ready to seize the new visibility and power that the patronage of the Société des Arts brought him. He offered a room in his palace for the Société's gatherings and obtained the title of "société académique" for the Société des Arts, which indicated the king's interest in at least the possibility of establishing an academy of the arts, though no further promise.[90] He also compiled new regulations for the Société, which he assumed the members would accept without further discussion. In fact, the Société spent the following year working on another new version of the regulations, which were eventually published in 1730. The first article of the official *Regulations* explained that the Société's main duty was "to conform to the orders of His Most Serene Highness," spreading a public image of the Société des Arts as Clermont's own project.[91]

The new Société des Arts took shape during the early years of Louis XV's reign in a changed political and cultural climate. After the financial instability that characterized the end of the Regency, the new minister of finances, Cardinal André-Hercule de Fleury, launched an era of reforms aimed at stabilizing the French economy, raising hopes for a new policy of encouragement for the arts and manufactures.[92] The members presented their association as functional to what James E. McClellan III and François Rigoud have termed the "Colonial Machine": a bureaucratic apparatus made up of state institutions that worked for the success of France's colonial enterprise.[93] In the first years of its existence, the Société des Arts examined inventions in the fields of geography, navigation, mechanics, metallurgy, and civil and military architecture—all sciences and arts that were not cultivated as primary subjects in other academies and that were essential to commercial and colonial expansion. The Compagnie des Indies, one of the main components of the Colonial Machine, was ready to capitalize on the expertise of the Société des Arts. Having received reports that the gold mines of Bambouk, in Senegal (a French colony), might be richer than commonly known, the Compagnie hired Jacques Louis Pelays, a mint chemist and one of the Société's founding members, for the task of assessing the richness of the mines and the cost of extracting gold.[94] Members of the Société, in their turn, were investors in the Compagnie des Indies.[95]

In fashioning itself as a new space for the improvement of the arts, the Société embraced the relationship among practice, theory, and improvement discussed in the *Three Letters* and in Sully's works. Theoretical knowledge offered principles that would inspire *artistes* to pursue unknown possibilities in their physical interactions with materials, but theoretical principles could be regarded as certain only when confirmed by the practical work of *artistes* committed to improving machines and instruments. This practical work would reveal the properties of natural materials more effectively "than the most sophisticated conjectures, founded upon the most learned theories."[96] This was in line with earlier writings by learned artisans, in particular Bernard Palissy's *Discours admirables*, which unabashedly attacked the value of erudition in the search for the secrets of nature, praising instead an approach based on learning by doing. Similarly, the members of the Société equated the pursuit of theoretical knowledge to idle curiosity. They were not, nor did they want to be, natural philosophers. Neither would they engage in the investigation of causes: they were producers of effects. But it was not practice per se that they intended to pursue. They regarded practice without theory as "skilled routine"; it was the province of workmen whose work, without the *artiste*'s supervision, would never lead to any improvement.[97] According to the Société's members, improvement could only occur through the "mutual assistance" of theory and practice. Just as Sully had actively sought the collaboration of geometers, mechanical physicists, *artistes*, and state administrators for improving his lever pendulum clock, so the Société would encourage the collaboration of *artistes*, savants, and men of state. Yet the "mutual assistance" between theory and practice as presented by the Société defined a hierarchical relationship between them, in which the former had value only in so far as it served the latter.

Between 1730 and the early 1736, the Société des Arts gathered regularly twice a week in the palace of the Count of Clermont. In the absence of detailed documentation, it is impossible to determine how many members belonged to the Société at any given time or even who or how many attended any specific meeting. The surviving minutes of the meetings record the topics that the Société discussed (mostly between 1732 and 1736) and the composition of the committees charged with evaluating reports. These documents show that the most active and loyal members through the life of the Société were clockmakers and surgeons. Lists of members, however, show that over two hundred people from a wide variety of arts, crafts, and social positions joined the Société during the fifteen or so years of its life, which indicates the wide appeal of the Société's cultural and political program (see appendix). The Société started to decline in 1736, when its meeting place moved from the count's palace to the houses of the clockmakers Julien Le Roy and Pierre Gaudron. The frequency of meetings was reduced to once every two or three weeks, and the Société ceased to exist by the early 1740s. The surviving materials on the Société des Arts indicate quite clearly that there was a significant difference between, on the one hand, the *artistes'* cultural and political ambitions as presented in the project they proposed to Bignon, in the published *Regulations*, and in their own publications, and, on the other hand, the Société's actual life and accomplishments. The Société des Arts presented itself as a new social model and political project through which improvement should be pursued. But, as we see in the next chapter, it remained entangled in Old Regime practices of patronage and privilege that ultimately undermined its existence.

Chapter 4

SOCIETY OF ARTS

"[The clock] is the rule of society," wrote the abbé Pluche in his widely read *The Spectacle of Nature*. Grounded in the visual and auditory experience of any Parisian of the time, his statement testified to the pervasiveness of clocks in everyday life:

> The coarsest and the most antique clock, were it even the old balance clock, with a ringing as dismal as that of the Samaritaine, yet never ceases, from the belfry which contains it, to speak to a whole people, and at equal distance to repeat the warning that it is appointed to give. It watches, and, from the beginning to the end of the night, informs every private man in the intervals of his sleep. It is the clock that gives the first signal of prayers, that causes the gates of towns to be opened, that calls assemblies together, and publishes all our works as they succeed one another.[1]

On top of manufacturing buildings, churches, towers, and private or public palaces, clocks set the rhythms of urban life. They marked the hour, the half-hour, or even the quarter-hour. The chimes of the "horloge du Palais," the oldest public clock in Paris, signaled to the city the times of the Parliament's meetings.[2] As a secular counterpoint to churches' bell towers, it also rang to proclaim extraordinary state events, such as royal weddings, the births of royal children, or peace treaties. In 1572, the sound of its bell launched the so-called St. Bartholomew's Day massacre, when Huguenots were mass murdered by a Catholic mob. The clock's reputation was such that the old quai Morfondu came to be renamed quai de l'Horloge because of its proximity to the clock tower.[3] Clocks marked the cadences of court life. The ceremony of the "levée du roi" (the rising of the king) commenced with the ringing of a clock. When the king died, the *horloger du roi* froze time by stopping all the clocks at court.[4]

Clocks were built on the notion that regularity was a fact of nature. The concept of an ordered universe in which celestial bodies repeated the same cycle after a certain amount of time underpinned cosmological systems and mechanical timekeeping devices alike. Whether one believed in a geocentric or a heliocentric universe, or in circular or elliptical planetary orbits, the succession of astronomical events in the universe repeated itself periodically, and therefore it could be reproduced by means of a mechanical device. Cosmic order had materialized in machines such as astrolabes and armillary spheres since antiquity.[5] These devices allowed the calculation of specific astronomical events, but they did not follow the motion of the heavens in real time. Mechanical astronomical clocks instead purported to do just that. With varying degrees of complexity and mechanical virtuosity, their dials displayed rising suns and moons, stars, eclipses, or other events, marking with their chimes the regularity of astronomical motions. Operated by a motive force that mechanized the turning of wheels, these automata replaced direct observations of the sun or other stars to tell the time of the day. In European cities, the most sophisticated public clocks displayed automated processions of humans or animals performing the same actions at the same time, mirroring simultaneously the periodic motions of celestial bodies as well as the ongoing organization of labor.[6]

The correspondence between cosmic order and the mechanical clock found its most powerful expression in René Descartes's conception of the clockwork universe. Put in motion once at the beginning of time by the divine clockmaker, Descartes's universe continued its regular operations forever, constantly preserving its quantity of motion. Paradoxically, however, when translated into materials, Descartes's theoretical conception of the universe resisted the eternal order that it intended to secure. The maker of astronomical instruments Jean Pigeon d'Osangis, later a member of the Société des Arts, designed for Louis XIV a mechanical armillary sphere that materialized the fundamental idea of Descartes's clockwork universe (fig. 4.1). Pigeon's instrument showed the earth and the known planets moving around the sun, animated by a working clock that towered above all spheres. Even if it was based on the best available astronomical observations, Pigeon's *sphère mouvante* lacked precision. By Pigeon's own admission, its clock mechanism did not run as regularly as the cosmological model it represented.[7] This was not just a problem of Pigeon's clock. Even the extraordinarily accurate repeaters that Louis XIV received from Charles II as diplomatic gifts lost their regularity after some time.[8]

As clockmakers knew all too well, Descartes's mechanical clock was an ideal more than a reality. Even with the introduction of the pendulum clock—which notably increased the clock's ability to keep time—the ideal of clockwork

Fig. 4.1. Jean Pigeon d'Osangis's automated armillary sphere. [Département Cartes et Plans, GE BB 565 (1, 15), Bibliothèque Nationale de France, Paris]

regularity did not correspond to the reality of material clocks. Public clocks required daily adjustment, and astronomers relied on a variety of observational instruments to regulate their clocks. Henry Sully's *The Artificial Regulation of Time* (1714) warned against unreasonable expectations of timekeeping born out of ignorance of the material limitations of clocks and especially watches. Even with the aid of correction tables, clocks and watches were intrinsically unable to tell astronomical time with unlimited precision. In the material world, the effects of friction were unavoidable, just like those due to changes in temperature or position, the smallest faults in the shape of gears, or other accidents related to the actual manufacture of timepieces. No eternal regularity was possible. Left to themselves, Sully warned, clocks and watches would irreversibly deteriorate.[9]

Like the Newtonian universe, which needed the intervention of an immanent providence to preserve its order, even the most perfect clock needed the intervention of a *régleur*, a regulator who, with delicate adjustments and careful maintenance, made clocks keep and tell reliable time.[10] As the clockmaker Julien Le Roy remarked, if it was desirable to have a good watch, it should be "equally desirable that it be in the hands of someone who can adjust it well [*qui la puisse bien régler*]."[11] It was no coincidence that Sully worked as a *régleur* while in the service of the Duke of Arenberg. The role of the *régleur* was what he and his colleagues at the Société des Arts wanted the *artiste* to play in French society. As the Newtonian system of the world became a model for a perfect system of government, so the Société des Arts articulated a political project that borrowed from the material life of clocks the emphasis on rule-makers (*régleurs*) as preservers of social order and guardians of the common good.[12] This chapter examines the social microcosm of the Société des Arts in comparison with contemporaneous forms of professional or polite sociability and argues that the Société was modeled on the working practices of *artistes*. I discuss the Société des Arts as an expression of the *artistes*' ambition to play an official role in the administration of the French state, and its crucial contradiction between a rhetoric of collaboration and cooperation among members and a practice of hierarchical relationships and search for patronage. I examine the reasons that led to the dissolution of the Société des Arts in the early 1740s and its legacy to the Académie des Sciences and the Académie de Chirurgie (Academy of Surgery).

RULES, REGULATIONS, AND RÉGLEURS

To paraphrase Pluche, if clocks were society's rule, *artistes* wanted to be its rule-makers. Rules, *règles* in French, were defining features of the arts. In the

Encyclopédie, Denis Diderot defined art as a "system of instruments and rules concurring toward the same end," and explained that rules and instruments were like "additional muscles for the arm" and "complementary springs for the muscles of the mind [*esprit*]."[13] François de La Peyronie, the secretary of the Académie de Chirurgie, compared the rules of the arts to the law. The rules of the law distinguished right from wrong; similarly, the rules of the arts taught how "to distinguish those who practice [them] properly from charlatans who dishonor [them]."[14] In other words, not only were rules defining features of the arts; they were also essential for preventing frauds.

Règles and *règlements* were at the heart of mercantilism's mathematics. In a world in which global wealth was finite, the mercantilist equation dictated that national exports should outnumber imports. Quality was the equation's crucial variable: it was by increasing the quality of nationally produced goods that a state could win the competition of the international marketplace.[15] Jacques Savary's *Dictionary of Commerce* explained that "quality" characterized the "good or bad nature of commodities," but it was not a subjective judgment.[16] Rather, quality was defined through written regulations (*règlements*) that codified all the manufacturing processes, including the type of machinery and raw materials to be used. The regulations were issued by the Bureau of Commerce, a government office that was reorganized in 1724 in order to regulate commerce and manufactures, mirroring the function that the Council of Finances served for all financial matters.[17] The bureau's members included intendants who resided in the French provinces, whose role was to inspect local industries and report to the bureau. Starting in 1730, the bureau established the positions of general inspectors of manufactures, who specialized in specific technical processes (textiles, dyeing, etc.) and inspected all manufactures under their supervision, as well as the provincial intendants.[18] This bureaucratic panopticon pursued what Philippe Minard has called a political economy of quality. It guaranteed the quality of French products by imposing and enforcing regulations on artisans and workers, and by policing the manufactures' compliance with such regulations.[19]

Regulations were not understood as obstacles to improvement. On the contrary, they were elaborated on the basis of what the bureau regarded as best practices and tools for each art, the very ingredients that made improvement possible. By imposing such practices and tools on all manufacturers, regulations would ensure that French commodities were produced according to the best standards available. Although they were written, regulations were subject to change, when altered circumstances demanded revisions of existing methods of production. To put it differently, when they were approved by the bureau, the regulations captured the bureau's current understanding of the state of the art. Following recurring requests for revisions, along with concrete proposals for changes, the bureau could and did issue new sets of rules.[20]

Regulations bridged the worlds of writing and making. Based on direct observation of artisanal work and expert assessment of manufacturing processes, they were similar, in this respect, to the encyclopedic projects discussed in previous chapters. Not surprisingly, as the editor of the Académie's *History of the Arts*, René Réaumur had suggested that the French state employ members of the Académie des Sciences as inspectors of manufactures.[21] Inspectors served the important function of persuading manufacturers to adopt new regulations: they spread technical information by offering public lectures and demonstrating new machines.[22] Réaumur's views were implemented during the reign of Louis XV, when the Bureau of Commerce built a strong relationship with the Académie des Sciences, drawing from it expertise on technical matters. In 1731, for example, the chemist and academician Charles de Cisternay Dufay was appointed general inspector of dyeing. Throughout the life of the bureau several other academicians served similar roles. The experts that offered assessments of manufacturing processes for the bureau were generally selected from the members of the Académie des Sciences. As noted by Minard, inspectors held a condescending attitude and sense of superiority toward artisans and workmen, whom they regarded as mere recipients of instruction, incapable of contributing autonomously to improvement. Dufay's work exemplified this attitude. Mirroring Réaumur's approach to the establishment of the Royal Manufacture at Cosne, he studied the best dyeing processes and compiled regulations that were imposed on workers and artisans, whom he instructed.[23]

It was this exclusion from the decision-making processes concerning the arts that *artistes* wanted to correct. If, as has been argued, artisans and *artistes* de facto participated in the bureau's networks of information and exchange, they were nonetheless excluded from formal positions because of their social status.[24] This exclusion was not limited to the Bureau of Commerce. The polymath and clockmaker Pierre-Augustin Caron de Beaumarchais was denied a position in the administration of Waterworks and Forests (*Eaux et Fôrets*) because of his upbringing as an artisan. The superintendents explained that "however famous one may be in this art [clock-making], such a condition is incompatible with the honors attached to superintendence."[25] The Académie des Sciences also excluded artisans from its membership.

Similar prejudices affected surgeons, regarded as men of practice and therefore excluded from the institutions of medical education. In 1607, the Parlement ruled against a Paris physician who wanted to teach physiology to surgeons, stating that "science is not for those who work only with their hands." More than a century later, in 1725, the rector of the University of Paris agreed with the medical faculty that surgeons should not be allowed to teach public courses in surgery.[26] Not surprisingly, the Société des Arts numbered several surgeons among its most committed members. As with other *artistes* in the Société, they were eager to establish an institution that would grant authority to their expertise without subjecting it to the more theoretically inclined physicians.

The Société des Arts was a manifestation of the *artistes*' ambitions to elevate their social status and to play an active role in the administration of the state. *Artistes* presented themselves as perfectly qualified for the task. They were, after all, *régleurs* in their fields. As learned artisans who excelled in their arts, they often performed as *jurés* or *guarde-visiteurs* in the governing bodies of their guilds. Prominent members of the Société des Arts, such as the clockmakers Pierre Gaudron, Nicolas Gourdain, Claude Raillard, and Julien and Pierre Le Roy, as well as the surgeon René-Jacques Croissant de Garengeot, held administrative roles in their guilds. As *jurés* or *garde-visiteurs*, *artistes* were guardians and arbiters of regulations: they articulated and enforced rules, inspected their colleagues' workshops, and interacted with state bureaus. They also proposed changes in regulations to pursue improvement. Julien Le Roy, for example, wrote a lengthy essay for the Bureau of Commerce in which he recommended a number of modifications to existing regulations in order to boost the national and international trade of French clocks and watches. He examined the licensing of clockmakers, the tax system connected to the trade of timekeeping devices, and the increasing number of low-quality Swiss watches that circulated in France. The significant revisions to regulations that he advocated were tightly connected to the ways of operating of the Société des Arts. He maintained that in order to improve the quality of French watches it was necessary to train apprentices in the theory as well as the practice of their art. They should study works of physics and mechanics in addition to carrying out practical sessions in the workshop. He also proposed the establishment of the position of "intendant supervisor of the progress of the commerce and art of clock and watchmaking," who would oversee a thorough reform of the statutes of the guild of clockmakers. In his plan, this position—which resembled that of an inspector of clock- and watchmaking—should be assigned to an *artiste*.[27]

In this attempt to constitute itself as an institutional body that would provide expertise on the arts, the Société des Arts de facto, if not deliberately, encroached on various established institutions. Its ambition to become a new state institution dedicated to the encouragement of the mechanical arts, for example, conflicted with Réaumur's vision for the Académie des Sciences, just as the surgeons' desire to be in control of surgical education conflicted with the status quo within the Faculty of Medicine. The Société, however, did not intend to challenge the administrative or social structures of the Old Regime. Its project was not revolutionary. The members of the Société des Arts subscribed to the Bureau of Commerce's political economy of quality. Improvement in their minds was to be achieved through regulations that would guarantee the high quality of finished goods. The state, through its apparatus of intendants, bureaus, guilds, and academies, was for them an unquestioned source of authority and power. They sought the privilege of working for it.

CLOCKWORK NETWORKS

There can be little doubt that clock-making offered a model for the Société des Arts. In 1732, when the Société discussed what to represent on the gold medal to be awarded for its annual prize, one of the possibilities that the members discussed was to engrave Vulcan holding a clock. Although this proposal was eventually rejected by the Count of Clermont as something that would be more appropriate to a clock-making society, the Société's *Regulations* of 1730 were de facto modeled on the practice of clock-making.[28] A clock was a microcosm of materials, mechanical skill, and astronomical knowledge, where each component contributed to the working of the whole. Clockmakers' workshops, just like the Société des Arts, functioned through networks of collaborations and exchanges involving *artistes*, savants, and patrons. The clockmaker—who understood himself as an *artiste*—served as the clock's designer and the workshop manager. He supervised any and all of the processes leading to the finished product and subcontracted other *artistes* or artisans. Subcontracting was required because of guild regulations that restricted the range of materials with which clockmakers could work, yet it was also an acknowledgment on the part of an *artiste* of the skill and expertise of another.[29]

The similarities between the practice of clock-making and the kind of relationships that constituted the microcosm of the Société des Arts are best exemplified by Julien le Roy's workshop. The most productive in eighteenth-century Paris, the workshop was populated by apprentices and other laborers who worked on the mechanical parts of clocks and watches. Le Roy frequently subcontracted components to former apprentices and, for components that by guild regulations could not be produced by clockmakers, he relied on the most highly regarded *artistes* in Paris.[30] Sometimes, though by no means always, these *artistes* became members of the Société des Arts. This was the case for William Blakey, an English maker of watch springs who had been a member of the Société des Arts of 1718 and also joined the new Société a few days after its establishment.³¹ Springs were essential components of watches, usually imported from England, where clockmakers jealously guarded the secrets of their art. Under the Regency, Blakey had been one of the key figures in John Law's plan for relaunching French manufactures, directing a manufacture of English watches in northern France. Following the collapse of the scheme, Blakey went back to Paris where he established another manufacture of watch springs and began his collaboration with local clockmakers, including Sully and Le Roy.³² Similarly, Nicolas Jullien, a painter who specialized in enameled dials and supplied Le Roy's workshop, became a member of the Société des Arts in 1730.³³ Unlike other clockmakers who produced only movements (the internal clock mechanism), Le Roy's workshop sold clocks complete with their cases, which he commissioned from the most celebrated *artistes* in Paris. Clock cases were highly ornamental and became the most distinctive feature of eighteenth-century French clocks. Le Roy's workshop could boast of collaborations with the royal goldsmith Thomas Germain (a member of the Société des Arts) and, later in the century, with the celebrated sculptor Jean-Jacques Caffieri.³⁴

These networks of artisanal collaboration characterized any reputable clockmaking workshop in Paris. In his article "Clockmaking," the *encyclopédiste* Ferdinand Berthoud listed fourteen different kinds of artisans who worked on the various parts of a clock.³⁵ His nephew's unpublished daybook reveals with methodical precision the range of specialized workmen who contributed to making clocks in his workshop: goldsmiths for the gilding of needles and pointers, enamellers and chasers for dials, sculptors and cabinet makers for cases, and so forth.³⁶ On Berthoud's lists of providers for various projects, the same names repeatedly recur for the same tasks, suggesting the stability of the clockmakers' professional networks.

Clockmakers' networks extended beyond the artisanal world, just as the Société des Arts included not only *artistes* but also savants, potential patrons, entrepreneurs, and state administrators. Clocks and watches were essential to a number of sciences, in particular astronomy and navigation, but in order to make good use of them it was essential to understand exactly what they measured. Although Johannes Kepler's laws of celestial motions were harmoniously regular, the apparent motion of the sun could make clocks seem whimsical.³⁷ This was due, Sully explained, to the fact that the "natural division of time," based on the apparent motion of the sun, consisted of intervals of unequal length (such as days and years), unlike the "artificial division of time" measured by clocks and watches, in which all time intervals were of equal duration. If a clock was regulated on the solar noon (when the sun crossed the local meridian) one day, after twenty-four hours it would strike noon a few minutes before or after the passage of the sun over the meridian. Sully devoted ample sections of his work to explaining the difference between "true" and "mean" time—notions that were essential for understanding that even when clocks seemed to run faster or slower than the sun, they were working well. Sully wanted watch and clock users to understand that the difference between the "true time" of astronomers and the "mean time" measured by clocks depended on the motion of the earth around the sun and on the inclination of its orbit.[38] Conventionally, a day was defined as the twenty-four hours that elapsed between two passages of the sun across the local meridian. However, the sun did not take exactly twenty-four hours to come back to the same position over the local meridian. The use of clocks and watches required mathematical and astronomical knowledge: civic life was still regulated on solar ("true") time, and users had to regulate their watches daily, using a variety of observational instruments (fig. 4.2).

Equation tables offered adjustments to convert mean into true time, but learned clockmakers strove to build clocks that could reliably tell true time. In this attempt, they collaborated with astronomers, geometers, physicists, and other savants. Le Roy, for example, worked closely with the mathematician and member of the Académie des Sciences Philippe de La Hire on a model of a pendulum clock that could measure true time.[39] Le Roy collaborated also with other instrument makers to improve horological devices. His partnership with fellow Société des Arts member Jacques Le Maire was crucial to his design of an improved version of the Butterfield horizontal dial.[40] This astronomical instrument was commonly used to find local noon, an essential operation for regulating watches and clocks. Le Roy intended to produce a new version that could be used at different latitudes, yet he was aware that ideal precision in this kind of measurement could not be achieved because of the imperfect horizontality of the dial's plate, the difference between magnetic and polar north, the thickness of the gnomon, and other such physical limitations. With Le Maire (who physically made the instrument), Le Roy worked on metals that allowed him to reduce the unwanted effects deriving from the very materiality of the instrument.[41] The resulting instrument materialized the kind of interactions that the Société des Arts intended to foster (fig. 4.3). Similar collaborations had occurred between Société des Arts members Henry Sully and Henri Enderlin in 1724. The famous collector Joseph Bonnier de la Mosson, himself a member of the Société des Arts, displayed both artifacts in his museum, a fact that highlights the iconic status of these instruments.[42]

Clock-making enabled astronomers to advance their field and contribute essential new knowledge to natural philosophy. This kind of collaboration

Fig. 4.2. The regulation of watches as a technique of the urban educated self: three gentlemen regulating their watches on the solar noon shown by a sundial built on a palace. The solar noon occurs when the bright spot within the elliptical shadow intersects the vertical line marked with XII. The gentlemen assemble around the instrument a little before the event and then regulate their watches. This was a daily routine that demonstrated the watch users' familiarity with basic astronomical instruments. There are still various examples of such sundials in European cities. From François Bedos de Celles, *La gnomonique pratique* (Paris: Briasson et al., 1760). [Beinecke Library, Yale University, New Haven, Conn.]

Fig. 4.3. Julien Le Roy's improved horizontal dial made by Jacques Le Maire. The screws allowed for horizontal adjustments, whereas a plumb line (missing) and the graduated scale below made it possible for the instrument to be used at different latitudes. [Whipple Museum of the History of Science, Cambridge, England]

between the sciences and the mechanical arts was foundational for the Société des Arts. As stated in the draft of regulations that the founding members submitted to Bignon, the most ambitious projects in the sciences, in particular navigation and geography, relied on the availability and quality of precision instruments.[43] Precision clocks proved crucial to determining the shape of the earth and played a key role in ending the philosophical controversy between Newtonians and Cartesians.[44] While supporters of the Newtonian and Cartesian systems of the world all agreed that the earth was not a perfect sphere, the former maintained that it was flattened at the poles whereas the latter believed that it was elongated. The Académie des Sciences assumed the task of finding an empirical answer based on precision measurements carried out in situ. Instruments were crucial to the success of the state-funded expeditions to Lapland and Peru. Pierre Louis Moreau de Maupertuis commissioned from Le Roy a clock to take to the Arctic Circle.[45] One of the *raisons d'être* of the Société des

Arts, the collaboration of *artistes* and savants, was a fact of life for Le Roy and other learned clockmakers.

The practice of clock-making likewise gives insight into the "honorary" membership that the Société des Arts reserved to aristocrats. The role of patrons in setting research agendas for early scientific institutions has been widely discussed.[46] With the establishment of the position of "honorary" members, however, the Société des Arts intended to subtly reverse the effects of the patron-client relationship. Aristocratic amateurs of the arts were inducted into the Société with the idea that they would provide funds or networking opportunities for other members.[47] This was nothing new for clockmakers such as Le Roy. In 1728, when he collaborated with Sully to reestablish the Société des Arts, Le Roy received the commission of a pendulum clock for the *levée* of the king. The clock he produced was a material representation of clockmakers' overlapping professional, intellectual, and patronage networks, as well as a material statement of the political ambitions of the Société des Arts (fig. 4.4).

Since the times of Louis XIV, the *levée du roi* had been a well-attended ceremony during which courtiers conducted business and formed alliances. Even though Louis XV was not as fond of court rituals as his predecessor, his day still started with the ringing of a pendulum clock.[48] The symbolic importance of the timepiece was clear to the Marquis of Beringhem, who commissioned the machine. Among his responsibilities as the first *écuyer* (squire), the marquis was in charge of the *levée* and was therefore eager to present an artifact that carried the marks of his own tastes and patronage of the arts. The marquis designed the clock's case himself, and he entrusted the sculptor Nicolas Le Sueur with its execution.[49] It was not unusual for a clockmaker to collaborate with an *artiste* outside his consolidated networks. Le Roy, though, was eager not to be overshadowed by Le Sueur or the Marquis of Beringhem. For him, this commission was a unique opportunity to directly interact with the king. As honorary members in the Société des Arts were expected to offer patronage to *artistes* and insert them into powerful networks, so the marquis's commission enabled Le Roy to achieve his goal of a royal audience. As Le Roy recounted, in restless anticipation of the meeting, he modified the usual clock's design by placing all the movements on the back plate, where his signature was visible. The king would thus be able "to remove the clock-face in order to see the clockworks revealed," never forgetting that they were the product of the hand and mind of the clockmaker.[50] By guiding the king's eyes toward the inside mechanism of the clock, Le Roy pointed not only to his own *esprit* but also to the microcosm of relationships, exchanges, and expertise that brought a clock into the world. It was this

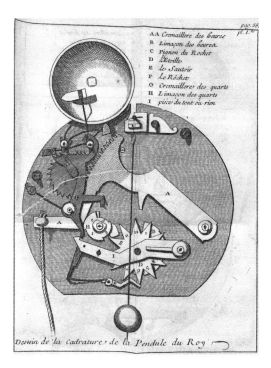

Fig. 4.4. Julien Le Roy's clock for the *levée du roi*. The clock's movement is on the back plate, where the maker would place his signature. From Henry Sully and Julien Le Roy, *Règle artificielle du temps* (Paris: Dupuis, 1737). [Museo Galileo, Florence, Italy]

microcosm that Le Roy wanted to institutionalize for all the arts through the constitution of the Société. As the king himself could appreciate by looking at the inner mechanism of the watch, the machine's order and regularity could only be preserved by the care of a *régleur*. This was a material metaphor for the state as conceived by *artistes*. Social order could be preserved, and the common good pursued, through a system of regulations.

By recruiting the best *artistes*, the Société des Arts presented itself to the public, and more specifically to the French government, as the most reliable author of new rules for the arts, crafts, and manufactures. The practice of clockmakers spectacularly illustrates the kind of collaborations among *artistes*, savants, potential patrons, and the world of entrepreneurship that the Société des Arts intended to institutionalize. The working practices of other *artistes* in Paris, such as mapmakers, painters, sculptors, architects, and mathematical instrument makers, similarly bridged the worlds of science, state offices, and manufactures. They readily enrolled in the Société des Arts.

THE SOCIÉTÉ DES ARTS'S SOCIAL MICROCOSM

The Société des Arts constituted itself as an arena where *artistes* collaborated with savants, entrepreneurs, and aristocrats toward the common goal of improving the arts. It was a social microcosm that mirrored the working practices of *artistes* and that shared important features with existing venues of artisanal or learned sociability—such as the guild, the academy, or the Republic of Letters—without replicating any of them. Like guilds, the Société intended to play a leading role in the decision-making processes related to regulations on the arts and crafts. Yet the systematic collaboration among *artistes* and savants that the Société enacted went beyond the possibilities offered by guilds. As Charlotte Guichard has shown in the case of painters and sculptors, guilds could be sites of legal and social conflict for *artistes*. In her analysis of the attempts to establish the Académie de Saint-Luc as a separate body within the guild of painters and sculptors, Guichard has focused on the legal battles between a group of masters and the guild's administrative body. She emphasizes that their conflict centered on the legal constraints within which guild members had to operate. In spite of the intellectual construction of painting and sculpture as liberal arts, which dated back to the Renaissance, the professional framework for painters and sculptors was still largely controlled by the guild. In their campaign for the independence of the Académie de Saint-Luc, a small group of master painters and sculptors, who consistently defined themselves as *artistes*, denounced the despotic power of *jurés* and *garde-visiteurs*, whom they defined as "simple artisans." In their views, *artistes* should be free of the regulatory power of the guild, just like members of academies or the *artistes* working at the Louvre and in other *lieux privilégiés*.[51]

Artistes such as clock- and watchmakers, surgeons, architects, painters, and sculptors hoped to elevate their social status through their association with academicians, savants, state officers, and aristocratic patrons. In Paris, several sites of royal patronage for select groups of *artistes* encouraged such ambition. The Louvre, the Académie de Peinture et Sculpture (Academy of Painting and Sculpture), and the Académie d'Architecture (Academy of Architecture) demonstrated that excellence in one's art, combined with the right connections, could bring convenient social returns.[52] Yet even such prestigious institutions did not provide the kind of systematic interactions among *artistes*, savants, entrepreneurs, administrators, and potential patrons that the Société des Arts intended to foster. Of the ten Parisian painters, sculptors, and architects who joined the Société between 1728 and 1736, only the architect Jean-Michel Chevotet was accepted to an academy (Académie d'Architecture) a few years after his admission to the Société, and only the architect Jean-Baptiste de Puisieux was never inducted into any academy. The others all belonged to either the Académie de Peinture et Sculpture or the Académie d'Architecture when they joined the Société.[53] The fact that these *artistes*, whose academic status placed them above their peers, joined the Société des Arts points to its prestige and broad appeal for practitioners committed to a collective project of improvement.

The changing world of the arts demanded new alliances that neither the old world of guilds nor the royal academies enabled. *Artistes* working in the beaux arts shared with clockmakers and surgeons the fundamental idea that their art would improve more efficiently through exchanges and collaboration with savants and other *artistes*. Excellence was the keyword of this shared project. The Société was the place where the best *artistes* shared their knowledge and joined forces to pursue the common goal of improving the arts. The need to share expertise derived from the *artistes*' working practices, which demonstrated a relationship of mutual dependence among the arts. Watches, clocks, maps, and other artifacts were produced through the collaboration of various artisans and *artistes*. An artifact's overall quality depended on the individual skill of the practitioners who contributed to its creation. This implied that in order to improve any art, it was essential to improve also the arts connected to it.

This changing world of the arts, where relationships of mutual dependence were frequently redefined, was reflected in the Société des Arts's social microcosm. The first six members—Henry Sully, the brothers Julien and Pierre Le Roy (clockmakers), the map engraver Henri Liébaux (the Société's secretary), the mathematicians Jean Baptiste Clairaut and his son Alexis—expanded the Société's membership by transferring into it preexisting networks created by their working practices. Liébaux was able to secure new members to the Société thanks to the connections he had established through his work as an engraver of maps. Since the quality of printed maps depended on the ability of the engravers that would prepare the copper plates for publication, cartographers were eager to collaborate with the best engravers.[54] The most celebrated cartographer in Paris, Guillaume Delisle, relied on the engraving skills of Liébaux and Desrosiers—the latter to become one of the first members of the Société des Arts.[55] It is possible that it was chez Delisle that Liébaux met Jean Paul Grandjean, who was one of Delisle's pupils. Grandjean, who served as the Société's secretary in 1732, became a member in October 1728, almost immediately after its constitution. Grandjean was extraordinarily active in recruiting new members. He inducted relatives and friends: his friend and former Delisle pupil Daniel Jousse; his uncle Hynault, a member of the Parlement of Paris; his brother-in-law Jean-Baptiste Laurent Hillerin de Boistissandeau, an inventor of calculating machines; the abbé de Pimaudan, who lived in Grandjean's premises; the wealthy financier Durey d'Harnoncourt; and the German chemist Jean Grosse.[56] Grosse lodged with the chemist Boulduc, a member of the Académie des Sciences who also hosted Jean Baptiste Clairaut and his sons, all members of the Société des Arts.[57]

Physical proximity and kinship similarly played a role in the expansion of the Société's membership, just as they contributed to defining the social world of artisans. Liébaux lodged at the Cour du Dragon, a site in the Saint-Germain area that was undergoing renovation under the direction of the architect Pierre Vigné de Vigny (for the Marquis de Crozat). The architect Puisieux, who worked for Crozat and knew de Vigny, also lived there. He offered his premises for the first meetings of the Société des Arts. Both de Vigny and Puisieux became members of the Société, along with Michel-Ange Slodtz, who sculpted the dragon that gave the Cour its name. As was common in the artisanal world, fathers introduced to the Société their sons and, at times, their brothers: this was the case with Julien Le Roy and his brother Pierre, with Clairaut father and sons, and with the painter François Lemoyne and his son. Masters and apprentices too found themselves together in the Société, as with Jean-Antoine Nollet and his master, the enameller Jean Raux, and with the architects Chevotet and his master Thomas Germain.[58] Working and familial relationships often overlapped. The wood engraver Jean-Michel Papillon, for example, was introduced to the Société by the calligrapher Honoré Sébastien Royllet, for whom he had engraved vignettes and capital letters. Within the Société, Royllet had joined forces with the painter Jean-Baptiste Oudry, who was Papillon's relative on his maternal side, to sponsor Papillon's candidature.[59]

It is of course likely that other members were recruited via connections established through common patrons. Parisian aristocrats delighted in the arts and in the sciences, and contributed to the fortunes of *philosophes* and *artistes* alike. Their salons facilitated polite sociability and were often open to *artistes*. The very idea of the Société des Arts emerged among the *artistes* that worked for Jean-Joseph Languet de Gergy, who acted as Sully's patron and who introduced to the Société the maker of navigation instruments Jacques Le Maire as well as the maker of mathematical instruments Jean Pigeon. *Artistes* who joined the Société over the years often worked for the court or for the most powerful aristocrats in Paris.

In November 1728, when Liébaux invited Bignon to take on the leadership of the Société des Arts, twenty-five members had worked together on the Société's regulations. Thirteen—more than half—were *artistes*. The majority of them were involved in clock- or watchmaking (four); the others were two surgeons, a goldsmith, the medal engraver for the mint, the assayer for the mint, a map engraver, a maker of navigational instruments, a calligrapher, and a carpenter and entrepreneur for the Office of Roads and Bridges. Of the remaining twelve, five were savants (three mathematicians and two astronomers), two composers, one a military officer, and one a teaching tutor; the treasurer of the mint, the director, and his brother completed the list. The diversity of the Société des Arts's membership was indeed the most distinctive feature that contemporaneous observers underscored. Newly elected associates enthusiastically described the mix of "professors of various sciences, among whom [there are] presidents and consultants of the parliament, as well as qualified people in different professions," while foreign visitors praised the heterogeneous expertise of the members who were assigned to different divisions.[60] The Swedish mathematician Jonas De Meldecreutz admired the fact that geometers, chemists, and mechanicians were allocated to various classes, "in accordance to the profit that crafts in the same class can draw from geometry, chemistry, or mechanics."[61] For him, as for many others, the strength of the Société consisted of its heterogeneity.

The publication of the *Regulations* of the Société des Arts in 1730 presented to the international public the innovative approach to knowledge that the Société represented. If the royal academies had become the institutions of the Republic of Letters, the Société des Arts presented itself as a new model of social and cultural interactions, a microcosm of an ideal polity that enacted the *artistes*' political project.[62] This ideal polity was a Society of Arts—the English here is meant to emphasize its ideal dimension—and differed in significant ways from the ideal of the Republic of Letters. While citizenship in the Republic of Letters was obtained through participation in the world of scholarship and learning—weaving networks through correspondence, personal acquaintance, or publication—membership in the Society of Arts was gained through sharing expertise and human capital, and it was maintained through commitment to useful knowledge. If the Republic of Letters was a world of scholarship and disinterested knowledge, the Society of Arts was a world of making, devoted to improvement. If the Republic of Letters reserved special positions for women, the Society of Arts remained a homosocial space.[63] Most importantly, the contrast between the Republic of Letters and the Society of Arts had to do with their relationship to the monarchy. Whereas the Republic of Letters contributed to the formation of a public sphere within which challenges to the authority of the French monarchy emerged, the Society of Arts was in the service of France's imperial and colonial expansion.

This ideal Society of Arts was not limited to *artistes*. Much like an ideal business partnership or *société*, it pooled together individuals who worked toward a common goal: *la perfection des arts*. In eighteenth-century France, *sociétés* were commercial partnerships formed through familial and/or professional relationships. Their organization reflected contemporaneous notions of sociability that merged the moral and the useful. Members of *sociétés* were expected to contribute their capital—whether in the form of money, if they were investors, or skill, if they were artisans—to achieve a result that would benefit each of them. The rhetoric on the pursuit of the common good was grounded in the Lockean idea of an orderly state of nature in which humans were naturally disposed toward collaboration.[64] *Sociétés* were not created on the premise of reciprocity, equality, or horizontal exchanges. Nor were they created on cosmopolitan ideals. They worked on the basis of different abilities put in the service of a common—often local—goal, sustained by order and politesse. Similarly, the Société des Arts was not a transnational republic of peers. Although international in its proclaims, it served local purposes and was deeply hierarchical. Its members held different roles according to their expertise, their social status, their provenance, and their commitment to the Société. The core members were divided into "assiduous," who were expected to attend each meeting, and "free," who were not obliged to do so. Those who did not reside in Paris were classified as "foreign," even if living in the French provinces. "Honorary" members were "people distinguished by birth, dignity, and love for the arts," and it was expected that they would support members of different rank by any means they were willing to offer. All members would share their secrets, disclosing details about materials, ingredients, machines, and tools. The assiduous and free members were selected among those who excelled in their field and could be savants or *artistes*.[65]

Mirroring the expert partnerships that took place in clockmakers' workshops, members would collaborate to develop nine focus areas: agriculture and economy; animal economy; manufactures (textile, dyeing, and leather tanning); military and civil architecture; the building of ports and ships; horology and mathematical instrument making; optical glassmaking; metallurgy; and *arts du goût*.[66] According to his expertise, each member (with the exception of the honorary) was assigned to one or more of the focus areas, so that each project would be carried out by a group of people with different skills. This rearrangement of the world of the arts reflected the project of improvement that the Société pursued. In the context of glassmaking for optical instruments, for example, two optical geometricians would work with a physicist, a mechanician, a glassmaker, two spectacle makers, and an enameller. In order to advance animal economy, the Société would assemble two physicians, three surgeon anatomists, two physicists, two geometers, and two mechanicians. Metallurgy fell under the care of two physicists, two chemists, and one mechanician. The *Regulations* of 1730 specified precisely how many and what kind of members would be assigned to each area. In addition to assiduous, free, foreign, and honorary members, the Société also included *répondans* (responding members), who were called on a case-by-case basis when expertise was needed from arts or crafts that were not represented in the Société.[67] They included: a cabinet maker, a candle maker, a founder, an artillerist, locksmith, a plumber, a potter, a spectacle maker, a gardener, an apothecary, a sculptor in marble, an accountant, and a painter in enamel. The printer of the Société's *Règlement*, Gabriel-François Quillau; the engraver and draftsman of the Société, Merne; and the maker of scientific instruments for the Société, Marc Mitouflet Thomin, were also *répondans*.

The social differences among members were codified in the *Regulations* and inscribed in the choreography of meetings: "In the meeting room, the honorary members shall be seated to the right and to the left of the patron; the officers will be placed all around the table, where the foreign members, the assiduous, and the free, will occupy the remaining places regardless of distinction of rank, according to the time of their arrival to the meeting. The *répondans* will stay behind the rows of the members of the first four classes."[68] The special placing of honorary members by the patron, and of *répondans* behind all other members, performed the hierarchical structure of the Société des Arts. The Société, and by extension the societal ideal it represented, was neither an egalitarian utopia nor a subversive political project. The social model it represented was grounded in Old Regime practices of patronage and privilege, where excelling in one's art and pleasing the right patron could bring appealing social returns. Obtaining letters of nobility was, for *artistes*, an occasional, though by no means rare, event. Several members of the Société des Arts had been close to the regent and John Law, while others held titles that highlighted their proximity to the court. Thomas Germain was the goldsmith to the king, Jean Faget the surgeon to the duchess dowager of Bourbon, Henri Liébaux the geographer to the Count of Clermont, Philippe Buache the geographer to the king, Pierre Gaudron and Jean Baptiste II Dutertre were clockmakers to the duc d'Orléans, Pierre de Vigny the architect to the king, and François Lemoyne first painter to the king. Several more members received commissions from the court, whereas others, such as the anatomist César Verdier, held institutional positions funded from the royal coffer.[69] The Société des Arts presented to the public a model of social relationships that consecrated the *artiste* as quintessentially distinguished from other practitioners of the arts and crafts.

The elitist attitude of the Société des Arts was particularly manifest in the creation of the class of *répondans*. Unlike other members, who could be *artistes* or *savants*, *répondans* were selected exclusively from the world of crafts. They were expected to submit reports on "all machines, instruments, devices and tools of their art, and the way to use them, together with the description of all matters and preparations they employ both for the preparation and for the making of their works."[70] They were obliged to attend at least two meetings each month but did not have any right to vote. Plans for this class, the only one exclusively composed of artisans with limited decision-making power, strikingly resembled the schemes for groups of artisans who would collect raw materials for the *History of the Arts* to be produced by the Académie des Sciences. Their labor would remain in the shadow, just as the labor of the many workers in artisanal workshops. In fact, the early draft of the *Regulations* of the Société des Arts did not include the class of *répondans*, which likely emerged from the negotiations with the Count of Clermont.[71] Nonetheless, the fact that the founding members of the Société accepted this class points to the widespread acceptance of hierarchy and social disparities within the artisanal world.

PRACTICE, THEORY, AND THE SOCIÉTÉ DES ARTS

Artistes were eager to distance themselves from the rough representation of craftsmen that circulated in the world of learning through works such as André Félibien's *Principles of Architecture*. They read philosophers such as Francis Bacon, René Descartes, John Locke, and Nicolas Malebranche; invested in sizeable libraries; authored books; and commissioned portraits in which they appeared in the same guise as gentlemen of letters.[72] Julien Le Roy had himself depicted with a book in his hand, without any tool or instrument indicating his membership in the world of craft (fig. 4.5). Similarly, surgeons publicly expressed their "pain in being confused with vile artisans" after their association with barbers in 1655. Reminding their readers of surgery's ancient history and of surgeons' humanistic training and knowledge of Greek and Latin, they denounced the "shame and humiliation" this confusion made them suffer.[73] These complaints and self-representations should not be interpreted too simplistically as the *artistes*' desire to be perceived as savants. No doubt, they wanted to present themselves as learned and educated. Although it was not required to practice their art, a few surgeons took master of arts degrees, while clockmakers' libraries reflected their desire to compensate for their lack of education in the

Fig. 4.5. Portrait of Julien Le Roy; engraved by Pierre-Etienne Moitte after a now lost portrait by Jean Baptiste Perroneau. [Chateaux de Versailles et Trianon, Versailles, France; © BnF. Dist. RMN-Grand-Palais/Art Resource, New York]

classics and liberal arts.[74] This was a manifestation of the *esprit* of the *artistes*. Yet they maintained the superiority of practical over theoretical knowledge, and they criticized the preference that state institutions such as the Académie des Sciences accorded to theoretical over practical subjects. When the mathematicians Alexis Clairaut and Joseph Saurin were nominated for the position of adjunct mechanician at the Académie, the inventor Du Quet remarked in his correspondence with the abbé Bignon that both lacked adequate qualifications, since neither had any background in mechanics or physics. He added that if the interest of the state was to perfect the arts, "it would be more advantageous to give the positions of geometers to mechanicians rather than let geometers take the positions of mechanicians."[75]

The founding members of the Société des Arts agreed on the epistemological superiority of practical over theoretical knowledge. The surviving minutes of the Société des Arts clearly indicate the members' commitment to the examination of practical inventions. In the course of their meetings, they offered each other feedback on watches, clocks, compasses, hydraulic and military machines, mathematical instruments, surgical procedures, bridges, fortifications, engraving techniques, and an "artificial shower."[76] Even members who were not *artistes*, such as Clairaut, Grandjean, and Jean Paul de Gua de Malves, subscribed to this understanding of the relationship between theory and practice, and presented to the Société works on practical matters.[77] In 1729, the Société established that theoretical works, even by its members, would not be read in full during its meetings. Consequently, Gua de Malves was allowed to read only a short abstract of his report on the geometry of curves, whereas he read in full his translation of a recipe for making vinegar.[78]

Prudence and skepticism about theoretical claims underpinned the Société's decision in the "Le Maire affair" of 1732. Jacques Le Maire was a mathematical instrument maker and an assiduous member of the Société who claimed to have contrived a declination-free compass. In addition to performing experiments with the compass for the Société, Le Maire had presented a report dedicated to the minister of the French Navy, the Count of Maurepas, which he wanted to publish with the formal approval the Société des Arts. In the report Le Maire revealed how to make such a compass and advanced a theoretical explanation of the observed effects. After careful consideration, the Société decided that it would certify the facts collectively observed, but it would resolutely refrain from expressing itself on Le Maire's theories. The episode was considered so exemplary of the Société's approach to theoretical and practical knowledge that the members planned to include it whenever a history of the Société was written.[79] Later, in 1733, the committee that examined the essay on agriculture by the botanist Pierre Deschiseaux scathingly defined it as a work of "misunderstood and misplaced erudition" that bore "no relationship with the views of the Society nor with the mechanical perfection of the useful arts, which is its only goal."[80] Similarly, the report of the surgeon Garengeot on a new method for extracting stones was regarded as "too savant and too theoretical for the Société des Arts" and therefore inappropriate for the Société's first public meeting.[81]

Members of the Société des Arts continued to engage in this kind of boundary work well after the Société's dissolution in the early 1740s. In 1765, when the Académie des Sciences voted for Jean d'Alembert to fill the position of *pensionnaire mécanicien*, his defeated rival, the former member of the Société des Arts Jacques de Vaucanson, commented: "The intelligent public understands without much effort that it is much easier to make meteorological observations, demonstrations on ice, lodestone or electricity than to invent and build a good machine. In the one case it is only a matter of explaining as one likes certain known effects; in the other one must produce new effects. This is why the great majority direct themselves towards theory rather than practice."[82] Writing to the director of the Bureau of Commerce, Daniel Trudaine, Vaucanson implicitly subverted the motif of the disinterested pursuit of truth that characterized the world of natural philosophy. It was thanks to the work and practical knowledge of the *artiste*, not the prattle of the savant, that improvement could be achieved. Not surprisingly, the perpetual secretary of the Académie des Sciences, the influential writer Bernard Le Bovier de Fontenelle, held the opposite point of view. He was convinced that theory advanced more speedily than practice because the former was "in the hands of intelligent people [*gens d'esprit*], who burn for the improvement of their sciences," whereas the latter was "handled by people who are ordinarily very vulgar and very indifferent to improvement."[83] The world of clock-making, however, offered evidence of the fact that theoretical results that promised great advancements on paper could in fact prove useless when translated into the world of materials.

This was the case with one of the pinnacles of modern horological knowledge, Christiaan Huygens's cycloid. Along this curve, Huygens demonstrated, pendulum oscillations were isochronous. The consequence was that a clock whose weight oscillated along such a curve would preserve its regularity for much longer. The clockmaker and member of the Société des Arts Henri Enderlin, as Sully had done before him, pointed out that the weight of a clock could be made to move along a cycloid only through the addition of several gears. Because of the usual interferences of friction, irregularities of the components, humidity, and similar problems, this addition de facto worsened the regularity of clocks built on Huygens's principle.[84] A theory might be good in principle, but practical advancements resulted only from accurate knowledge of materials. Without the latter, a good theory could be useless in practice. This was a defining feature of the difference between the Académie des Sciences and the Société des Arts: an institution devoted to the advancement of theoretical learning could not by itself contribute to the economic advancement of the state. That should be the role of the Société des Arts.

The world of entrepreneurship was sensitive to the important role that the Société could play in advancing technical innovations and overcoming guild resistance. The introduction of the English method for rolling large sheets of lead presented to the French audience an exemplary case in which the expert opinion of the Académie des Sciences proved insufficient to assess the value of a new technology. As Pierre Rémond de Sainte-Albine, who reported on the episode, explained, "In a matter concerning exclusively a learned and delicate theory, one would not conceive of any other authority than that of the Académie des Sciences. But in a matter such as we are dealing with, its judgement acquires new strength by being confirmed by people of the art."[85] Large sheets of lead were extensively employed in architecture, especially in the covering of churches and large buildings, and to create water reservoirs. In Paris, lead covering was regarded as a luxury feature, and it was applied to the most fashionable buildings. The English method for making large sheets of lead centered on a horse-operated rolling machine, which flattened lead on a large table (fig. 4.6), whereas the traditional French process consisted of pouring melted lead into a flat iron mold. In 1728, an engineer named Fayolle presented the rolling machine to the Académie des Sciences, which evaluated it favorably. One year later, a group of entrepreneurs—among them the royal architect Germain Boffrand—bought the invention from Fayolle and obtained a privilege for a Royal Manufacture of Rolled Lead in the faubourg St. Antoine in Paris, an area that was outside guild supervision. In spite of the Académie's positive judgment, the guild of plumbers questioned the quality of large sheets of lead produced by the rolling machine. Their numerous objections prompted the Bureau of Commerce to request that the French ambassador in London

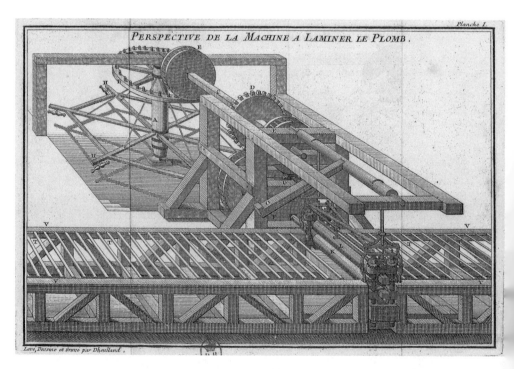

Fig. 4.6. Machine for rolling lead sheets; from Rémond de Sainte-Albine, *Mémoire sur le laminage du plomb* (Paris: Guerin, 1746). [Bibliothèque Nationale de France, Paris]

report on the cost effectiveness of the machine and the quality of rolled sheets of lead. Although the report was favorable again, the entrepreneurs of the Royal Manufacture resorted to the Société des Arts for a final assessment.[86]

The discussion on the best method for making large sheets of lead was not a dull, technical affair. Rather, it was a fashionable debate on an expensive, desirable commodity that interested not just plumbers and entrepreneurs but also the Parisian elite. The vicar of Saint-Sulpice and honorary member of the Société des Arts, Jean-Joseph Languet de Gergy, was obviously interested in the issue, given the grandiose renovation of the church he was organizing. The Count of Clermont himself, with his interests in metals, was also intrigued. He joined the committee appointed by the Société in their visits to the manufacture and witnessed all the tests. He made sure to be present at the meetings when the Société discussed the committee's report and cast his own vote, joined by four honorary members.[87] The debate on the quality of rolled sheets of lead cast practitioners (the plumbers) versus savants (academicians) but with their request to the Société des Arts, the entrepreneurs made clear that either party lacked substantial expertise. In order to evaluate the Royal Manufacture, the Société formed a committee that included reputed *artistes* as well as savants.[88] They examined both the machine and the sheets of lead it produced—just as the Académie des Sciences had done—and expressed a favorable opinion. The Société authorized Rémond de Sainte-Albine to publish an account of the entire controversy and its resolution. The booklet, *An Essay on the Rolling of Lead Sheets*, went through three editions in the course of fifteen years, a success that bore testimony to the popularity of the topic among the reading public. Sainte-Albine's publication presented the way in which the Société operated, showing that its assessments were based on both practical and theoretical knowledge. The Société could easily arbitrate between practitioners unable to change their routine and savants who did not understand the technical aspects of the affair. For entrepreneurs, this was a highly valuable source of expertise. Boffrand, one of the entrepreneurs involved in the manufacture and the architect to the Count of Clermont, was won over by the approach. He joined the Société des Arts soon afterward, in 1731.[89]

THE SOCIÉTÉ DES ARTS AND THE ACADÉMIE DES SCIENCES

The Société des Arts modeled its ritualized practices on those of the Académie des Sciences. Admissions of new members, the structure of meetings, and the review of reports followed the same protocols that were in place at the

Académie.⁹⁰ The meeting days of the Société (Thursdays and Sundays) did not overlap with those of the Académie des Sciences (Wednesday and Saturday), which allowed for double membership. In the course of two hours, members discussed new candidates, voted on procedural issues, and read one or more reports according to the agenda decided by the secretary. The reports were subsequently reviewed by ad hoc committees. In the course of 1732, the Société started planning annual meetings that would be open to the public, the publication of its "history" (summaries of reports that were read or presented), and the award of prizes to nonmembers.⁹¹ All of these were routine activities at the Académie des Sciences.

Several members of the Société des Arts had interacted with the Académie des Sciences before the foundation of the Société. The clockmakers Sully, Julien Le Roy, and Jean Baptiste Dutertre, as well as the inventor Edme Guyot, the instrument maker Le Maire, the mathematicians Clairaut (father and son), and the surgeon Garengeot had presented reports to the Académie before 1728. They had collaborated with academicians, who had publicly acknowledged their contributions.⁹² Relationships between Société members and academicians were not occasional. Sully and Le Roy interacted frequently with the Académie on problems related to astronomy and navigation, and Jacques Le Maire's inventions were known to Réaumur through Bignon.⁹³ With the foundation of the Société these interactions continued. Société members presented their work to the Académie and experimented together with academicians.⁹⁴ The Count of Clermont himself interacted with the Académie des Sciences. In 1729, he joined Réaumur in a series of experiments with noble metals, aimed at finding methods for gilding and silver plating.⁹⁵ In the same year, the count became interested in the various techniques for making porcelain; he set up a workshop on his premises and entrusted its direction to a craftsman that Réaumur had procured. In November 1729, Réaumur read a report to the Académie on the making of porcelain and revealed that the count had been directly involved in this research.⁹⁶

The exchanges between the Société and the Académie des Sciences were not unidirectional. The Société's innovative approach to knowledge-making constituted a model that the Académie strove to appropriate by creating new positions, initiating new projects inspired by the Société's activities, and recruiting Société members. In 1731, the Académie established the new position of adjunct geographer, geography being one of the areas of practical knowledge that the Société des Arts intended to promote. The two candidates for the position, the cartographers Philippe Buache and Henry Liébaux, were both members of the Société des Arts.⁹⁷ In 1729, the Académie started the project of a multivolume publication that would showcase all the machines it had approved since its foundation, the models of which were stored at the Royal Observatory. This project originated with a member of the Société des Arts, Jean-Gaffin Gallon. Until then, the approved machines were only quickly mentioned in the Académie's periodical publication, without any images. During his visit to the observatory, Gallon realized that "countless people who take pleasure in machines, by having them under their eyes could therewith draw ideas able to improve them, or be inspired to invent new ones." He proposed to the abbé Bignon the publication of a multivolume work in the tradition of the theaters of machines, which would showcase the collection at the Royal Observatory by presenting illustrations of each machine accompanied by a short description. He believed that even "people who don't have any mechanical knowledge, such as most artisans and workers, could contribute to the improvement of these machines, or more generally to the art of machines, by means of such publication."[98]

Gallon had been brought to Bignon's attention in 1728, and Bignon had immediately recommended him to Réaumur as someone who, having "a great talent for drawing" and good mathematical knowledge, could prove useful should the project of a description of the arts and crafts resume.[99] When Gallon proposed his project for a collection of machines and inventions approved by the Académie des Sciences, Bignon encouraged Réaumur to sponsor it. Within a few months, Gallon was entrusted with the compilation of the collection and the drawings for the plates, while Réaumur was appointed a member of the committee in charge of supervising the work.[100] Gallon's proposal did not immediately earn him a position within the Académie.[101] During the four years that it took him to complete the multivolume *Machines and Inventions Approved by the Académie des Sciences*, he presented his inventions both to the Académie and the Société, being particularly active in the latter. Eventually he obtained a position as a corresponding member (not a full-fledged one) of the Académie in 1735. Several other Société members were offered positions at the Académie des Sciences: Alexis Clairaut (1731), Charles Marie de La Condamine (1730), Jean Paul Grandjean (1731), Jean Grosse (1731), abbé Nollet (1739), Jean Paul de Gua de Malves (1741), Jacques Vaucanson (1746), and François Quesnay (1751).[102] Société member Jean Charles Habert was considered together with Grosse for the position of adjunct chemist, which eventually went to the latter.

Historians have interpreted the offering of positions at the Académie des Sciences to the Société's members as the main cause for the dissolution of the Société. In particular, they have interpreted these recruitments as Réaumur's planned strategy. He offered membership in the Académie to a few key members of the Société under the condition that they withdraw from it. Since Réaumur believed that the Académie des Sciences should serve the state by providing expert consultants as well as knowledgeable inspectors of manufactures,[103] he opposed the Société because its goals overlapped with those of the Académie.[104] This point of view finds partial support in the admission of La Condamine—director of the Société des Arts, later member of the Académie des Sciences—who in 1765 wrote to his fellow member Grandjean:

> You witnessed personally, Sir, the shame I suffered upon my admission to the Académie, 35 years ago, when they [members of the Académie] demanded that I resign from the Société des Arts, without incurring the disfavor of a prince of the blood, who was then my colonel, and who wrote a letter in which he stated that he saw no incompatibility between my role as an academician and that as the director of the Société of which he was the patron.[105]

The induction of a few members of the Société des Arts into the Académie, however, was not a sufficient reason for the former's dissolution in the early 1740s, some fifteen years after its establishment.[106] No doubt Réaumur's vision for the Académie des Sciences had much in common with the goals of the Société des Arts, and even contemporary observers noticed the overlap in the activities of the two organizations.[107] Yet membership in the two associations was not always mutually exclusive. In 1732, when the news of the foundation of the Société des Arts was spreading throughout Europe, the French ambassador in Copenhagen, the Count of Plélo, wrote to Guillaume Delisle that the Société's members included several famous French *artistes* as well as "five or six of your fellows at the Académie."[108] Grandjean served as the Société's secretary in 1732 while he was a member of the Académie, and Grosse was active in the Société after his admission to the Académie in 1731.[109] The abbé Privat de Molières became a member of the Société des Arts in 1732, when he was already a member of the Académie des Sciences.[110] Furthermore, members of the Académie des Sciences who had been members of the Société des Arts continued to interact with their former colleagues, privileging their practical expertise over that of fellow academicians. In 1732, La Condamine asked the Société to provide a thorough evaluation of two samples of steel. In a diplomatic move, since Réaumur had published on the subject in 1722, he explained to his former colleagues that the assessment would have to remain informal.[111] The collaborations between Réaumur and Clermont make it unlikely that he would have opposed an association that enjoyed the patronage of a prince of the blood, one whom Réaumur had publicly praised as a patron of the sciences and the arts and with whom he had repeatedly worked.[112] Rather, it seems that relationships among key members of the two associations were strong enough so as to make membership in the Société a sort of proving ground for future academicians, or a consolation prize for those who were not granted the acknowledgment they sought at the Académie. This was the case of the Italian Carlo Taglini, who, in spite of his aspiration to become a corresponding member of the Académie, through the mediation of Bignon and Réaumur was offered instead membership in the Société des Arts.[113]

THE SOCIÉTÉ DES ARTS AND THE ACADÉMIE DE CHIRURGIE

If the Académie des Sciences was a model for the Société des Arts, the Société des Arts was a model for the Académie de Chirurgie (Academy of Surgery), founded in 1731.[114] The group that worked for the constitution of the Académie de Chirurgie included two active members of the Société des Arts, Croissant de Garengeot (treasurer for the Société des Arts) and Henri François Le Dran, who were selected by the king as two of the six officers of the academy. Quesnay, a foreign member of the Société des Arts, would become secretary of the Académie de Chirurgie in 1748. Five other members of the Société des Arts became members of the Académie de Chirurgie: César Verdier, Jean Faget, Jacques-François-Marie Du Verney (consultants to the Perpetual Committee), Pierre Bassuel (adjunct to the consultants to the Perpetual Committee), and Daniel Medalon (corresponding member, 1742).[115] Their involvement in the new association did not conflict with their membership in the Société des Arts. On the contrary, they continued within the Académie de Chirurgie the project of improvement pursued by the Société des Arts.

The similarities between the goals of the two groups are striking. The first regulations of the Académie de Chirurgie stated that its primary goal was to improve the practice of surgery. In order to accomplish this, the Académie committed itself to the compilation of a complete history of surgery, "which not only includes all ancient practices, but also the origin of those that have replaced them, and the reasons for the preference that have led us to adopt them." Aware of the complexity of the task, the founding members decided to start with a "catalogue of all the books, ancient and modern, whose abstracts could contribute to the execution of this project."[116] This was fully in line with the Société des Arts's *Regulations*, which stated that the members would work toward the compilation of individual catalogues (collections of abstracts) of the arts.[117] The person responsible for codifying such abstracts at the Académie de Chirurgie was Le Dran, a member of the Société des Arts. The *Regulations* of the Académie de Chirurgie included an article that, as the secretary Sauveur-François Morand highlighted, had "escaped the other academies" and was instead one of the articles in the *Regulations of the Société des Arts*.[118]

Surgeons were particularly fond of the theme of the "mutual assistance" of theoretical and practical knowledge, which the Société des Arts advocated for all arts. They had to withstand the contempt of theory-inclined physicians, who regarded surgery as purely manual work that was pursued by ignorant barbers and that should be subjected to the scrutiny of the medical faculty.[119] The establishment of the Académie de Chirurgie aimed to transform the public perception of surgery from manual labor to useful knowledge. Surgery contributed to moral progress and the economic advancement of the state by creating healthy subjects and monarchs:

> The rich and the poor become indistinguishable in their need of our help. The fate of the prince and that of the artisan are equally in our hands: our skill preserves for the state, as our incompetence steals from it, some portion of its prosperity in the person of the laborer, of its wealth in that of the merchant, of its ornament in that of the savant, of its force and splendor in those of the warrior and the nobleman, of its support and happiness in that of the very same monarch who governs it.[120]

Part and parcel to this makeover of the public perception of surgery was the creation of a prize to be awarded annually to the best essay on a surgical topic. The prize was a gold medal worth two hundred livres, designed in collaboration with the Académie of Inscriptions and Literature. It was a visual endorsement of the combination of theory, practice, and patronage pursued by the Société des Arts. The medal represented Louis XV as Apollo Salutaris (Apollo the savior) who, "having next to him, on the one side, the main instruments of practical surgery, and, on the other, the symbols of the theory of the same art, such as books, skeletons, furnaces, balm urns, etc., seems to dictate to Minerva Hygeia notes on the usages of the one and the other kind" (fig. 4.7).[121] As the ancient god of poetry and medicine, Apollo embodied the surgeons' combination of erudition and practical knowledge. The allegory of the royal patronage of the Académie de Chirurgie represented on the medal reassured polite audiences that surgery did not consist merely of the handling of pincers and saws, but that it was a useful branch of natural knowledge, rooted in erudition and learning, and devoted to the pursuit of the public good.

The cultivation of both theory and practice implied that surgeons would collaborate with physicians, overcoming traditional competition, just as the

Fig. 4.7. Apollo Salutaris, project of a medal for the Académie de Chirurgie, 1731; from *Mercure de France*, June 1733. [Bibliothèque Nationale de France, Paris]

Société des Arts intended to foster interactions and exchanges between savants and *artistes*. This collaboration was a matter of fact within the Société des Arts, where surgeons and physicians worked together in the "anatomy and animal economy" division. In 1732, the director of the Société des Arts, the would-be physician Medalon, proposed the election of the reputed physician Michel Procope (the son of the owner of the famous Café Procope) and was supported by the surgeons Le Dran and Garengeot. In 1732, soon after the foundation of the Académie de Chirurgie, the latter chose the Société des Arts as the arena for discussing his new surgical procedure for extracting stones. Garengeot revised it in light of the criticism Société members had advanced, and the essay was chosen as one of three to be read during the Société's first public meeting.[122] Medalon participated in the Académie de Chirurgie's annual competition with an essay in which he advocated for the use of surgical interventions in certain kinds of tumors. The Académie awarded the prize to him.[123]

FRICTIONS

What caused, then, the dissolution of the Société des Arts in the early 1740s? In 1753—upon his admission to the Académie Française—the Count of Clermont took responsibility for the Société's failure: "I have always loved the sciences and the arts. During my early youth, I formed a society whose goal was to cultivate them. The war, which the King made me the honor of enrolling me in, prevented me from taking care so as to maintain this institution."[124] This statement overemphasized Clermont's role in both the creation and dissolution of the Société des Arts, and his contemporaries echoed that distortion. This was due, in part, to the published *Regulations*, which presented conformity to Clermont's orders as the Société's "principal duty," and that led several contemporaries, including Voltaire, to believe that the count was the founder of the Société.[125] Clermont's patronage was, no doubt, a double-edged sword for the Société. If, on the one hand, it contributed to building the Société's legitimacy and reputation, on the other hand, it left the future of the association to the whims of an aristocrat who had several other priorities. In his initial enthusiasm for the Société, the count made an annual donation of 1,200 livres for funding two prizes: a gold medal worth 300 livres that the Société would award to the author of the best essay on a subject to be determined each year by the Société, and six gold medals, worth 100 livres each, to be awarded to *artistes* who had distinguished themselves for their contribution to the perfection of the arts.[126] The call for essays and inventions to be evaluated for the prizes was published in various magazines, thus spreading simultaneously the Société's self-presentation as an assembly of experts in the arts and the role of the count as its patron.[127]

Clermont, however, hardly ever attended the weekly meetings. His presence, as well as that of honorary members, was meant to provide opportunities to the rest of the Société. Their repeated absences disillusioned several members and discouraged them from participating in the ordinary work of the Société. Members' absences became such a matter of concern as to require an extraordinary meeting, on 25 January 1733, which resulted in a formal request to the count to join the meetings more often and to invite the honorary members to do the same.[128] The request did not lead to any change in the count's attitude to the Société. In the 1730s, Clermont obtained a special permission to join the French Army in the War of the Polish Succession, a commitment that took him away from Paris and distracted him from his commitments to the Société.[129] Nonetheless, he found time to install the comte de Morville, the strenuous opponent of the abbé de Coyer in the debate on the *noblesse commerçante*, as an honorary member of the Société in 1731.[130] This was a quite despotic appointment, given that several members had embraced Law's apology of the *noblesse commerçante*, but it did not have any major consequence as Morville died only one year later, without ever attending the Société's meetings. The Count of Clermont's absences, on the contrary, had the major consequence of delaying, and ultimately halting, the preparations for the Société's first public meeting.

Like the Académie des Sciences's *séances publiques*, the Société's public meeting was planned as an event open to foreign visitors and local cultivated elites. It would showcase the members' expertise, in the sense expressed in contemporaneous dictionaries: their excellence in each art, leading to their ability to report to state authorities about that art.[131] The secretary would initiate the event by giving a short history of the Société and an overview of its goals, followed by a select group of members who would each read an essay on an innovative topic. Artworks by other members would be displayed. The committee charged with the organization of the event selected Gallon's essay on methods for pulling gold threads, Julien Le Roy's essay on his improved horizontal quadrant, Garengeot's essay on a new surgical procedure for extracting stones, and artworks by Oudry, Germain, the engraver of precious stones for the royal cabinet François-Julien Barrier, and the wood engraver Jean-Michel Papillon.[132] This ceremony also was meant to celebrate the Société des Arts's authority in all technical matters with the award of prizes funded by the count.[133] Clermont's military commitments, however, kept delaying the event. Aware that the Société's public meeting would make a firm impression of his patronage of the arts, the count wished to be present and denied his permission to hold the event in his absence. Ignoring multiple petitions by the director, Claude Bottée, Clermont ordered the Société to wait for his return but was unable to commit to any firm date, repeatedly postponing his availability over the course of two years.[134] The failure to host a public meeting, whose preparation took a full year, dampened the enthusiasm of the active members. The director was dismayed by a decision that he regarded as superficial and outraged by the fact "that one wants to be obeyed without reply."[135] The Société had advertised the event in the most popular magazines and the prospect of having to postpone it indefinitely was a cause of great embarrassment. Moreover, they communicated with the count via his secretary, the writer François-Augustin de Paradis de Moncrif (soon to become one of the forty "immortals" of the Académie Française), whose "despotic style" did not ease interactions. The realization that Clermont's patronage paradoxically impeded the Société's activities resulted in the decision to terminate the relationship. In 1737, the Société resolved to ask the count's permission to look for another place for its gatherings.[136] There is scattered evidence that the Société continued activities in the houses of the clockmakers Julien Le Roy and Pierre Gaudron until the very early 1740s.[137]

Although the unreliable patronage of the count hindered the Société des Arts, it primarily exacerbated what was an already precarious situation. There were several sources of friction within the Société. The *Regulations* of 1730 fixed the number of free and assiduous members at one hundred, to be distributed within the nine focus areas. Each area could only be constituted of a fixed number of *artistes* and savants. This entailed a large yet structured expansion of the Société from its original twenty-five members. However, before the *Regulations* of 1730, new members had been inducted through individual connections, and it was not easy to retrospectively fit them into the nine areas. Julien Le Roy's recruits, for example, were supernumerary in the class of clock- and watchmaking, and were given different roles (mechanicians, physicists, etc.) or temporarily assigned to the class of foreign members. Even if during the first two years of its existence the Société was "purged twice" of members that did not live up to expectations,[138] the "distribution of the arts"—the attribution of each member to the Société's focus areas—appeared in the minutes only a couple of times in 1732, and it was referred to as a task that was not progressing as rapidly as it should.[139] The quality of the Société's members remained an issue that the various directors monitored. An undated draft on "the improvement of the Société des Arts" explained that only those who excelled in their art were to be admitted to the Société, and their selection ought to be "extremely scrupulous."[140]

The Société des Arts consisted of overlapping networks. Outside the Hôtel de Clermont subsets of members met at the Académie de Chirurgie; the Académie d'Architecture; the Académie de Peinture et Sculpture; and in guilds, workshops, and aristocratic circles. Even if they all shared the main agenda of perfecting the arts and agreed on the epistemological superiority of practical over theoretical knowledge, their actual interactions within the Société were not always collaborative. With Clermont's patronage becoming more nominal than factual, only a few *artistes* within the Société continued to believe in the collaboration and constructive criticism envisioned by the founding members. In 1733, when the schedule of the first public meeting was finally decided, the Société agreed that the secretary should carefully select excerpts from the members' essays so as not to give any sense of internal disagreement.[141] Garengeot, in particular, was formally invited to revise his essay in order to eliminate his harsh statements on alleged plagiarists.

Unlike in the Republic of Letters, politeness and civility were not constitutive of the Société des Arts.[142] Bitter disagreement occurred, especially among members in similar businesses. In 1731, the economist Nicolas Dutot, a former collaborator of John Law, presented to the Société an analysis of the technical processes related to the minting of species. He examined in particular the "remedy," a technical term that indicated the tolerance that the king granted to the Royal Mint when it came to the difference between the actual quantity of gold and silver in each coin and the legal standard. Dutot harshly criticized the edict signed by the regent in 1719, which exonerated the directors of all French mints from the charge of illegally profiting from the remedy. He also denounced their abuse in appropriating more than their due share of (precious) wastes.[143] Renard Du Tasta, the director of the Paris Mint, must not have been too pleased. Surgeons and clockmakers too engaged in confrontations. In 1734, the Société des Arts director Claude Bottée wrote to Moncrif that he felt often unable to reconcile the heterogeneous spirits of the members.[144] Contemporary observers did not fail to notice the Société's internal tensions, implicitly evoking stereotypes of the litigious artisanal world.

Savants' post-mortem analyses cast the Société des Arts as an association doomed to failure because of its artisanal constituency. In his obituary for Jean Paul Grandjean de Fouchy, the marquis de Condorcet exalted the cultural experiment represented by the Société des Arts but remarked that "prejudices and petty jealousies made this useful institution fall."[145] D'Alembert explicitly attributed the failure of the Société to the poor nature of its members: "The count of Clermont formed a literary society, whose meetings he sometimes attended, and which had taken the title of Société des Arts. This sort of academy intended to put together the sciences, the liberal, and the mechanical arts. The project was grandiose, but too vast, and for that matter it was too badly organized by those whom the prince charged with its execution."[146] Making a caricature of the intellectual goals of the Société des Arts, D'Alembert went on to discuss the "strange idea" conceived by "the editors of its regulations":

> Not only did they want, which is reasonable, to wed, so to speak, each mechanical art to the science from which this art could draw some enlightenment, such as clockmaking to astronomy, the making of lights to optiks; they also wanted, if you let me say so, to join each of these arts to the kind of writing [belles-lettres] that according to them had more to do with it; the embroiderer to the historian, the dyer to the poet, and so on with the others.

His conclusions pointed even more sharply to the members of the Société des Arts—not to the idea of the Société itself—as the cause of its failure. The prince's confidence "was ill served . . . by those whom he had honored." His "laudable vision for the progress of the sciences, letters, and arts, remained ineffective because he was not fortunate enough to find collaborators worthy of seconding his views."[147]

D'Alembert's dismissal of the members of the Société des Arts reinforced stereotypical visions of artisans as vulgar men of the hand, unable to advance knowledge or even the arts they practiced, expressed decades earlier by fashionable savants such as Bernard de Fontenelle. These visions subsumed the heterogeneity of the artisanal world as represented in the Société des Arts into the universalizing category of the intellectually limited artisan. The Société des Arts, instead, was a space in which distinctions between artisans and *artistes*, and even among *artistes*—as evidenced by the class of *répondans*—were quite clear. The Société existed by virtue of these distinctions, which enabled the collaboration of people with different expertise, united by the common goal of improving the arts.

Later *artistes*, such as the clockmaker Ferdinand Berthoud and the surgeon Antoine Louis, offered an alternative interpretation of the reasons leading to the dissolution of the Société des Arts. Berthoud admitted the existence of "pettiness" and "mean envy" among the Société's members, which betrayed the collaborative spirit on which it was based. He was convinced, however, that all frictions would have disappeared if the government had granted to the Société the financial support it deserved. Louis similarly pointed out that the Société's "lethal vice" was lack of financial support. As the perpetual secretary of the Académie de Chirurgie, which existed thanks to the unrelenting patronage of the first surgeon to the king, Louis readily acknowledged the importance of a patron, whether in the person of a powerful man or the state.[148]

I suggest that the main reason for the failure of the Société des Arts had to do above all with the inherent paradox of its political project. While the Société presented itself as an association devoted to improvement through collaboration and collective work, its *artiste* members pursued an ideal of upward mobility restricted to a select few such as themselves. This point is clearly illustrated by the creation of the class of *répondans*. *Répondans* were *artistes*: they were ingenious, learned, and excelled in their craft. Unlike other *artistes* in the Société, however, they were excluded from any decisional process. This distinction undermined the Lockean conception of human nature upon which the Société was founded. As the calligrapher Royllet—the author of *New Principles of the Art of Writing* and a *répondant* of the Société des Arts—explained, it was "unnatural" for any *artiste* to be confined to a class that did not offer any expectation of promotion.[149] This understanding stemmed from the working practices of the *artiste*: unlike merchants, who simply circulated already-made commodities, *artistes* manipulated natural materials in order to produce luxury items or useful objects. This transformative act demonstrated that art helped nature achieve a more perfect, orderly status. This reconfiguration of the natural world by art corresponded to the social ascent *artistes* wished for themselves, and which they often obtained.

The possibility of upward mobility was grounded in the *artistes'* understanding of nature, and human nature, as capable of change and improvement. This was a notion that started to circulate in France around this time. Between 1722 and 1735, the French diplomat and natural historian Benoît de Maillet produced *Telliamed*, an irreverent work on the age of the earth that was based on the notion that changes occurred to the crust of the earth as well as to living creatures over long periods of time. De Maillet conceived of life as a natural phenomenon that originated in water and gradually conquered land. This idea challenged the immutability of the chain of being (and the creation story of Genesis), offering instead a new vision of the natural world in which living creatures constantly underwent processes of transformations. Henry Liébaux, the first secretary of the Société des Arts, knew de Maillet and owned the manuscript.[150]

Members of the Société des Arts experienced firsthand the opportunities of social metamorphoses that Old Regime practices of patronage offered. To take only a few exemplary cases: François Quesnay spectacularly climbed the social ladder from his apprenticeship as an engraver in the French provinces to his position as the secretary of the Académie de Chirurgie, and, finally, as Madame de Pompadour's personal physician. The abbé Nollet started his trajectory as the apprentice of the enameller Jean Raux and, thanks to his secret service to the state, obtained the chair of physics at the Collège de Navarre. Henri Liébaux started as a map engraver and, thanks to his leadership of the Société des Arts, became the geographer to the Count of Clermont.[151]

As Royllet pointed out, lack of clear advantages deterred *artistes* from joining the Société. The number of *répondans* in the Société had spectacularly shrunk from twenty-four at the end of 1730 to zero by the end of 1733. Royllet's counter-proposal was that all new members should start as *répondans*, with the best among them gradually promoted to the class of foreign members and then to the class of assiduous or free.[152] The fact that this was never discussed shows that the project of the Société des Arts was neither meritocratic nor democratic. In contrast to the Lockean understanding of human beings as naturally collaborative, it was through the establishment of a state institution, or through the support of a committed, powerful patron, that the Société des Arts hoped to keep all the members together. The societal microcosm that the Société des

Arts represented was intrinsically paradoxical. If, on the one hand, its project relied on natural law conceptions of human relationships based on collaboration and the pursuit of the common good, without the authority of a patron, or the status of a state institution, its members found themselves competing against one another in pursuit of distinctions and privileges. If Locke was the ideal, Hobbes was the reality. As in a clock without an effective *régleur*, internal friction, without the regulating power of the state, caused the Société des Arts to fall into disarray.

Part 3

Writing and Making

Chapter 5

THE POLITICS OF WRITING ABOUT MAKING

The practice of the arts depended on embodied skill, the know-how or tacit knowledge that practitioners acquired in the workshop, generally under the guidance of a master. As a nonverbal ability, embodied skill could not be fully captured in textual or visual representations. Pre- and early modern practitioners repeatedly complained about what they termed the "insufficiency of words," or the inability of the written medium to accurately record all the information that was necessary to reproduce artisanal processes or recipes. Nonetheless, they authored a vast range of how-to manuals, recipe books, technical dictionaries, pedagogical treatises, and encyclopedias that detailed various aspects of their art. Several scholars have discussed the reasons that compelled artisans to write, in spite of their acknowledgment of the limitations of words. Pamela Long has emphasized the social benefits that practitioner-authors obtained as a result of their writings on technical matters. Hélène Vérin has underscored the normative and pedagogical functions of such treatises. Pamela Smith has shown that writing down practical knowledge was a cognitive exercise, a "meditation" on materials that resulted in the articulation of a "vernacular science of matter and transformation." The encyclopedic craze of the seventeenth and eighteenth centuries did not sweep away the idea that words were inadequate to fully represent artisanal knowledge. On the contrary, as Ann Blair and Richard Yeo have argued, projects of "total works" on human knowledge, such as encyclopedias and dictionaries, were responses to the overflow of information that derived from the cornucopia of publications that overwhelmed readers. By selecting, abridging, and organizing relevant bits of information, encyclopedias and dictionaries attempted to offer functional pathways through the labyrinth of knowledge that the "multitude of books" had generated.[1]

This chapter builds on such studies to argue that, for *artistes*, writing about the arts was a political act. I discuss writing about making as a process of intersemiotic translation, predicated on the impossibility of fully communicating everything. Intersemiotic translation is a concept introduced by the linguist Roman Jakobson to discuss translations between different systems of signs, for example from books into films or vice versa. All translations entail processes of selection, interpretation, and adaptation, but the category of intersemiotic translation invites reflections on the active process of recreation that is associated with translating across different media.[2] The emphasis on the awareness of the impossibility of a simple transfer of knowledge from the workshop to the book is particularly useful in the analysis of the writings by *artistes*, as it calls attention to the strategic choices of practitioner-authors. In other words, it brings the politics of writing about making to center stage.

By emphasizing the political dimension of writing on the arts, I do not mean to suggest that *artistes* were taking collective positions in favor of or against the political institutions of the Old Regime. I argue that their advocacy for the relevance of the mechanical arts was intended to position them, and not the savants, as the experts that should help the French state in making decisions about the encouragement of trade and manufactures, and about technical innovations. This vision implicitly called for a reconfiguration of existing social hierarchies that failed to differentiate between *artistes* and other artisans. It was a political project that was consistent with, not in opposition to, the structures and institutions of the Old Regime. As in intersemiotic translations, *artistes* made strategic choices about what to present in their published writings when they recreated their art in textual and visual media.

I argue that writing about the arts was a political act that concerned the adjudication of expertise and the articulation of rules. *Artistes* advocated that they, not savants, should write about the theory and practice of their arts, and that only they could assert the rules for practicing their art at its best. These rules constituted what the surgeon François Quesnay—a member of the Société des Arts who would later become a leading physiocrat—called the "theory" of the art, a process that historians of technology have discussed as the beginning of a normative discourse on the "science of the arts," or "technology," which occurred in the late eighteenth century.[3] Writing in the 1740s, Quesnay emphasized that the theoretical aspects of the arts differed from speculative knowledge because they emerged from practice and experience. The "true theory" of surgery, he remarked, "must be formed upon diseases, upon operations, upon their success: there it is the only source of the principles that must guide us." This true theory consisted in the articulation of rules for the arts, the formalization of the surgeon's *modus operandi*, which could only be done by surgeons. At a time of heated disputes between surgeons and physicians about who should be responsible for surgical education, Quesnay did not want his readers to miss the full implications of his statement: the theory of surgery could only be the work of surgeons.[4] Rules and principles were not speculations. They resulted from practical experience, not from abstract notions grounded in interpretations of ancient texts.

This chapter discusses the politics of writing about making by examining publication plans (such as the Société des Arts's "History and Description" of the arts), several works that various members of the Société des Arts published under the banner of the Société, and the publications that *artistes* outside the Société, or former members, authored after the dissolution of the Société. I show that *artistes* within and outside the Société des Arts used writing as a means to present themselves in opposition to other artisans and craftsmen, whom they presented as devoid of *esprit*. In their writings, *artistes* articulated a theory of cognition that celebrated the role of embodied knowledge. Their theory differed from the earlier "artisanal epistemology" discussed by Pamela Smith because *artistes* did not claim that there was an approach to knowledge-making that was distinctive to all artisans. On the contrary, they were eager to be perceived as essentially different from other practitioners of the arts and emphasized the features that distinguished their own way of operating. Through writing, and above all publishing, *artistes* circulated a discourse on their own distinctive way of knowing. It was an inherently political discourse, as it concerned the *artistes*' relevance to the French state and the role they could play in projects of improvement.

AUTHORSHIP, ENCYCLOPEDIAS, AND PROGRESS

In response to the increasing number of artisan-authors, the Republic of Letters debated whether practitioners or savants were more qualified to write on the arts. In fact, the abundance of writings on the arts prompted the publication of satirical books, such as *L'art de peter* ("The art of farting," 1751) and *L'art de plumer la poule sans crier* (literally, "The art of plucking a hen with no screaming," or, less literally, "The art of fleecing someone discreetly," 1710), which ridiculed the emphasis on "art" as a set of rules for properly performing certain functions.[5] From the pages of the *Journal de Trévoux*, the anonymous reviewer of Antoine Thiout's *Treatise on Mechanical and Practical Horology* (1741) voiced widespread skepticism about "men of the hand" who picked up the quill, rooting his remarks in the "simple observation" that "the mind is one thing, the hand is another." He explained that even though one could occasionally find among artisans and *artistes* "tempers of mind able of reflection, and superior to their craft," their written accounts needed the supervision and the additional work of professional writers.[6] This view was grounded in the assumption that artisans lacked clarity of expression and the ability to present technical notions in a pleasing way. The reviewer referred to the dispute sparked by the publication of the chevalier de Folard's *New Discoveries on the Art of War* (1724), which had occupied numerous pages in the *Journal de Trévoux* for almost two decades. De Folard maintained that war was a science, an art, and a craft, and, as such, it evolved as much in the field as in the cabinet. But, he warned, it was detrimental to write about it without the sound knowledge that emerged from firsthand experience.[7] His supporters sharply criticized men of letters, such as Machiavelli, who "believed themselves able to write [on the art of war] as well as the greatest Captains." They had seen armies only in paintings and made war only in their reading rooms.[8] The greatest captains, instead, based their writings on their valiant experience, their genius, and their passion for war. As "masters of the art," they offered general rules grounded in successful experience, which could usefully serve as models to their readers.[9] De Folard's opponents countered that if only military men could authoritatively write on the art of war, then "one would need to have been a writer for his entire life in order to write on any subject . . . because we write with the quill, and make war with the sword."[10]

The debate between people of the arts and crafts on one side, and literary writers and historians on the other, divided authors and the reading public, and constituted the context within which the Société des Arts articulated its encyclopedic projects. In his *Artificial Regulation of Time*, Henry Sully anticipated the objections to practitioners of the arts and crafts who picked up the quill. Although he conceded that "it would be bad for a craftsman to claim the title of *bel esprit*," he countered with an example from warfare that someone assigned to the trench might yet be able to excel in leadership roles: "If a man glories himself with the title of Soldier, he does not limit himself, by virtue of that, to only carry a musket."[11] In other words, the status of artisan did not preclude the possibility of writing about one's own craft. Indeed, it was the *artiste*'s task to offer certain rules for practicing the various arts. The reviewer of Thiout's *Treatise* proposed a compromise between the two parties, recommending a division of labor between practitioners and savants that mirrored the Académie des Sciences's approach to the *History of the Arts*. As Thiout's work exemplified, he explained, artisans and *artistes* should provide raw materials that professional writers would transform into a superior and better-organized work. The notion of the *enchaînement* of the arts, which underpinned the encyclopedic projects of the Académie des Sciences, was central to the reviewer's critique of Thiout's work. Whereas Thiout had presented a compendium of individual contributions to the art, professional writers could "compare reports, merge them in bodies of work, nestle them within great plans, deduct principles, and connect [enchaîne] all circumstances and norms."[12]

What was at stake in this debate? As Quesnay and Sully explained, it was crucial for *artistes* to be authors, to speak for their own art. The articulation of a "theory of the arts" had important normative and pedagogical aspects that *artistes* could not delegate to anyone else. Through writing, *artistes* made the case for their expertise and their ability to lead projects of improvement. The Société des Arts approached the debate, and the mounting skepticism within the Republic of Letters toward *artiste*-authors, by partially embracing the compromise that the reviewer of Thiout's work would later propose. The Société entrusted the position of *indicateur*, or indexer, to a member with demonstrated writing skills, who would compile annual catalogues of "all new books, journals, theses, and any other work that will appear in Europe concerning the arts as well as the sciences related to them."[13] This position was first held by the abbé Jean Paul de Gua de Malves, a mathematician who worked as a translator from the English language and who would be the first editor of the *Encyclopédie*. The next *indicateur* was Pierre Rémond de Sainte-Albine, who later became the editor of the *Mercure de France* and a celebrated playwright.

If these selections can be seen as a concession to the argument that privileged professional writers over practitioner-authors, the Société des Arts's approach to writing on the arts considerably diverged from the proposal of Thiout's reviewer. In the first place, the ideal of the *enchaînement* of the arts that the reviewer evoked, and that fascinated the savants at the Académie des Sciences, was not as appealing to the *artistes* in the Société. They knew from their own experience that the ways in which the arts related to one another changed through time. The arts were linked together by a relationship of mutual dependence, but this relationship was not as permanent as the *enchaînement* of the arts implied. The variable nature of artisanal collaboration was at odds with the notion of a timeless order that underpinned the encyclopedic project of the Académie des Sciences and with the notion of the chain of being upon which the project was founded. The Société des Arts itself was an expression of the plasticity of the relationships among the arts, and among the arts and the sciences. The Société was a sort of living encyclopedia where connections were created according to the demands that emerged from common projects. Its nine focus areas, based on the collaboration of *artistes* and savants with different expertise, identified new fields of practical knowledge that eluded any permanent classification.

The writing projects of the Société des Arts were grounded in the radically different belief that in the world of the arts progress occurred frequently. Accordingly, the Société prioritized short-term publications over an encyclopedic project, which, though present in the draft proposed to Jean-Paul Bignon in 1728, disappeared from the *Regulations* of 1730. In the style of literary journals of the time, such as the *Nouvelles de la république des lettres* or the *Journal des sçavans*, the Société planned to publish monthly compilations of technical news that would serve the broader community of people interested in the arts by providing information on the latest inventions or discoveries. Like the Académie des Sciences, the Société would also publish its own "History," a periodical publication that detailed the works in progress of its members. All these publications were intended as a representation of the collective expertise that the Société offered to the public and to the state, and of its commitment to improvement and the public good. In contrast to the encyclopedic projects that intended to record best practices and tools as models to be emulated, the Société's publication projects intended to make the public aware that the world of the arts was a moving target, a forward-looking context where improvement occurred both frequently and unexpectedly.[14]

The other crucial difference between the writing projects of the Société des Arts and the model proposed by Thiout's reviewer concerned the division of labor between practitioners and savants. The idea that savants would reorganize the raw material provided by practitioners fundamentally contradicted the goals of the Société. This is best illustrated by the plans for the publication of an encyclopedia on the arts that the director, Daniel Medalon, resumed in 1732. The Société established a committee of nine members, one for each focus area, who would organize and direct the "History and Description" of the various arts practiced in Paris. These members were not exclusively savants: they included the writer François-Augustin de Paradis de Moncrif (also secretary to the Count of Clermont); the notary and royal secretary Louis Auvray; the lawyer and secretary to the Société des Arts, Hynault; the director of the mint, Mathieu Renard Du Tasta; the surgeons Henri François Le Dran and René-Jacques Croissant de Garengeot; and the clockmaker Pierre Gaudron. Although the minutes of the Société meetings do not explain how the work would have been organized, they make it clear that the "distribution of the arts," in other words the allocation of the Société members within the nine focus areas, was functional to the completion of any writing project. Likely, the nine members would commission and gather various memoirs from the areas under their supervision. This was a radical departure from the encyclopedic projects of the Commission des Arts or the Académie des Sciences. Through its encyclopedic project, the Société des Arts intended to translate in written form the kind of cross-professional relationships that motivated its existence. The ordering principle that organized the work was no longer an abstract notion of interconnectedness among the arts that stemmed from a Baconian understanding of the relationship between nature and the mechanical arts. It emerged from the project-oriented approach to improvement that the creation of the Société's focus areas made visible. These focus areas (agriculture and economy; animal economy; textile and dyeing manufactures; military and civil architecture; the building of ports and ships; horology and mathematical instrument making; optical glassmaking; metallurgy; *arts du goût*) reconfigured existing skills and literacies in the service of France's commercial and imperial expansion. An encyclopedic project structured along these lines was a strong political statement about the relevance of the Société des Arts. If the Ancients celebrated the past, and the Moderns the present, the Société des Arts was interested in France's future.

The "History and Description" of the arts shared the destiny of all of the publishing projects presented in the *Regulations*: it never reached completion. In 1733, when it became clear that the Count of Clermont was more committed to his military career than to the Société, and therefore that the much awaited first public meeting would not take place as planned, absences from regular meetings became more and more noticeable. General disaffection for the Société was evident. Nevertheless, a few members did produce works that might have been part of the "History and Description." In 1733, Jean-Gaffin Gallon presented a report on "the art of drawing and spinning gold threads," which was then selected as one of the essays to be read during the Société's public meeting and, the following year, the clockmaker Claude Raillard discussed during four meetings his work-in-progress on the history of clock-making.[15]

Although we do not have direct sources to examine what the "History and Description" of the arts coordinated by the Société would have looked like, the prospectus of an encyclopedia on horology that Sully outlined in the same year as the establishment of the Société des Arts casts light on an *artiste*'s vision of a "total work" on his art, as well as on the difficulties that such a project encountered. The most salient difference between Sully's and the encyclopedic projects discussed earlier concerned the understanding of how often improvement in the arts occurred through time. Sully's encyclopedia, which was ended by his sudden death, intended to present horology as a constantly evolving branch of human knowledge that relied equally on materials, practice, and theory. Its aim was to educate the public and stimulate emulation and collaboration in other clock- and watchmakers. Its six volumes traced a progressive path of learning leading to a full understanding of Sully's most recent inventions, the lever pendulum and the marine watch. As the most advanced innovations in the field, Sully's discoveries would constitute the starting point for new developments in the art, which would result in an increased quality of clocks and watches, as well as new contributions to horological methods for finding longitude. Sully's encyclopedia was addressed to different audiences, ranging from leaders of trading companies and the navy to *artistes* and savants actively engaged in horology.[16] To the nonprofessional reader, the work would offer relevant knowledge to make informed decisions about which clocks and watches to purchase, or in which projects to invest. To Sully's fellow makers, it could serve as a reference book that informed them about the best techniques and the most advanced discoveries. To savants, it would present the state of the art to which they could contribute by offering criticism or suggestions, which would in turn stimulate the best *artistes* to revise and advance their own work.[17] Such encyclopedic work, as Sully acknowledged, could not be the work of one man only, but it could be accomplished only through collaboration with other *artistes*. Completeness, the most distinctive quality of an encyclopedia, was a particularly difficult target at a time of great innovations. Gathering as much updated information as possible was crucial. Consequently, Sully invited his colleagues in France and abroad to send him reports that he would subsequently publish. However, the creation of a network of collaborators was not easy to obtain, and without it, the task of capturing in writing the ever-changing world of the arts was simply impossible. Sully's vision of the Société des Arts as an association of savants and *artistes* that agreed to work on focus areas was obviously related to his understanding of encyclopedias of the arts as collaborative projects that contributed to improvement.

As an existing network of experts, the Société des Arts seemed poised to complete the encyclopedic task it had set for itself. So why did it not succeed? A comparison with the quite formidable success of the *Encyclopédie* sheds light on the reasons for the Société's failure. The *société de gens des lettres* that Denis Diderot and Jean Le Rond d'Alembert constituted was brought together not only by an ideological commitment to the project; it was also maintained by an entrepreneurial plan that rewarded its contributors, subscribers, and printers. When Diderot and D'Alembert stepped in as the new editors of the *Encyclopédie*, their contract stated that they would receive, respectively, three thousand livres and seven thousand livres.[18] The contract also allocated funds to collaborators.[19] The number of subscribers to the *Encyclopédie* increased with the publication of the first volumes in 1751, and it escalated when the work was placed on the Index of Forbidden Books of the Catholic Church and eventually suppressed in 1759. The *Encyclopédie* was big business. In 1771, Diderot estimated that it earned the printers a total of 2 million livres.[20] With the weak support of the Count of Clermont, the Société des Arts lacked the financial resources to carry out a similar project. It also lacked the status of an academy, which might have provided the institutional authority for accomplishing the project. The Regent's Survey, which similarly required the collaboration of a great number of artisans and practitioners in the provinces, had been successful thanks to the networks of inspectors that enforced royal authority at the local level.[21] Internal divisions and the pressure to publish quickly under one's own name in order to secure prestige and reputation in the public arena did not allow for the community-building ethos that kept the *encyclopédistes* together.

THE *ARTISTE*: A SELF-PORTRAIT

Writing about their own art gave *artistes* the opportunity to respond to common literary representations that portrayed artisanal work as purely rote. *Artistes* differentiated themselves from other practitioners of the arts and crafts, whom they discussed as mindless "artisans" or "workmen." The most common theme in the *artistes*' rhetoric about other artisans was their blind attachment to routine. Sully attributed the slow progress in clock- and watchmaking to the workmen's adherence to repetitive tasks: "They [workmen] only do what they have seen their masters do, without ever thinking."[22] Decades later, his fellow clockmaker Jean-André Lepaute denounced the overwhelming abundance of workmen, especially striking when compared to the scarcity of *artistes* capable of "reasoning about this art." This imbalance, according to Lepaute, continued to hinder the progress of clock-making.[23] This point of view was common among members of the Société des Arts. While defending surgery as an art that required as much intelligence as manual skill, Quesnay disparaged surgical empirics as "mere workmen bound to a miserable routine, often pernicious to patients and always harmful to the progress of the art."[24] He highlighted instead the industry, research, experience, dedication, and skill that characterized surgeons' work. He then asked, "Are all these resources, so precious for human life, worthy only of the vile name of artisan?"[25] Authors of works that required memory, eyes, hands, and diligent imitation were only "worthy of the title of skilled workmen." But the arts that displayed "intelligence [*esprit*] and invention," those that originated from ingenuity, those that operated according to "the most enlightened reason," were the domain of those whom he defined as *artistes*. He included in this category geometers, architects, sculptors, painters, and chemists, together with surgeons. Their practical work, a combination of senses and *esprit*, was "more estimable than the sterile speculations of philosophers."[26]

The definition of the *artiste* as quintessentially different from other practitioners was central to the *Essay on Horology* that the watch- and clockmaker Ferdinand Berthoud published in 1763. Berthoud, who, like Sully, was interested in the problem of longitude and hoping to revive the Société des Arts, underscored that horology should not be mistaken for the craft of making clocks or watches by "servile imitation." That was the job of manual workers. Horology, instead, was "the art of disposing a machine after some principles, after the laws of motion, and by employing the simplest and most solid means." As such, it was "the work of the men of genius."[27] So, even though all those who practiced the craft of watch- or clock-making were called *horlogers*, Berthoud contended that a distinction should be made between the "*artiste* who masters the principles of the art" and those who obtained their license simply by demonstrating their manual ability to make a clock movement. The *artiste horloger* was like an architect: he designed and planned the machine, and then selected the workmen who would build it.[28] The *artiste horloger* studied the theoretical and practical principles of his art, engaged in experiments and trials, and supervised mere workmen. In order to make his readers appreciate the kind of expertise required to carry out the task of directing workmen, Berthoud detailed the various professional figures who collaborated on the making of clocks and watches. This impressive list, which amounted to, respectively, sixteen and twenty-one different kinds of craftsmen, highlighted the complexity of clock- and watchmaking. It offered a glimpse of the large and heterogeneous workforce employed in the manufacture of each piece: enamellers, engravers, gilders, wheel cutters, needle makers, chain makers, spring makers, and so on. Berthoud then identified the *artiste horloger* as the masterful coordinator of these heterogeneous workmen, superior to any one of them because of his ability to "execute by himself, when needed, all the components that constitute clocks or watches." Berthoud cast the *artiste horloger* against an undifferentiated crowd of "workmen who are always occupied with making the same thing." The *artiste horloger*, instead, was able to supervise and intervene in each stage of production when he detected problems.[29]

The surgeons and Société des Arts members Le Dran and Garengeot also warned against the "danger" of mindless workmen, which was more immediately evident and acquired new urgency in the context of surgery. Both Le Dran and Garengeot underscored that the surgeon should have an open mind when selecting which method to employ: cutting open a body could always produce unexpected surprises. The intelligent surgeon should always be in the moment, ready to change his strategy according to unpredictable circumstances. Surgeons who operated mechanically, without thinking, were dangerous assassins, a threat to both human life and the art itself.[30]

Artistes wanted the state to understand how they stood apart from—and above—"those who [we]re merely workmen," an argument based on their new notion of ongoing progress in the arts.[31] Whereas the encyclopedic projects on the arts planned by the Académie des Sciences rested on the idea that the arts progressed only slowly, *artistes* emphasized that new discoveries occurred frequently in the workshop. However, such discoveries went often unnoticed because routine-bound artisans were blind to potential improvements.[32] Not only did their habit of performing the same tasks in the same way slow down progress; it also made them "dangerous" since they constantly missed various useful discoveries. Echoing the argument of the master painters and sculptors who battled for the independence of the Académie de Saint-Luc from the rest of the guild, Berthoud proposed that only "the most intelligent *artistes*" should be entrusted with the administration of guilds. The "miserable workmen" who typically held positions of leadership in the guilds' regulatory bodies hindered the progress of the arts when they did not "endeavor to destroy it."[33] The royal enameller Jacques-Philippe Ferrand put forward a similar recommendation, inviting state administrators to bestow their favors only to practitioners who demonstrated a "superior genius for the arts, and a natural talent for their practice."[34]

Another aspect of the *artistes*' self-presentation was the campaign in the name of openness and the public good, aimed at distancing themselves from the stereotypical representations of secretive, mischievous artisans that savants had circulated at the end of the previous century. The opposition between self-interest and public good had been a typical theme in the savants' discourse on natural knowledge. In his *Principles of Architecture* (1676), André Félibien described the difficulties that he encountered due to artisans' reluctance to reveal craft secrets. In order to obtain information about their practice and tools, he had to pay for secrets without ever being sure that artisans would not hide crucial details, or even lie, in order to make extra profit.[35] The search for truth was at odds with the pursuit of profit, and this was the reason why merchants and tradespeople had been excluded from learned institutions.[36] As Pamela Long has shown, however, in the artisanal world the boundary between openness and secrecy was never clear cut, and technical writings that detailed processes and machines related to various mechanical arts had been produced even before the introduction of the press.[37] In the sixteenth and seventeenth centuries, "secrets" became a best-selling keyword in the titles of a variety of books offering medical, culinary, cosmetic, or reproductive recipes.[38] All these texts shared with the encyclopedic projects of the late seventeenth and early eighteenth centuries the concern with the preservation of practical knowledge.

Learned artisans, and savants alike, were familiar with this literature. By taking a position against secrecy, then, *artistes* were appropriating the stereotypical critique of craft secrecy formulated by academicians and savants. In doing so, they highlighted their own investment in the public good and their ability to partake of the gentlemanly codes of the Republic of Letters. The abbé Nollet, for example, praised the generosity of the Fellow of the Royal Society and lecturer of experimental philosophy Jean Theophilus Desaguliers who, during Nollet's visit to England in 1734, had instructed him "freely" about instrument making. At that time, Gregorian telescopes were sought-after novelties in France, but very few Parisian artisans were able to make the essential metal alloy for obtaining clear mirrors. Those who succeeded, Nollet remarked, "made a mystery of their discoveries." Desaguliers, instead, had asked the best workman in London to reveal his recipe to Nollet. Emulating the attitude of the learned Desaguliers, Nollet now offered the recipe to his readers.[39] In a similar vein, the surgeon Le Dran declared that hiding useful knowledge was "contrary to the good of society."[40] These were no longer the times when Ambroise Paré, who published illustrations of surgical instruments, was attacked by the Surgeons of Saint Cosme for violating the guild's norms of secrecy.[41] Garengeot, who served as *juré* in the same guild, regarded open discussions of surgical tools as essential for fostering much-needed collaboration between surgeons and instrument makers. He advised young surgeons to beware of those "misers" who, because of their "petty jealousy," kept everybody else in the darkness of ignorance.[42]

Artistes emphasized that sharing knowledge was for them a moral imperative. By highlighting the pettiness of other artisans and craftsmen, they cast themselves as champions of openness whose personal interest was one and the same with the public good. They wrote about the arts because the circulation of knowledge, which secrecy impeded, was a precondition for technical progress.[43] Not only was secrecy as detrimental to projects of improvement as the loss of knowledge; it also inhibited emulation, the healthy competition among practitioners that triggered the ambition to excel and therefore resulted in the increased quality of commodities and goods. So the clockmaker Le Roy offered textual and visual descriptions of his new inventions while also publicly disparaging the secretiveness of his London colleagues. The makers of the repeater clocks that king Charles II sent to Louis XIV as diplomatic gifts had sealed the clock cases so as "to deprive others of the sight of their machines," a narrow-minded approach that kept Parisian clockmakers from the benefits of mutual collaboration. Le Roy, instead, offered visual representations of his new inventions.[44]

Artistes offered accounts of inventive processes in terms that evoked ecstatic raptures. Similar to the mystic who experienced the presence of the divine physically, the *artiste* received inspired visions of possibilities never attempted before. This process of material revelation distinguished the *artiste* from his artisan peers. When discussing the innovative design of the clock he made for the ceremony of the *levée du roi*, Le Roy related that the Marquis of Beringhem, who was in charge of starting the ceremony, had commissioned the mechanism from him after choosing an unusually small case for the clock. The honor of working for the king had "inflamed" Le Roy's imagination and guided his reflections about a new design that would fit in the case. This meditative process reached its climax when an idea presented itself to his intellect (*esprit*): Le Roy felt initially struck, then "transported with pleasure," as he envisioned all the advantages of the new design that formed in his mind.[45] The clockmaker Pierre Gaudron similarly stressed the astonishment he felt when an idea suddenly presented itself to his *esprit* during his first conversation with the regent, soon to become his patron. When the regent, who was knowledgeable in clock-making, addressed some key issues on the regularity of clocks, the idea of a remounting weight "caught [Gaudron] all of a sudden." The "genius" that inspired him in that "happy moment" also guided him during the production of a new clock, which he later presented to the Société des Arts on his admission.[46]

In their descriptions of the inventive process as a revelatory event during which ideas presented themselves to the *esprit*, a number of *artistes* gestured to the medieval tradition of the arts as a path to salvation. At least since the Middle Ages, the mechanical arts had been described as means for human redemption and salvation. The ninth-century scholar John Scotus Erigena had discussed them as human practices that led to spiritual and material improvement, a theme that was variously revived and rearticulated in the following centuries. The mechanical arts restored humans' prelapsarian state of grace and achieved human dominion over nature.[47] *Artistes* constructed representations of themselves as geniuses enlightened by such transcendental forces. Le Roy declared, "In the very moment that this idea presented itself to my intellect, I was surprised that it had escaped all my colleagues."[48] He underscored the simplicity of his improvement to the mechanical parts of tower clocks, expressing surprise that it had "escaped skilled clockmakers who had meditated on this matter before me."[49] In order for his readers to judge for themselves the ingenious simplicity of his innovation, he described his invention after offering descriptions of traditional models.

Similarly, in 1721, the enameller Ferrand employed a metaphor from the world of esoteric knowledge to discuss the master-apprentice relationship. In *Art of Fire*, his treatise on enameling, he declared that he wanted to be regarded as a solitary pelican "who gives blood from his breast to enable his little ones to survive by themselves, when they are sufficiently strong." The self-sacrificing pelican, like the crucified Christ, symbolized the master's generosity toward "the good disciple who is chosen for the practice of this science."[50] Mastery of the arts required initiation and a "natural disposition." Although Le Roy and Gaudron did not suggest that there was something innate that distinguished *artistes* from other practitioners, other authors did so. The abbé Nollet underscored that his success in the art of instrument making was facilitated by a "natural dexterity" with manual work, which he had cultivated since childhood.[51] Ferrand reminded his colleagues to keep in mind, when selecting apprentices, that not everyone could succeed in the art of enameling: "natural talent" was as essential as ingenuity, genius, and intelligence.[52] Similarly, Berthoud explained that the improvement of existing machines could not be achieved by any practitioner. The *artiste horloger* produced new inventions thanks to a "talent" that could not be acquired. The contributions of a talented *artiste* were essential for the project of improvement, and so his work should be rewarded by the state.[53]

SENSORIAL INTELLIGENCE

Artistes wrote about their art to call the readers' attention to the cognitive processes that underpinned their work. Materials presented numerous constraints to the human imagination, and *artistes* conceived clever expedients to overcome them. It was not just their minds that *artistes* wanted readers to appreciate. They wanted to challenge the simplistic distinction between the mind and the hand that informed the savants' approach to the arts, and especially the association of artisanal work with the latter. Not that they disdained the hand: on the contrary, the hand was one of the cognitive tools through which *artistes* approached the natural world. But it was not the only one.

Quesnay was one of the most vocal authors to counter the association of his art with the hand. As a surgeon, he was particularly sensitive to arguments built on the etymology of the word "surgery" as the work of the hand (from the Greek words *kheir* for hand and *ergon* for work), which physicians articulated to limit surgeons' academic ambitions. Objecting to the value of etymology for the definition of professional boundaries, Quesnay reminded the "miserable etymologists" that their way of reasoning implied that geometers would have to limit themselves to the measure of lands. Surgery, he explained, could not be reduced to the work of the hands. A good surgeon was somebody who mastered not only manual operations but also "the diligent use of the eyes, and of the other senses," and the acquired ability "to see, to touch, and to feel the objects of this knowledge." His senses worked together with his *esprit* in order to articulate the rules and the elements of surgical knowledge.[54]

If in the classical tradition the art shaped the body, and by extension the mind, of the artisan, the *artiste* now defined himself by virtue of distinctive features that, in other contexts, could be seen as flaws. Marc Mitouflet Thomin, a maker of optical instruments and a *répondant* member of the Société des Arts, remarked that some conditions that were perceived as defective in ordinary people became "natural advantages" for certain arts. Near-sightedness, for example, had allowed celebrated engravers to work without magnifying lenses. The most famous miniature painters were also all near-sighted. There was an important lesson there for Thomin's readers to learn: parents of near-sighted children should direct them "to occupations that are compatible with this disposition, which would place them in a position to excel in the genre of their choice."[55]

These kinds of reflections emphasized the impossibility for the written medium to fully convey the *artistes*' knowledge. Reading works by *artistes* was certainly important for other practitioners, yet it could not replace the experience of learning by doing, nor could it make up for lack of talent or bodily fitness. The fact that *artistes* were nonetheless eager to write about their art points to the political agenda that was inherent in their commitment to publishing. In the process of translating their practices into words, *artistes* made choices about what to highlight. They made clear that the master-apprentice relationship, consisting of verbal as well as nonverbal communication, was essential to the transmission of technical knowledge. They also made clear that their way of knowing was grounded in sensorial intelligence: as *artistes* confronted practical problems, their senses created knowledge in synergy with their minds. The mind by itself did not suffice for the work of the *artiste*; rather, the *artiste*'s entire body was an intelligent medium through which knowledge was acquired.

Writings by *artistes* discussed cognition as an embodied process, grounded in sensorial, open-ended interactions with natural materials.[56] Sully explained how senses and intellect worked together in his own practice. Echoing Nicolas Malebranche's *Research on Truth*, he emphasized that new ideas were constantly generated by what the senses experienced in the processes of selection, combination, separation, construction, and demolition in which *artistes* engaged. In order to become new knowledge, these ideas had to be screened by both senses and *esprit*. Since in the practice of any art, experimentation and improvisation occurred constantly, new discoveries were always possible, but it was important to seize the moment. For this reason, Ferrand recommended that dedicated practitioners should always keep paper, quill, and ink by their side in order to record any relevant circumstance that brought an unexpectedly good result.[57] It was this kind of written record that, at a later stage, would allow for the evaluation of new ideas by the senses and the intellect. Making and remaking were necessary to the acquisition of new knowledge, and writing down their notes helped *artistes* keep track of their progress.[58]

Sensorial intelligence had to be stimulated, and *artistes* detailed methods for doing so. They read works by their colleagues or by predecessors in the arts, they visited cabinets of mechanical devices, and they constantly looked for inspiration in the world of nature and the arts. Jean-André Lepaute underscored the importance of reading anything even vaguely related to horology. But interactions with machines, whether real, projected, or just imagined, were even more important. Masterpieces such as Jacques de Vaucanson's automata, or the mechanical extravaganzas in the cabinet of the Lyon inventor Nicolas Grollier de Servière, were significant sources of inspiration for the *artiste*'s imagination.[59] Jean-Gaffin Gallon shared with his readers the deep impression made on him by his visit to the models of machines kept at the Royal Observatory. He proposed the publication of the *Machines and Inventions Approved by the Royal Academy of Sciences* as a means to make this kind of inspiration more widely available. Collecting was another recommended activity for the distinguished *artiste*. Julien Le Roy was an eager collector of timepieces and of any item related to the arts of watch- and clock-making. He even acquired prototypes from his colleagues.[60] The process of making, or even repairing, was also conducive to new ideas. The instrument maker and Société des Arts member Jean-Antoine Nollet declared that nothing had been more useful in the process of learning how to make philosophical instruments than disassembling and reassembling machines and apparatus together with the most famous makers in Holland and England.[61]

INVENTION AND USEFUL KNOWLEDGE

Unlike the ordinary practice of the arts and crafts, the open-ended nature of the *artistes*' cognitive process could not be "reduced to art," that is to say, it could not be formalized in a set of rules.[62] The entry "Invention" in Antoine Furetière's *Universal Dictionary* underscored this point. He quoted the popular verses: "There is no art for invention, it does not depend on us; it is a gift of heaven, like a pension that one cannot touch when one wishes."[63] Similarly, Sully emphasized the improvisational nature of inventions: "Nothing is rarer in the land of invention than being able to know what to look for, to pursue research in a natural way, and to finally find what one wishes. The path to these happy inventions is very difficult, and for every one aspirant who gets there, ten thousand get lost."[64]

This discussion of the inventive process referred to the sensitive issue of the role of chance as the supreme cause of progress in the arts. Francis Bacon advanced this notion to explain why the mechanical arts had been so slow in their path to perfection. In 1734, Voltaire revived this idea. In his *Philosophical Letters*, he digressed on Bacon's *Novum organum*, reiterating the point that the most notable inventions of the past, including the compass, the press, engraving, gunpowder, optical glassmaking, and the discovery of America, had been the result of chance, not the work of the greatest philosophers. Voltaire added that all arts were born from a "mechanical instinct" common to most men, not from "good philosophy."[65] His statements, quickly spreading through the Republic of Letters, reiterated the savants' views of artisans as replaceable pieces of machinery. Even if he seemingly blamed other philosophers for the lack of contributions to such epoch-making discoveries and inventions, de facto his analysis erased *artistes* from the scene of mechanical inventions.

Writing one year after the publication of Voltaire's *Philosophical Letters*, Julien Le Roy countered Voltaire's notion of a common mechanical instinct with a different version of the inventive process:

> When we consider a machine with the plan of improving it, it is not in our power to form ideas in order to succeed in the project and its execution; this is because we cannot represent to ourselves ideas that we have never had. On the contrary, it is quite easy to recall those of which we already have some knowledge. From this it follows, that it is so easy to copy most machines, and that it is so difficult and rare to invent new ones. Since our imagination and knowledge [*lumières*] are not always able to give us new and useful ideas, and since it is chance that makes us perceive them, we should not be surprised that the progress of the arts has been so slow, because the very same progress has often been the work of chance, which reflection puts to work.[66]

While conceding that chance did play a role in the production of new knowledge on the arts, Le Roy emphasized here the inability of most practitioners to reflect on the opportunities that chance offered. A mechanical instinct could well be common among men, but only *artistes* were able to develop the possibilities that chance offered and transform them into useful devices.

Sully made a similar point about the role of chance in creating connections that *artistes* were able to identify and exploit when he referred to a German satirical work entitled *Wise Nonsense, Nonsensical Wisdom*. The book, which dealt with "the limited genius of alleged inventors," described failed inventions into which a lot of effort, work, ingenuity, and money had been invested (wise nonsense), as well as serendipitous inventions and discoveries that proved unexpectedly useful (nonsensical wisdom).⁶⁷ Serendipitous discoveries demonstrated the *artiste*'s unique qualities: he mastered the rules and was simultaneously open to new possibilities. He did not exercise his mind at leisure in order to surprise and astonish audiences with the display of fanciful mechanical wonders that would never work. His was no longer the time of baroque technological dreams. Through his experiments and open mind, the *artiste* was ready to catch the fleeting moment in which new ideas arrived. It was serendipity, not chance, that governed progress in the *artiste*'s workshop. Chance, in other words, was not a completely blind force. By reading, observing, and making, *artistes* trained their sensorial intelligence so as to be ready to seize the opportunities that chance offered. As Sully explained, chance helped a well-prepared mind: a suspended wheel that turned a millstone in the countryside had inspired him to adapt that design to clock-making.⁶⁸

Artistes contested the notion of invention that was used by other practitioners in search of public recognition, and in so doing they defined their own expertise in terms of its usefulness. The clockmaker and member of the Société des Arts Pierre Gaudron explained that to invent, from the Latin *invenire*, simply meant "to find." The principles upon which machines were conceived could hardly be considered new as they derived from the imitation of nature or from well-known practices in the arts. Self-proclaimed "inventors" simply ignored the history of the arts, which demonstrated that inventions occurred frequently in artisanal workshops, and claimed rewards for something that might never work. Inventions were valuable only in so far as they were useful: *artistes* could consider their work accomplished when they were able to build "a useful machine, which is solid, simple, and known for that use."⁶⁹ Sully agreed that an invention "that has only novelty to boast . . . has very little merit." He added that the ability to make machines that worked distinguished useful inventions from the combination "of things very ingeniously imagined, if considered individually, and whose assemblage is worth nothing with respect to the goal that one had proposed."⁷⁰

The *artistes*' conceptualization of the inventive process marked a clear difference from the earlier tradition of technical writing known as "theaters of machines." That genre, which had inspired Gilles Filleau Des Billettes and the early work on the *History of the Arts* carried out at the Académie des Sciences, emphasized the ingenuity of design over the possibility of execution. Most machines in those treatises were masterpieces of mechanical imagination that often failed to perform when built. *Artistes* wanted polite audiences and state administrators to realize that improving a machine or inventing a useful device were not exclusively thinking processes. Improvement required manual dexter-

The Politics of Writing About Making 163

ity, knowledge of materials, as well as the ability to organize, maintain, and direct a workshop. In other words, it required an *artiste*'s useful knowledge.

IMAGES AND AUDIENCES

Before standardized production, the choice of what artifact to depict and how to represent it on a bookplate required no less thinking than the decision about what and how to write about the arts. In the Old Regime, artifacts created in artisanal workshops were unique pieces that always differed from one another in at least a few details, from decoration to size, from materials to design. Manufacturing processes based on the use of interchangeable parts were introduced by military engineers only in the late eighteenth century. Before then, recreating artifacts from verbal and visual descriptions always entailed some degree of interpretation and imagination. As Ken Alder has argued, the emergence of standardized production was intertwined with the introduction of various tools for preventing disagreement over artifacts' defining features, such as gauges and machine tools. It also depended on the introduction of a series of drawing techniques that standardized the intersemiotic translation from image to artifact, and therefore minimized the processes of interpretation and imagination that characterized the earlier system of production.[71] Diderot underscored the limitation of textual description with an effective anecdote. He told the story of a man who commissioned the portrait of his lover to a hundred painters, but, being unable to show her to them, he sent a detailed written description of her looks and character. When he received all the portraits he noticed that they all accurately represented his description, yet they were all different, and none resembled his lover.[72]

Diderot and D'Alembert invested heavily in the visual apparatus of the *Encyclopédie* in order to overcome the limits of words, yet the plates of the *Encyclopédie*, just like those of the *Descriptions des Arts et Métiers* and their unpublished predecessors, were highly negotiated representations.[73] Geraldine Sheridan, who has analyzed the visual vocabulary developed by the Commission des Arts at the turn of the eighteenth century to represent the various processes and tools involved in each craft, has shown that each vignette was the result of numerous reflections, corrections, and adaptations that involved draftsmen, writers, and academicians. The final representations were as much informed by the rules of geometry and perspective as by rhetoric and the stylistic codes of the Académie de Peinture and Sculpture.[74] These negotiations involved also *artistes*, who sometimes were disappointed with the outcome.[75] This was the case with the wood engraver and member of the Société des Arts

Jean-Michel Papillon, who contributed several articles on wallpaper and wood engraving to the *Encyclopédie*. In 1765, he published a long list of corrections to be added to his *Encyclopédie* articles in which he explained that, while he accepted Diderot's abridgements to his texts, the philosophe's arbitrary modifications of the plates offered inaccurate and misleading views of wood engravers' work, misrepresenting roles, tools, and gestures (fig. 5.1).[76]

The idea that words and images could fully convey the necessary information to learn an art did circulate in published works. In his widely influential *The Art of Turning*, first published in 1701 and then translated into various languages over the course of the eighteenth century, the Minim friar Charles

Fig. 5.1. Detail from the *Encyclopédie* showing a wood engraver's workshop, including several artisans working: sketching plates (at left), quenching tools (lower right), annealing them in sunlight (at center table), and several people engraving on wood (at right). Jean-Michel Papillon, who contributed textual and visual materials on wood engraving to the *Encyclopédie*, criticized this representation as too distant from a simpler one he had submitted. He explained that wood engravers do not sketch with mallet and gouge, as shown at left; this was a technique employed by wood sculptors.

He also criticized the choice of showing tools common to other arts that were only rarely used by wood engravers; these tools were employed by cabinetmakers, to whom wood engravers usually commissioned blank plates. [Museo Galileo, Florence, Italy]

Plumier declared that his goal was to offer readers of any status or profession all of the information they would need in order to learn the art of turning at the ornamental lathe without a master. Plumier, who was no *artiste* but mastered the art, cultivated the ambition to detail "all that a person who wishes to distinguish himself from vulgar artisans needs to know."[77] Accordingly, his text, both in Latin and French, began with an erudite history of the art of turning, with references to the Bible, up to the most secret techniques developed around the world, which he had learned over the course of many years and after several journeys and conversations with the most skilled turners in the world. In line with his plan to fully disclose all information needed to learn the art without a master, Plumier added numerous plates as visual complements to his written text. As a skilled naturalist, well known for his *Description of American Plants* (Paris, 1693), a work of natural history that resulted from his travels to the French Antilles, Plumier was fully aware of the role of images in conveying information about something that readers could not see with their own eyes. In order to enable readers to build the machines he represented in *The Art of Turning*, he gave several perspectival views of each machine, as well as various projective views of each component. Since perspectival views gave a good sense of what the machines looked like but failed to offer useful information on how to actually make them because of the altered proportions among the parts, his readers would get the missing data from projective views and the proportion ratio at the bottom of each plate, which would enable them to calculate the real dimensions (figs. 5.2–5.3). In contrast to his repeated declarations on the possibility of learning the art without a master, however, Plumier acknowledged that particularly complicated techniques could only be learned through long practice: "Because, to tell the truth, words by themselves are insufficient to teach thoroughly somebody on this matter."[78]

The emphasis on the limitation of words called attention to the role of nonverbal transmission between master and apprentice, something that was grounded in practices of making, and that words and images, as accurate as they could be, were unable to replace. Because apprenticed artisans acquired this nonverbal know-how, it was possible for them to fill the informational gaps that derived from the intersemiotic translation of practice into text and images. In other words, artisans might be able to recreate artifacts described in text and images, even if not exactly the same, because they were accustomed to interpreting and reimagining the information contained in verbal and visual media. This would not be the case for readers who did not have the tacit knowledge acquired during the processes of making and remaking that characterized artisanal practice. In the third edition of his *Essay on the Rolling of Lead Sheets*, former Société

Fig. 5.2. Perspectival view of a lathe for turning wood; plate from Charles Plumier, *L'art de tourner en perfection* (Lyon: Jombert, 1701). [Bibliothèque Nationale de France, Paris]

des Arts member Pierre Rémond de Sainte-Albine implicitly pointed to this difference in the understanding of visual representations. He boasted of having finally obtained permission for including plates representing the machine for rolling lead sheets and explained that the entrepreneurs of the Manufacture for Rolled Lead Sheets had previously withheld permission for fear of imitators, an argument that he found unsound.[79] If people like Rémond de Sainte-Albine, who had no direct involvement in making, believed that words and images did not suffice to successfully translate them into a functioning machine, the entrepreneurs realized that visual representations might add crucial information

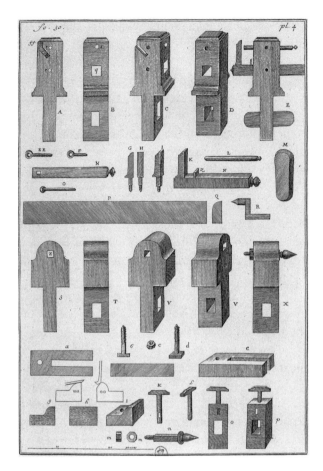

Fig. 5.3. Projective views of the shifting heads of lathes for turning wood (above) and iron (below); plate from Charles Plumier, *L'art de tourner en perfection* (Lyon: Jombert, 1701). A proportion bar appears at the bottom. [Bibliothèque Nationale de France, Paris]

which, combined with textual descriptions, would enable people familiar with the art to fill in the gaps with their own tacit knowledge and know-how.

Artistes made strategic choices when it came to what kind of visual representations to include in their publications. The clockmaker Le Roy included full-page plates of his inventions that did not give any information about the components' size, dimensions, or materials—the kind of details that makers needed to build the clock—attracting instead the reader's attention to the novelty and ingenuity of his design (fig. 4.4). The intersemiotic translation that he operated was meant to call attention to his own *esprit*, not to enable readers to build exactly the same machine.

Similarly, the abbé Nollet selected representational techniques based on the audience he wished to address. In his widely successful *Lectures on Experimental Physics* (1743), which presented the lectures he had given to the French dauphin and was addressed to the fashionable elites, he used exclusively perspectival representations of his demonstration apparatus. This technique allowed viewers to form a three-dimensional impression of the artifact they saw, but it did not convey any information about the relative proportions of individual components. In his last work, *The Art of Experiments* (1770), he employed a variety of alternative techniques. Addressed to amateur demonstrators and instrument makers, *The Art of Experiments* offered rules and principles on how to make scientific instruments, with the goal to enable amateurs in the provinces to build their own *cabinets de physique*. The numerous plates that accompanied the work represented the very same instruments illustrated in *Lectures on Experimental Physics*, but they were markedly different in style (figs. 5.4–5.5). They erased the aristocratic settings that provided the background for the instruments in *Lectures* and offered a variety of points of view.

Translating text and images into working artifacts was not a straightforward process. In the absence of a standardized method of production, the translation depended on the ability of local workmen and the availability of proper materials. *Artistes* presented themselves as experts in the management of workmen and materials, thereby creating a new readership among provincial *artistes* who wanted to contribute to projects of improvement but lacked materials or reliable workmen. Just as the *artiste horloger* was able to manage and supervise a diverse range of workmen, so the abbé Nollet detailed the various strategies he had devised to make glassblowers understand his demands, persuading them to modify their routine when they produced the glass components of his instruments. Since making demonstration apparatus required assembling pieces made by various craftsmen, his readers needed to learn how to interact with them. The text and images of *The Art of Experiments* would enable this ideal reader to make sure that routine-bound artisans would understand how to properly work. Nollet explained that an essential moment of this interaction consisted in cutting out real-size paper models of the desired components and writing detailed instructions on each side.[80]

Similarly, the surgeon and member of the Société des Arts René-Jacques Croissant de Garengeot presented himself as a mediator for aspiring *artistes* in the provinces. In 1727, he published an illustrated treatise on surgical instruments with the intention to encourage further research in the field. The cutlers who

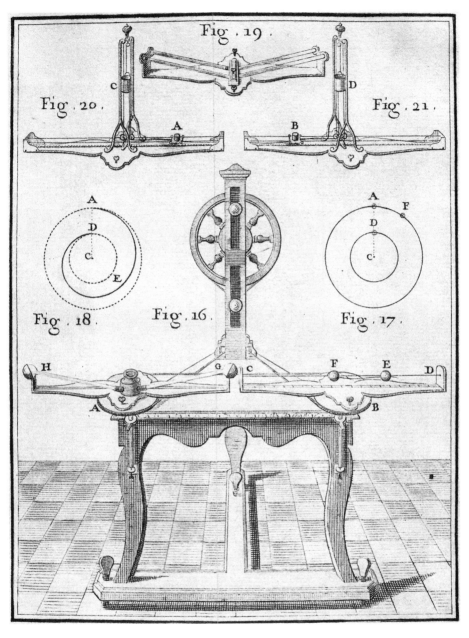

Fig. 5.4. A machine for demonstrating centrifugal forces, illustrated in perspectival view on a patterned floor; a plate from abbé Nollet, *Leçons de physique expérimentale* (Paris: Guerin, 1743–64). [Museo Galileo, Florence, Italy]

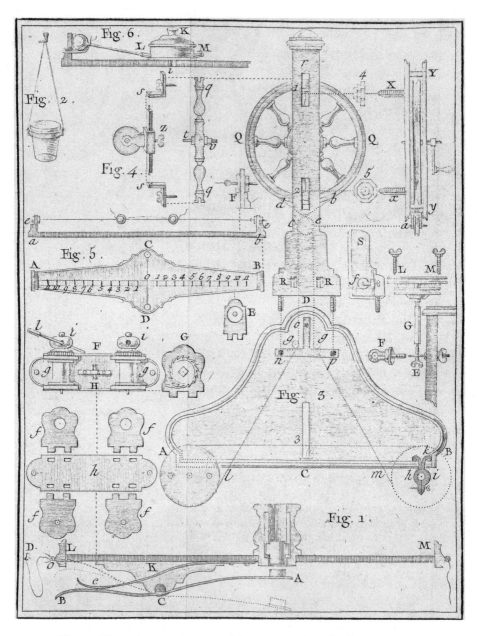

Fig. 5.5. The various components of the same machine for demonstrating centrifugal forces shown in the previous figure, all disassembled and without any context; plate from abbé Nollet, *L'art des expériences; ou, Avis aux amateurs de la physique* (Paris: Durand, 1770). [Museo Galileo, Florence, Italy]

made surgical instruments were unable to understand the surgeons' needs, and they often made pieces that proved uncomfortable to handle—leading, in some cases, to procedural mistakes.[81] According to Garengeot, new surgical instruments should be designed on the basis of knowledge that emerged from the practice of surgery. Surgical work revealed new details about internal anatomy, and instruments should be modified accordingly in order to optimize procedures.

The interaction with local makers was a crucial step for obtaining high-quality tools. In Paris, the goldsmith Lequin Jr. had learned how to adapt his manual skills to the surgeons' demands. Surgeons in the provinces could use Garengeot's text and images as a guide to direct and supervise local artisans.[82] The plates in his book were visual intermediaries between workmen and *artistes*, enabling the latter to guide the former. The illustration of scissors, for example, called the surgeon's attention to the specific components that should be adapted to the size of his hands. This was not a model to be mechanically imitated, which is what cutlers would do without instructions. It was a diagram that attracted the reader's attention to the parts of the instruments that should be adapted to the individual (fig. 5.6).

Artistes were authors and readers. They read works on natural philosophy as well as books on the mechanical arts. While in Vienna, Sully was a regular in the library of Prince Eugene of Savoy, a reading hub for the local and international intellectual elite. There, Sully compiled a reference notebook for his own use, in which he copied all that was relevant to his art from the *Memoirs of the Academy of Sciences* as well as from several texts on the arts in the various languages he could read. The information he gathered, together with his interactions with old and new timepieces, proved crucial for his resolution to improve the regularity of watches and clocks.[83] Le Roy too was an eager reader who availed himself of his network of friends and colleagues to obtain the most recent books. He read Isaac Newton soon after the publication of Voltaire's *Philosophical Letters*, when it became fashionable in Paris to do so, and Madame Du Châtelet's *Institutions of Physics* (1740) as soon as it became available.[84] Papillon detailed his searches in various Parisian libraries for books related to his art, and the inventor Jean-Gaffin Gallon also digressed on the relevance of technical literature in his *Machines and Inventions*.[85] The inventory at death of the instrument maker abbé Nicolas Noel indicates that he owned a large number of books on the mechanical arts, from the theaters of machines to the most recent works.[86]

These examples forcefully indicate that *artistes* wrote for other *artistes*, just as they wrote for the broader public that they wished to educate on the relevance of the mechanical arts. In the former case, their works read as how-to manuals,

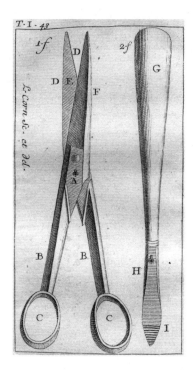

Fig. 5.6. Plate from René-Jacques Croissant de Garengeot, *Nouveau traité des instrumens de chirurgie* (Paris: Cavelier, 1727). [Yale University, Harvey Cushing/John Hay Whitney Medical Library, New Haven, Conn.]

which served as sources of documentation and inspiration, and as references for the processes of adaptation, interpretation, and recreation that characterized Old Regime manufacturing processes. In the latter case, their works read as treatises that showcased their expertise, the importance of the mechanical arts to the projects of improvement, and the valuable contributions that *artistes* were able to offer to the French state.

EDUCATION, ENLIGHTENMENT, AND HISTORY

The political relevance of writing was not lost on the members of the Société des Arts, even after its dissolution. In an essay on the methods for promoting international and domestic trade of French clocks and watches, which he addressed to state administrators, Julien Le Roy advocated for a number of reforms, including the establishment of a new position of "Inspector of Watch- and Clock-Making." This, he recommended, should be entrusted to one of the best *artistes* operating in Paris. In addition to reforming clockmakers' regulations, the inspector should publish articles in the most fashionable literary journals. Le Roy discussed writing as strategic to the promotion of French trade. By spreading authoritative and reliable knowledge about the art of clockmaking, these articles would enable audiences to assess new inventions more critically. Provincial clockmakers would be informed of new discoveries occurring in Paris, and amateurs considering the purchase of a new watch would be warned about the faulty foreign watches that circulated in France.[87] Le Roy's publication plans resembled the monthly "catalogues" that the Société des Arts intended to produce. They were expression of the watch- and clockmakers' needs to be up-to-date with the numerous innovations and improvements that their colleagues continued to produce. For *artistes* who wanted to drive innovation in their field, keeping track of the new was more important than recording the old. Publishing frequently in the *Mercure de France*, as Julien Le Roy did during and after his membership in the Société des Arts, was more relevant than planning encyclopedias on the arts. As the *Mercure*'s editor, Rémond de Sainte-Albine facilitated Le Roy's publications, which strengthened the public presence of the Société des Arts.

The education of the general public was a project in useful knowledge that *artistes* particularly cherished. Their expertise, the material difficulties they overcame, their original contributions and skills could only be appreciated by an educated audience. Appearances, in art as in nature, deceived the untrained eye. With the increasing fashionableness of highly decorated watches and clocks, the number of faulty mechanisms and even fakes circulating in France was on the rise. Sumptuous decorations designed to catch the eye of uninformed customers often hid poor movements and undermined the reputation of the art. Sully's publications were motivated by his desire to instruct readers about the principles of watchmaking so that they would understand what to reasonably expect from a watch. Too many people, he complained, nourished "unreasonable, even impossible expectations" about their watches because they lacked knowledge of what a watch could in fact accomplish. These people were potential prey to deceitful merchants who sold them the impossible dream of "imaginary perfection."[88] Similarly, medical imposters fed on unreasonable expectations born out of ignorance or professional neglect. Because surgical students erroneously did not regard certain disorders (such as eye or tooth problems) as pertaining to their area of study, patients resorted to "charlatans whose only merits are insolence and bravado."[89]

Writing, then, served crucial educational functions. It calibrated expectations, helping readers to avoid deceit and fraud. It also stimulated emulation among practitioners, contributing to raising the standards of each art. The search for a horological solution to the problem of longitude, for example, had been regarded by many as chimerical because changes in temperature caused irregularities to clock and watch movements. The fact that this limitation was widely deemed to be insurmountable was, according to Sully, a theoretical bias, which some grounded in "common opinion, without further knowledge [*autres lumières*]." Other people, who thought "more philosophically," were even more biased against improvement because they did not consider something as "possible, unless they could conceive in their imagination the way in which it can be."[90] According to Sully, the antidote to the "vanity and narrow-mindedness [*petitesse d'esprit*] that produced this arrogant condescension" was familiarity with the *artiste*'s working practice:

> How many surprising effects of human ingenuity and industry do we see all the time, indeed so commonly that we do not pay attention to them, and that the great men of past centuries would have regarded as impossible? If somebody had asked one of the world's most learned men of the fourteenth century if he believed it possible that one day in the future one would find some machine for telling the time with more accuracy than the sun, very likely his answer would have been that he did not believe so. Only the *rare modesty* that always accompanies *real knowledge* would have prevented him from declaring the thing absolutely impossible.[91]

No matter how promising a theory seemed, only the "learned and discerning *artiste*" was able to put together all elements, theoretical and material, needed to improve an existing machine or to build a new one.[92] With proper knowledge, directors and entrepreneurs involved in international trade companies would stop regarding the solution as chimerical. They would easily realize that horological developments would bring forth useful developments in navigation, geography, and cartography—all areas of strategic relevance to commercial and colonial expansion.[93]

If, as Immanuel Kant would later claim, Enlightenment was human beings' emergence from their self-incurred immaturity, educating audiences was an enlightenment project in which the *artiste* was a protagonist. *Artistes* offered the public a landscape of limits and possibilities within which improvement could realistically occur. They encouraged well-informed assessments of inventions and welcomed criticism from their peers. They campaigned against "imaginary perfection," pernicious to the pursuit of useful innovations, and against unsound skepticism, which prevented patrons and entrepreneurs from investing in promising innovations. Most importantly, by improving the arts, *artistes* were committed to bettering humanity. Thanks to their work, humanity would acquire the practical tools for emerging from its infancy.

Indeed, it was the history of the arts that demonstrated how human progress could be achieved. Writing such a history was to engage in propaganda in the name of improvement. As the astronomer Jean Paul Grandjean explained to his fellow members of the Société des Arts in 1728, the history of the sciences and arts did not require the "rigorous accuracy" of political or religious history. The "History of Astronomy" he was preparing would offer modified accounts of historical circumstances and facts "whenever they do not seem convenient to me, or fail to support the order and organization of my plan," because "in these matters it is more important that the work be intelligible rather than its history accurate." Through a selection of key inventions and authors, Grandjean intended to present events "as they should have been rather than as they have been."[94] Even Diderot echoed Grandjean's views when he stated that, so conceived, a history of the arts would present "the progress of an art . . . in a clearer and more instructive manner than by its true history. . . . The difficulties that had to be overcome to improve the art would occur in an entirely natural order, the synthetic explanation of its successive steps would render it comprehensible even for very average minds, and this would divert *artistes* onto the path leading to perfection."[95]

This retrospectively constructed history of linear progress distinguished the arts and sciences from other branches of human knowledge. It depicted an ideal landscape in which improvement had occurred without impediments deriving from human imperfection. It was a process of selective memory that reflected the *artistes*' understanding of progress as essentially connected to advances in the mechanical arts and their role as agents of improvement. If their *esprit* elevated them above other practitioners, the *artistes*' manual abilities distinguished them also from philosophers and savants. Their "artisanal enlightenment" was predicated on the combination of these two key ingredients. If the *esprit philosophique* defined the modern, enlightened age for the *philosophes*, the *esprit* of the *artistes* revealed itself in machines that performed the useful function for which they were designed.[96] As the next chapter shows, *artistes* believed that it was thanks to these machines that the modern age could truly become enlightened.

Chapter 6

L'*ESPRIT* IN THE MACHINE

"The philosopher is a human machine like another man, but he is a machine that, by its mechanical constitution, reflects on its movements.... He is, so to speak, a clock that sometimes winds itself." Thus wrote the philosopher and grammarian César Du Marsais in 1730 in a highly influential and subversive pamphlet that circulated in Paris in manuscript form for several years until it was published anonymously in 1743. Du Marsais's *The Philosopher* was republished several times throughout the eighteenth century, including by Voltaire, who added a revised and abridged version at the end of his *Laws of Minos* (1773). Most notably, Du Marsais's text constituted the bulk of the article "Philosopher" of the *Encyclopédie* (1765).[1] By placing thought within the physical human machine, Du Marsais's definition broke with the Cartesian divide between thinking and corporeal matter, a philosophical departure that also characterized Julien de La Mettrie's *Man, a Machine*, published in 1747. Unlike La Mettrie, however, the machine metaphor served Du Marsais for marking a clear distinction between the philosopher and other humans. While all humans were machines, the faculty that distinguished the philosopher-machine was its ability to reflect on the mechanism that constituted its nature—an ability that, in turn, rendered the philosopher a superior kind of machine, a self-winding clock that did not depend on external forces to form its own thoughts. Free from the yokes of religion or other forms of established authority, the philosopher-machine reflected on itself in virtue of its mechanical constitution. What we may find surprising—the fact that the thinking philosopher should be likened to a machine—was instead at the core of Du Marsais's vision, which, embracing Baruch Spinoza's determinism, equated rationality with the order and organization of machines. Unlike other humans, Du Marsais's philosopher knew the causes that prompted him to act, and so he was able to keep away from irrational passions.

Not surprisingly, the machine that represented the philosopher was a clock, the most sophisticated mechanical device that, by making sensible through auditory and visual means the regularity of nature, was an iconic symbol of order and organization. Analogous to the movements of a clock, rationality moved the philosopher's actions. The machine analogy also evinced important implications for the philosopher's moral rectitude, or *honnêteté*. Since his actions derived uniquely from reason and sense of order (*esprit d'ordre*), those actions were motivated by neither fear of hell nor hope of paradise. Because of his disinterested pursuit of virtue and his veneration for the "rules of probity," the philosopher venerated civil society as his only deity.[2]

Du Marsais considered mechanical operations as the apotheosis of organized microcosms, a perspective he shared with many of his contemporaries. This view had gained traction in Paris thanks to an increasing public appreciation for mechanical devices. In addition to numerous publications on the arts with impressive visual representations of machines, mechanical masterpieces were displayed in exhibits that attracted crowds. Popular magazines such as the *Mercure de France* hosted discussions on the latest innovations in manufacturing, and even the Académie des Sciences dedicated its public meetings to the presentation of mechanical innovations.[3]

Scholars have argued that later in the century, gloomier visions of the analogies between humans and machines became more widespread; yet, as I show in this chapter, the notion of the machine as a sophisticated, orderly system continued to underpin not only the self-identity of the philosopher but also educational programs as well as political and economic plans.[4] Building on scholarship that has pointed to the politics of artifacts, this chapter focuses on machines that shared one crucial feature: they insinuated tools, techniques, and gestures of artisanal workshops into the palaces, intellectual cravings, and political visions of the French elites.[5] I argue that machines of this kind, which became iconic in the latter part of the century, were an essential manifestation of the "artisanal enlightenment." This enlightenment, which drew from and contributed to the Enlightenment of the *philosophes*, expressed itself in particular through educational projects that emphasized the epistemological primacy and edifying nature of the mechanical arts. This chapter employs three case studies to demonstrate that however different these projects may have been, they embodied the belief that machines designed with *esprit* would contribute to humanity's moral and material improvement.

THE *BUREAU TYPOGRAPHIQUE*

In 1728, around the time that Du Marsais composed *The Philosopher*, the pedagogue Louis Dumas invented a machine for teaching young children to read and write. His *bureau typographique*, or printer's desk, was a wooden cabinet with six rows of thirty small boxes, each containing playing cards that displayed the letters of the alphabet (fig. 6.1). At the printer's desk, two-year-old children learned as they played: with the assistance of their tutor, they identified letters, then started to compose words, and, as they grew older, advanced to create their own books, which they stored in a separate box called "the printer's library." The *bureau typographique* could also be used with Latin and foreign languages, mathematics, mechanics, natural history, and other branches of knowledge. Dumas's aim was to introduce a new educational method based on the Lockean idea that knowledge is produced through the senses. Children learned to read and write through their touch, sight, and hearing. The pedagogical philosophy that underpinned this project was similar to the "rational recreations" that engaged both men and women who were interested in learning the basics of the new experimental philosophy. It also rested on a new conception of the learning child as responsive to teaching methods based on movement and play,

Fig. 6.1. The *bureau typographique*; from Louis Dumas, *La bibliothèque des enfants* (Paris: Simon, 1732). [Houghton Library, Harvard University, Cambridge, Mass.]

in contrast to a tradition that imposed immobility and rigor. Conceived of as an aid for tutors working in aristocratic households, as well as for teachers in public schools, the *bureau typographique* challenged traditional pedagogical methods and sparked a heated controversy that lasted over two decades. Detractors and supporters confronted each other from the pages of the *Mercure de France*, and several inventors put their skills to work in order to improve the machine, making it simpler and more affordable.[6]

In 1730, at the peak of the pedagogical controversy sparked by his creation, Dumas strategically sought the support of the Société des Arts. The *bureau typographique* was a mechanical invention that complemented a new pedagogical method based on a phonetic (rather than a standard) alphabet. Evaluating the invention therefore required the consideration of multiple elements, ranging from the mechanical to the pedagogical. This range posed a challenge to the narrowly specialized expertise of existing academies but was ideally suited to the diverse proficiencies among the Société des Arts. Several members of the Société worked as instructors in private households, public schools, or military colleges, and they too had devised instruments and machines for educational purposes. Not surprisingly, the *bureau typographique* elicited great interest among Société members. The Société issued a certificate of approval, and two of its members, Jean-Baptiste Clairaut and Daniel Medalon, found the machine so intriguing that a few years later each proposed his own modified, improved version.[7] The Société commended Dumas's invention for its ability to instruct children through the stimulation of their senses and imaginations. The certificate it issued, published by Dumas in 1733 and discussed in several popular journals, attributed the success of the machine to its "easy and totally mechanical practice," which was "nonetheless founded on the most exact and followed theory." This theory-informed mechanical practice introduced children to a "habit of order and labor" that would serve them for the rest of their lives.[8]

Like Du Marsais, the Société des Arts associated orderly and rational thinking with the mechanical nature of the operations performed at the printer's desk. The *bureau typographique*, however, was not just a metaphor: it was a material quotation of artisanal knowledge. It enacted a defining feature of artisanal apprenticeship by providing children with bodily training that simultaneously shaped their thinking. It honed hands and mind. At the same time, as a machine that enhanced children's learning abilities, it transfigured sweaty workshop gestures into an educational pantomime worthy of the court. The "child typographers," as the children trained at the printer's desk were called, mimicked the real-life gestures of typographers at the composition table. From lining up letters on the desk to laying out books for the library, all of the movements that children performed at the *bureau typographique* replicated the actions of real artisans. In their workshops, artisans gained new understandings of nature and its laws through their sensory engagement with materials, just as children advanced their own cognitive development through bodily interaction with the machine. On the educational stage created by the *bureau typographique*, repetitive artisanal choreographies became sure means to achieve superior reading and writing skills. In the context of a new pedagogical system based on forward-looking philosophical theories, these gestures and operations became educational techniques that the elites eagerly sought. The abbé Alary, tutor to the royal children, acquired a printer's desk for the education of the young princes.[9] To a society still fascinated with *enfants prodiges*, child typographers suggested that, with the right method, any child, male or female, could acquire superior reading skills.[10] After the adoption of the machine by the royal children, testimonials from across the social spectrum multiplied all over Paris. The three-year-old daughter of a silk merchant and a young boy who lived with one of the brothers Procope were among several child typographers who made a show of their surprising reading skills for curious visitors.[11]

The appreciation for mechanical operations redesigned as educational games was far from unanimous. Opponents of the *bureau typographique* criticized its elevation of repetitive routines as instruments of learning. A scathing review published in 1732 compared the child typographers to carnivalesque caricatures who were not even as well trained as domesticated animals. At the fair, the review claimed, "we saw a dog who had been so dressed up and so well directed by her master, that she played cards and won, she assembled letters and composed words spelled much better than those usually composed according to the principles of the supporters of the bureau [*buralistes*]." This review challenged the Société des Arts's competence in these matters with sarcastic remarks about the expertise of the "mechanicians" who endorsed the printer's desk, echoing the ongoing debate over whether practitioners or savants were better qualified to write on the arts. The fact that people of practice claimed authority in anything related to the world of learning, of which education was an important aspect, challenged established conventions. After the positive evaluation by the Société des Arts, observed the reviewer, could an opponent of the *bureau typographique* seek recourse from the Académie Française or the Académie des Sciences without insulting these venerable institutions?[12]

Debates on the education of children were debates on the future of society. Pedagogy was crucial to societal improvement and to the project of the Enlightenment, but precisely how to train future generations was a matter of great contention. The increased visibility of *artistes* in the public sphere, and their estimation of the mechanical arts as worthy topics for the enlightened, generated anxiety about the moral future of humanity. Famously, in 1750 the Académie de Dijon launched a contest for the best essay on the question, Has the restoration of the sciences and arts contributed to the purification of morals? If Jean-Jacques Rousseau had no doubt that the arts and sciences were means of moral corruption, *philosophes* such as Voltaire and Diderot maintained that the sciences and arts were in fact tools of enlightenment.[13] The latter view had been at the core of the Société des Arts's activities. Even after the Société's dissolution, former members were active in educating the public on the collective advantages that would derive from deeper knowledge of the mechanical arts. Not only did they author texts that explained to the reading public the usefulness of their inventions; they also engaged in educational programs that presented the mechanical arts as foundational in any project of improvement. Such programs, of which I discuss two exemplary cases below, centered on newly invented machines that embodied their designers' visions of artisanal work and its moral worth.

EDUCATING *L'ESPRIT* THROUGH THE MECHANICAL ARTS: THE ABBÉ NOLLET

Born in Picardie in 1700, Jean-Antoine Nollet obtained a degree in theology from the University of Paris in 1724 and became a deacon soon afterward. He did not advance any further in the ecclesiastical career; the title of *abbé*, not to be confused with that of abbot, was commonly used for people who took the minor orders. Nollet initially worked as the tutor to the children of the Taitbout family, whose male members had worked for generations as registrars at the Hôtel de Ville. It was at the Hôtel de Ville that Nollet set up a laboratory and started an informal apprenticeship under the guidance of the royal enameller Jean Raux (who joined the Société des Arts in 1729).[14] Enamellers worked on varnishes, lacquers, and painted glass, and built the glass components of various scientific instruments, particularly thermometers and barometers. It is likely that his skills in enameling and, more generally, in instrument making earned Nollet his admission to the Société des Arts. Few records of his membership survive. Admitted in December 1728 in the class of "free" members as a mechanician, Nollet presented a terrestrial globe upon admission and, two years later, a celestial globe dedicated to the Count of Clermont (fig. 6.2).[15] His skills in the art of enameling connected him to Charles de Cisternay Dufay, a member of the Académie des Sciences who had translated Filippo Buonanni's *Treatise on the Varnish Commonly Called Chinese*, and who was researching methods to produce Chinese lacquer with European materials.[16] Dufay's interests intertwined

Fig. 6.2. Celestial globe designed and assembled by Jean-Antoine Nollet, presented to the Société des Arts in 1730. [Bibliothèque Nationale de France, Paris]

in several ways with the art of enameling. Lacquers were employed for coloring glass and making porcelain, and Dufay's most recent work concerned the mercurial glow that was visible, under certain circumstances, in barometric tubes, instruments typically made by enamellers. Practical knowledge of materials connected Nollet not only with Dufay but also with René Réaumur, who was interested in the manufacture of porcelain and was Dufay's patron within the Académie des Sciences. In 1732, Nollet started to work for Réaumur, and it is likely that he broke his formal relationship with the Société des Arts around that time.[17] Réaumur entrusted Nollet with the direction of his laboratory and relied on Nollet's skills for his work on thermometry and natural history. After Nollet's election to the Académie des Sciences in 1739, Réaumur assigned to him the compilation of the volume on the art of enameling, glassmaking, and glazing for the *History of the Arts*.[18]

In the years between his work for Réaumur and his election to the Académie, Nollet completed a self-directed training in instrument making and opened the most popular school of experimental physics in France. In 1734, he accompanied Dufay during his travels to England and, two years later, to the Netherlands. During these journeys, Nollet met John Theophilus Desaguliers and Pieter van Musschenbroek, two of the most famous Newtonians of the time, with whom he perfected some of the techniques for making demonstration apparatus for experimental natural philosophy.[19] On his return to Paris, Nollet trained several workers and started a successful business as an instrument maker. Aristocratic amateurs and *philosophes* alike appreciated the distinctive decorations of his artifacts: the wooden parts of his instruments were styled on fashionable chinoiserie furniture, finished with gold leaf on colored lacquer, a material quotation of his collaboration with Dufay. Voltaire spent the exorbitant amount of ten thousand livres on a *cabinet de physique* entirely equipped by Nollet. As he waited for the completion of the work, Voltaire oscillated between adoring compliments ("he is a real philosopher, the only one who can furnish my cabinet"), and apprehension ("L'abbé Nollet makes me bankrupt"). His anxieties illuminate the double standards that practitioners of the arts faced in their interactions with the world of learning: now they were regarded as enlightening peers, now as self-interested merchants.[20]

As Voltaire acknowledged, however, instrument making was not just business for Nollet; it constituted the lens through which he understood and reinterpreted the experimental sciences. Like most natural philosophers, Nollet deemed instruments essential to experimental physics, but he also believed that the ability to build instruments should be part of the experimental physicist's knowledge. As he pointed out, the accuracy of instruments (or lack thereof) determined "the certainty of what [physics] can teach us."[21] Friction, small imperfections, changes in humidity and temperature, usage, and other unpredictable factors always interfered with the correct functioning of machines. It was in the very nature of machines to eventually break down, and it was the task of the experimental physicist to restore order. In order to master the art of performing experimental demonstrations it was crucial to cultivate both "sagacity of mind [*esprit*]" and dexterity of hands. Intelligence alone, Nollet noted, did not suffice. Even the most intelligent of men would come to a standstill without the manual skills to repair his apparatus.[22] Performing experiments was an art (which he aptly called "the art of experiments"), and it centered on making, repairing, and maintaining instruments.[23]

The program of the Société des Arts was never lost on Nollet. The epistemological superiority of practice over theory shaped his approach to experimental physics, and it was at the core of the school of experimental physics he opened in 1735, in his house in rue du Mouton, near Hôtel de Ville.[24] His educational project was timely. Demonstrations of experimental philosophy staged as theatrical performances were popular, particularly in England and the Netherlands, where public lecturers promulgated the principles of Newtonian philosophy by means of mechanical devices that illustrated the laws of nature without any mathematics. Speaking directly to the senses, these demonstrations became fashionable spectacles and endowed the fortune of a variety of demonstrators, ranging from university professors to itinerant lecturers and instrument makers.[25] In Paris, the tradition of experimental demonstrations predated both the debates on Newton's natural philosophy, which started in the 1730s, and the Anglomania that boomed in the latter part of the century. Already in the 1650s, Jacques Rohault's lectures on Cartesian philosophy illustrated by mechanical devices enthralled aristocratic audiences, and in the first decades of the eighteenth century, Pierre Polinière, later a member of the Société des Arts, replicated this kind of educational performance during his lectures at the university and in his private courses.[26] Nollet's school, however, differed in significant ways from these precedents, and not because he embraced the newer Newtonian natural philosophy. Opening his school one year after the publication of Voltaire's *Philosophical Letters*, Nollet resisted the mounting craze of the fashionable Parisian for anything English and declared that he was no Newtonian, no Cartesian, and no Leibnizian. His school introduced his students to a theme that was central to the Société des Arts: the subordination of theoretical claims to the principles that emerged from the practice of the mechanical arts.

By emphasizing the role of experimental demonstrations, Nollet elevated the methods and procedures of the mechanical arts to the role of providers of evidence. In the years of the "Newton wars," he condemned the disputes that pivoted around theoretical differences, emphatically scorning those (such as Voltaire, whom he did not name) who affected to be Newtonians in Paris or Cartesians in London.[27] The controversies that such polemical positions sparked overlooked the basic processes through which certainty about the operations of nature should be established. Practice, not theory, according to Nollet, should guide the mind toward the contemplation of truth. Demonstration devices rendered his lectures entertaining, but Nollet's students were not offered a spectacle of "pure amusement, where one sees a great number of experiments being replicated without plan or choice, able only to busy the eyes."[28] Similar to the *bureau typographique*, amusement at Nollet's school of "sensible" physics was a means to the end of introducing the French public, in particular the elites, to the methods of the mechanical arts.[29] Nollet contrasted his school with Rohault's weekly demonstrations and Polinière's lectures, explaining that it neither provided "puerile recreations" nor stimulated "forbidden curiosity"; rather, it offered instead a course of lectures that revealed the mechanisms of nature's operations through experimental demonstrations. Even the most abstract principles, those which one would normally learn through "painful application," penetrated the mind more easily through demonstrations "that compel us to acknowledge their usefulness and truth."[30] Each topic discussed in the lectures was presented in connection not only to natural phenomena but also to "the practices of the arts, and to the machines most commonly used for the convenience of civil life."[31] By attending Nollet's lectures, audiences did not simply learn the laws of nature; they understood the principles necessary for evaluating machines and the techniques employed in manufacturing, agriculture, mining, and other mechanical arts. These principles would prove valuable when it came to assessing inventions, investing in manufacturing, and promoting building or mining projects.

Experimental physics, according to Nollet, was public capital, or, in his words, a "property whose ownership is common to all." The visual and tactile evidence offered by experimental demonstrations made natural philosophy accessible to people of both sexes, all ages, and all social conditions.[32] Like a financial advisor, Nollet aimed to educate his audiences on how to best invest that common capital. The first step was for the French elites, and the public at large, to become knowledgeable in the mechanical arts. It was no coincidence that Nollet praised *The Spectacle of Nature*, a multivolume work by the abbé Noël Antoine Pluche, the tutor to the son of the intendant of Normandie. Published in 1732, *The Spectacle of Nature* was a small encyclopedia of natural history that, in Baconian fashion, encompassed the mechanical arts. Pluche had collected information on the arts by visiting workshops and interviewing artisans. Nollet added to Pluche's descriptive approach a pedagogical method that drew from the culture of artisanal workshops an emphasis on observation, intuition, and improvisation. During Nollet's lectures, the "exposition of facts" by means of demonstration devices often took over the use of words. Students could thus appreciate the immediacy of practical learning and put themselves in the shoes of workshop apprentices who learned by doing. In order to highlight the difference from choreographed spectacles, Nollet did not read his lectures from written texts because he believed that unexpected events during demonstrations could offer unscripted opportunities. Demonstrations, like artisanal processes, were never exactly the same, and students were welcome to attend the same course more than once.[33] By replicating workshop techniques in his school, Nollet's goal was to dispel prejudices and ignorance about the mechanical arts that were widespread in French society and that made it vulnerable to fraudulent schemes. His education program was an enlightenment project intended to contribute to economic advancement and, more broadly, the public good.

Contemporaneous debates over Newtonian natural philosophy fed a public appetite for natural philosophy and sanctioned Nollet's success, in spite of his proclaimed theoretical neutrality. Yet what Nollet offered to his audience was more similar to a course on the methods of the mechanical arts than to Newtonian philosophy. The learning process that Nollet designed for his students illustrated principles that, as François Quesnay and others were forcefully arguing in those years, emerged from the practice of the arts, not from the application of theory to practice. Properties of matter that might appear too abstract to grasp acquired sensorial immediacy when presented in the context of artisanal practices. So Nollet explained the ductility of metals by demonstrating how artisans drew thin wires or plates from gold or silver. He taught the porosity of solid matter through the action of solvents, pigments, and varnishes commonly used in the arts of dyeing and of engraving. He demonstrated the incompressibility of fluids by showing how pumps were used in forges and mines. He discussed the communication of motion through impact by illustrating various building techniques in military architecture. In short, he introduced no theoretical notion without an example from the world of the mechanical arts, with mills, pumps, cranes, hydraulic machines, cannons, fortifications, and even dancers presented as real-life testimonials to the truth of the laws of physics.[34] His students quickly appreciated that knowledge of the laws of nature resided in artisans' workshops at least as often as in books of natural philosophy.

Nollet's educational program was a project in useful knowledge and social improvement addressed to a broad public and, in particular, to children. The education of children was the key to effect positive changes. In order to instill in future generations an informed interest in the study of nature and the mechanical arts, Nollet believed it essential to initiate the younger generations in these subjects. They would learn while playing, and their intellect would grow accustomed to such topics. As the members of the Société des Arts had remarked about the *bureau typographique*, mechanical, repetitive operations could prove instructive when mindfully choreographed. By observing experimental demonstrations, children would notice that from the same causes followed the same effects. So, their "nascent reason" would learn to appreciate the uniformity and regularity of nature, if not the full implications of Newton's or Descartes's physics. By means of their familiarity with machines and experimental apparatus, children—and therefore future society—would be freed from "popular mistakes," "ridiculous beliefs," "false marvels," and other charlatanries that affected the wealthy and poor alike.[35] For Nollet, just as for the Société des Arts, the pedagogical value of machines rested in the fact that they materialized rational thinking and thus they were tools of enlightenment. They offered students of any age the opportunity to discipline their minds and appreciate the importance of the mechanical arts for any project of improvement.

The marquise du Châtelet reported that Nollet's lectures had become *à la mode* among the Parisian aristocracy, and that "carriages of duchesses, peers, and pretty ladies" lined up in front of the school's entrance.[36] Foreign visitors seeking the most fashionable events in the city also flocked to his lectures, and Nollet's celebrity reached well beyond Paris. In 1739, he was invited to Piedmont by the king of Sardinia to teach physics to the young prince, and five years later he was summoned to Versailles as the physics tutor to the French dauphin. His lectures soon enthralled the entire court and earned him monetary rewards, the position of *maître des enfants de France*, and lodgings at the Louvre. In 1743, Nollet published the lectures he gave to the dauphin in a multivolume work, *Lectures on Experimental Physics*, which became an international bestseller. A few years later, in 1753, the king established a new chair of experimental physics at the Collège de Navarre and entrusted it to Nollet.[37]

Nollet's commitment to educating the public, in particular the French elites, on the mechanical arts continued as his career escalated. In the 1740s, he dedicated himself to the new science of electricity. The science of sparks and shocks was a hot topic in polite conversations, with the media of the time extolling instruments and inventors. Not only savants but even "ladies and people of quality, who never regard natural philosophy but when it works miracles," became interested in electrical effects. Glass tubes rubbed by hand revealed the otherwise invisible "electric fire" to the eyes, the ears, and even the nose: livid lights, crackling noises, and sulfurous smells were distinctive features of performances carried out by itinerant demonstrators all over Europe. In darkened salons, spectators admired spirals of sparks glittering inside glass tubes, luminescent words flashing onto wooden boards, eerie glows filling glass vessels.[38] Friction, the very natural phenomenon that made machines break down, was responsible for making electrical phenomena sensible. Dufay, the first member of the Académie des Sciences who devoted himself to the study of electrical phenomena, produced electric effects by rubbing a glass tube. Nollet, instead, designed a large machine, known as "Nollet's electrical machine," that became his signature instrument. People in the audience operated the instrument by turning the large wooden wheel that was connected to the glass globe and by placing their hands on the rotating glass globe (figs. 6.3–6.4).

In amusing themselves with the electrical machines, Nollet's audiences, like the child typographers, replicated artisans' quotidian gestures. Operating a process of material translation, Nollet's electrical machine metamorphosed one of the most common artisanal tools, the turning lathe, into the most fashionable

Fig. 6.3. Nollet's electrical machine; from Jean-Antoine Nollet, *Essai sur l'electricité des corps* (Paris: Guerin, 1746). [Yale University, Harvey Cushing/John Hay Whitney Medical Library, New Haven, Conn.]

instrument of experimental physics. With a few modifications, a lathe, such as those that were frequently found in the workshops of pin-makers, cutlers, and many other artisans, could be transmuted into a wonder-making machine worthy of the most exclusive salons (figs. 6.5–6.7). With well-dressed aristocratic ladies replacing sweaty artisans, Nollet's electrical machine enabled audiences to perceive at will what nature displayed only under constraints. A philosophical novelty that literally inflamed the spirits of the elites (burning alcohol by sparks was a common electrical experiment), Nollet's electrical machine embodied the *artiste*'s combination of dexterity and intelligence. Through the tangible presence of Nollet's *esprit*, a common mechanical tool such as the lathe became an instrument of enlightenment.

The translation of workshop culture to the sites of polite sociability also characterized Nollet's work for French educational institutions. The inaugural lecture he gave at the Collège de Navarre in 1753 was a manifesto of the relevance of the mechanical arts to experimental physics. Rehearsing the main themes that he had presented to the French public when he opened his school almost thirty years earlier, Nollet profiled the "intelligent physicist" as somebody who was "as initiated in the mechanical arts as in the knowledge of natural effects," and lectured on the fragile nature of machines. The most sophisticated of

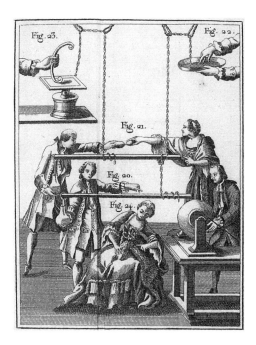

Fig. 6.4. Electrical experiments with Nollet's electrical machine (the glass globe is shown on the right); from Jean-Antoine Nollet, *Leçons de physique expérimentale* (Paris: Guerin, 1743–64). [Galileo Museum, Florence, Italy]

clocks, Nollet explained, or even the best apparatus, if left to itself, would break down and fail to perform as expected. The intelligent physicist was capable of acting on his instruments as the *régleur* operated on clocks and watches. Machines lived a material life that made them susceptible to external and internal changes: excessive cold rendered the best thermometers useless, just as excessive weight made the best balances inaccurate. The unpredictability of such changes required that the intelligent physicist understood instrument making, which constituted the core of "the art of experiments."[39]

Without mastering this art, physicists depended on the skill of available artisans and could never be in the moment when something unexpected happened. Chance, in experimental physics as in the arts, offered opportunities that only intelligence could put to work. Physicists without intelligence were to experimental physics as artisans without *esprit* were to the mechanical arts. By contrast, the intelligent physicist cultivated knowledge of materials as a dedicated naturalist studied the productions of nature. He was able to repair and maintain machines, and did not fall prey to the "dangerous" seductions of the *esprit géométrique*, which instilled a love for precision and full certainty that could not be found in the world of experiments.[40] Nollet's tirade against abstract

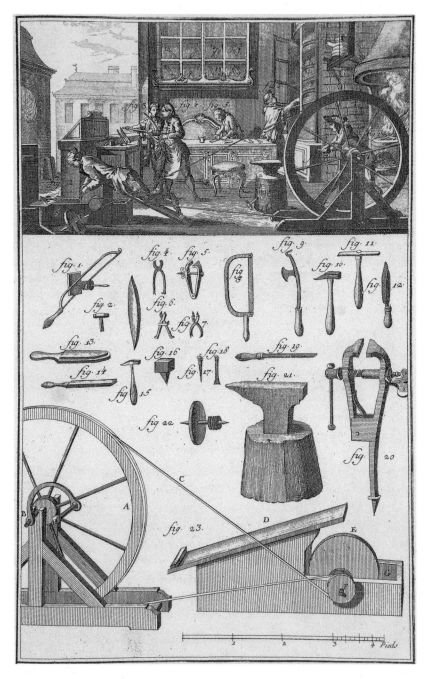

Fig. 6.5. Plate representing the workshop of a cutler, from the *Encyclopédie*. The equipment shown appears remarkably similar to Nollet's electrical machine portrayed just previously. [Museo Galileo, Florence, Italy]

Fig. 6.6. Plate representing the workshop of a turner, from the *Encyclopédie*, again showing a marked similarity to Nollet's electrical machine. [Museo Galileo, Florence, Italy]

mathematical speculations, combined with his diffidence toward philosophical systems, continued the Société des Arts's campaign against the authority of theoretical speculations and in favor of practical knowledge. In his private school, at the Collège de Navarre, at the École du Génie, and finally at the École Royale d'Artillerie at La Fère, Nollet continued to propagate the notion that the mechanical arts were key to social improvement and technical progress.

In his celebration of artisanal knowledge, Nollet endorsed the notion of the *artiste* as essentially different from other practitioners. In his bestselling *Lectures on Experimental Physics*, he included a portrait of the royal enameller Jean Raux, who had introduced him to the art of working glass at the lamp, as a sign of everlasting gratitude (fig. 6.8). In this rare representation of a real-life *artiste* at work, the light emanating from the lamp makes Raux's facial traits particularly vivid, marking a stark contrast to the empty chair that only alluded to the *artiste*'s presence in André Félibien's *Principles of Architecture* and to the anonymous workers that populated the plates of the *Encyclopédie*. By highlighting the individuality of Raux, the portrait stated his unique relevance, an argument that *artistes* variously articulated, within and beyond the Société des Arts. At the same time, the anonymous traits of the assistant sitting in front of Raux reaffirmed the *artiste*'s self-understanding as essentially superior to any other practitioner, as well as his appropriation of the savants' demeaning representations

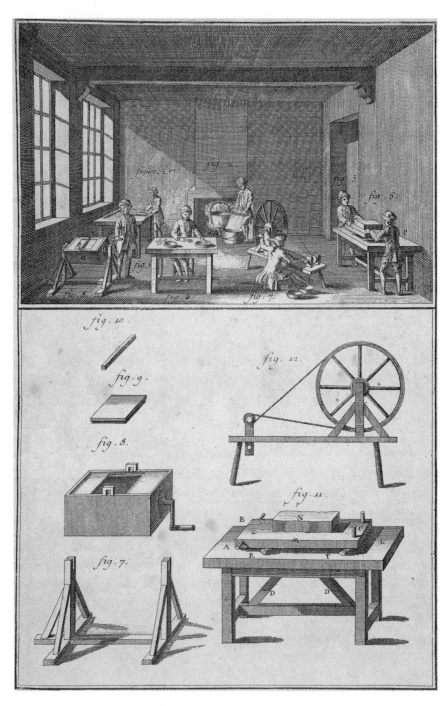

Fig. 6.7. Plate representing the workshop of a pin-maker, from the *Encyclopédie*, also quite similar to Nollet's electrical machine. [Museo Galileo, Florence, Italy]

Fig. 6.8. Portrait of the royal enameller Jean Raux, from Jean-Antoine Nollet, *Leçons de physique expérimentale* (Paris: Guerin, 1743–64). [Museo Galileo, Florence, Italy]

of other artisans. This elitist attitude toward other practitioners characterized most strikingly the mechanical inventions of Jacques de Vaucanson, who, like Nollet, designed machines based on his observation and study of artisanal gestures and operations. Unlike Nollet's machines or educational program, however, Vaucanson's inventions degraded artisanal labor, pushing the *artistes*' ambition to distinguish themselves from other practitioners to the extreme.

MECHANIZED LABOR AND VIRTUOSO MACHINES: JACQUES DE VAUCANSON

In 1735, when Nollet opened his school of experimental physics, Jacques de Vaucanson was admitted to the Société des Arts in the class of the assiduous members. A native of Grenoble (born in 1709), Vaucanson worked for some time in Lyon as a mechanician and, after touring France as a maker of automata, he settled in Paris in 1734. There he soon entered the social orbit of Alexandre Le Riche de La Poupelinière, a wealthy tax farmer who hosted one of the most vibrant salons of the Enlightenment, popular with Denis Diderot, Jean d'Alembert, Jean-Philippe Rameau, Jean-François Marmontel, and Madame de Genlis.[41] Vaucanson's mechanical inventions ranged from automata to machines for spinning and weaving silk. However disconnected they may now appear, these inventions were all grounded in his study of the human body and its functions. During his first years in Paris, Vaucanson attended anatomy lectures at the Jardin du Roi and began working on "moving anatomies," a term that indicated automata shaped as human bodies. Unlike wax or other anatomical models, moving anatomies simulated physiological processes that could not be investigated by dissections of dead bodies, such as blood circulation, muscular motion, digestion, evacuation, and respiration.

Vaucanson's project of creating an *homme automate* was also cultivated by surgeons such as Claude-Nicolas Le Cat and François Quesnay, and it was relevant to the focus area on animal economy of the Société des Arts. It is possible that it was through Quesnay that Vaucanson was introduced to the Société. Since 1730, when Quesnay published his *Observations on Bloodletting* under the auspices of the Société des Arts, he had advocated an experimental approach to this ancient practice. The discovery of the circulation of the blood challenged the original theoretical premises of bloodletting, which the medical profession continued to defend on the basis of new theoretical principles. In order to respond to the speculative approach of the medical profession, Quesnay had conceived of a hydraulic machine that simulated the circulatory system, through which he intended to study quantitatively the effects of bloodletting. The project was halted by the significant distortions introduced by available materials; as Quesnay realized, the rigidity of glass or metal tubes could not account for the action of elastic vessels such as veins and arteries within which blood circulated.[42]

Vaucanson cultivated the project of building a human-like automaton throughout his life. In 1741, he presented to the Société Royale de Lyon a project of an automaton in the shape of a human body that "will imitate in its movements animal operations such as the circulation of blood, respiration, digestion, and the motion of muscles, tendons, nerves." The machine was intended as a tool for medical research: "By means of this automaton we will be able to make experiments on animal functions and draw inductions to understand the various states of human health and heal our ills."[43] There is no evidence that such an automaton was ever built, but we know that in later years Vaucanson circumscribed his project to a human-like machine that would simulate the circulatory system. In 1762, he obtained the support of the king, the collaboration of Henri Léonard Jean Baptiste Bertin, the general controller of finances, and that of the former member of the Société des Arts, Charles Marie de La Condamine, now Vaucanson's colleague at the Académie des Sciences. Bertin and La Condamine offered intelligence on *caoutchouc*, a natural gum found in Guyana, Brazil, and Peru, which seemed promising for making artificial blood vessels.[44] Scattered sources suggest that Vaucanson did build a version of this automaton, though no further information survives.[45]

The imitation of bodily functions through the study of anatomy and the observation of human gestures was at the heart of Vaucanson's mechanical inventions. His automata have been widely discussed. Simon Schaffer and Jessica Riskin, in particular, have convincingly argued that these machines redefined, simultaneously, intelligence and human labor.[46] I wish to underscore that Vaucanson's mechanical inventions embodied the political project of the *artiste* in two of its most evident aspects: the ambition to contribute to the state's decision-making processes about manufactures and the arts, and the statement of one's own expertise through the representation of other artisans as replaceable pieces of machinery. Vaucanson's automata and his inventions for the silk industry expressed his conviction that perfectibility, or improvement, had to do more with machines and mechanical devices than with human beings. For Vaucanson, the automatic, repetitive operations carried out by machines could potentially outperform those carried out by human beings. Artisanal work was based on improvisation and embodied skill, and so its outcome was never exactly the same. Conversely, when properly designed, machines could perform orderly and regularly, just like the mind of the philosopher in Du Marsais's definition. These "rational" machines were brought into existence by the *esprit* of the *artiste*. They redefined artisans' embodied skill as subjective and imprecise, and therefore undesirable. At the same time, they presented improvement and progress as something that should be pursued by investing in technical innovations and by a selective focus on the education of the elites in the mechanical arts.

First displayed in Paris in 1738, Vaucanson's automata (flute player, digesting duck, and tambourine player) became sensations throughout Europe. Spectators especially admired the ability of the automatic flute player to produce sounds by actually blowing in the flute, unlike other musical automata that produced sounds by a separate mechanism. In his presentation to the Académie des Sciences, Vaucanson underscored that the automaton resulted from his knowledge of mechanics and acoustics, as well as from his observation of how human flute players positioned their lips when they blew into the instrument.[47] With their uncanny resemblance to living beings, Vaucanson's automata called attention to the ingenuity of their maker and, in 1741, earned him the position of inspector of silk manufacture for the Bureau of Commerce. This was the same year in which he announced his project of an artificial man to the academy in Lyon. Vaucanson's study of physiological functions mirrored the observation of artisanal gestures and embodied skill that constituted the foundation for his work on silk manufacturing. Although no written source addresses directly the links between his automata and the manufacture of silk, one of his contemporaries remarked that "while Mr. Vaucanson employed his talents to make some automata play or digest, he was already applying himself to improving silk."[48] Most likely, during his years in Lyon Vaucanson had realized the strategic relevance of silk to the French economy. As with other members of the Société des Arts, Vaucanson was eager to put his mechanical talents in the service of the Bureau of Commerce.

Silk was big business in the eighteenth century. With the introduction of annual fashions, Lyon had become the unrivaled capital for the production and trade of silk fabrics. Yet the Lyon merchant-producers imported threads from the neighboring state of Piedmont, due to the perceived superior quality of the foreign product.[49] Since the times of Jean-Baptiste Colbert, the Bureau of Commerce had unsuccessfully encouraged large-scale plans aimed at improving the quality of French-made silk threads. In his position as the inspector of silk manufacture, Vaucanson traveled to Piedmont and Lyon in order to acquaint himself with machines and techniques for weaving fabrics and making threads. Just as his automata were grounded in his anatomical and physiological studies, so his machines for the manufacture of silk stemmed from his observations of artisans at work. He came back to Paris with a plan that introduced several mechanical inventions in the context of a larger vision for industrial production. In Vaucanson's analysis, the high quality of Piedmont's threads was ensured by a highly regulated system of production, which the state enforced strictly and that concerned all aspects of silk manufacture, from the breeding of cocoons to the spinners' apprenticeship and the composition of the mechanical apparatus. Because of its specificity, this system could not be reproduced in France. Vaucanson's alternative proposal consisted of a "great plan" according to which the French state would supervise the creation of five great Royal Manufactures, funded by a large company made up of the wealthiest Lyon merchants, who had an interest in locally produced, less expensive threads.[50] The Royal Manufactures would operate according to regulations that Vaucanson prepared, and they would employ silk mills and spinning machines that he designed. The plan failed spectacularly.

The reasons for this debacle illuminate the politics that underpinned Vaucanson's mechanical inventions. His top-down approach favored the 250 wealthiest silk merchant-producers in Lyon and neglected to consider the negative repercussions of the new regulations on approximately 3,000 master workers who ran their own shops. In 1744, the violent riots they organized to oppose the establishment of the Royal Manufactures obliged Vaucanson to flee Lyon disguised as a monk to save his life. The fierce repression organized by the king, followed by reconciliatory amnesty for the rioters, reveals the firm support that Vaucanson obtained within the Bureau of Commerce.[51] The manufacturer and economist Jean-Marie Roland de la Platière remarked that Vaucanson had "worked more as a mechanician who seeks the savants' admiration than as an *artiste* who must be useful to manufactures."[52]

Indeed, one year after the failure of the "great plan," the minister of the navy, the Count of Maurepas, proposed (and obtained) Vaucanson's election at the Académie des Sciences.[53] The Académie provided not only a public arena for Vaucanson's inventions; it also offered human resources to be recruited to his new plans. In 1749, the Bureau of Commerce asked the abbé Nollet's collaboration in its efforts to gather technical intelligence on the manufacture of silk in Piedmont. Thanks to his public reputation in experimental physics, Nollet could and did easily disguise the real reason for his travel. Taking advantage of his reputation within the Italian community of experimental philosophers, he presented his new journey through the peninsula as motivated by a controversy over the medical virtues of electricity, and successfully completed the mission. His reports on the manufacture of silk, intended as fact-checking on information the bureau had accumulated since the late seventeenth century, were secretly passed on to Vaucanson, who was preparing a new plan for improving the manufacture of French silk threads.[54]

Why did the bureau invest so conspicuously in Vaucanson's inventions, even after spectacular failures? In order to answer this question, we need to consider that the design of manufacturing machines was not a purely technical endeavor. It was informed by conceptions of human labor and social hierarchy that were visible to the members of the Bureau of Commerce. As social studies of technology have pointed out, in order to understand what it meant that a machine "worked," it is essential to move beyond determinist notions of technical progress and success.[55] Although they have been interpreted as innovations that proved unsuccessful because they were ahead of their time, Vaucanson's machines in fact materialized his understanding of workers as dehumanized operators who should have no say in the organization of the manufacturing process. The bureau's unrelenting support of Vaucanson's machines indicates that its members saw and shared the social vision embodied in Vaucanson's projects. His plan for improving the manufacture of silk introduced a new system of production that mechanized human labor and silenced potentially riotous workers.

Taking advantage of the ongoing fascination of Parisian society with manufacturing and mechanical ingenuity, Vaucanson presented his new plan at a public meeting of the Académie des Sciences. Making silk had become a fashionable domestic amusement for those who wished to savor countryside practices in an urban context: among the many goods that could be purchased in the high-end rue S. Honoré, there were prepackaged boxes containing silkworm grains, mulberry tree leaves, and instructions on how to breed silkworms and spin silk from cocoons at home.[56] In his presentation at the Académie des Sciences, Vaucanson attributed the French backwardness in the manufacture of silk to widespread lack of education in the mechanical arts, a theme that resonated with the projects of enlightened reforms that the new leadership of the Bureau of Commerce intended to pursue.[57] In line with the emphasis on education and improvement of the bureau, Vaucanson launched a campaign against the ignorance that pervaded the manufacture of silk in the southern provinces and solicited the intervention of the central government to eradicate what he saw as the faulty procedures of narrow-minded artisans-turned-entrepreneurs.[58]

This new campaign pivoted around Vaucanson's conviction that the path to improvement went through the design of more efficient machinery and systems of production. For him, it was easier to design improved machinery than to educate rural workers. The manufacture of silk, Vaucanson declared, was in the hands of "people from the countryside, unable to correct themselves, and normally little inclined to allow others to instruct them." Given the impossibility of making these workers comply with state-imposed regulations, Vaucanson's solution was to limit their role, deskilling their labor, and reducing them to mere providers of motive force. His new spinning machine would accomplish just that. He explained that the high quality of Piedmont's threads depended on a delicate process, called *croisade*, which rested entirely on the skill of Piedmontese spinners. The *croisade* consisted in the twisting of two threads around each other several times as they were wound on a spinning reel. Since the twists gave the threads uniformity and resistance, the greater the number of twists, the finer the thread would become. However, the spinner had to be cautious, as each additional twist increased the risk of breaking the thread, and when a thread broke, all operations had to start again. Piedmont's spinners knew in their own hands how many twists each combination of threads could tolerate. They acquired this embodied skill, which was virtually absent in France, in the course of a long apprenticeship. Vaucanson's solution for the absence of such embodied expertise was to replace the action of the spinners' hands with a machine that could be operated by anyone (fig. 6.9). This spinning machine, Vaucanson underscored, would actually perform better than the ablest spinner.[59]

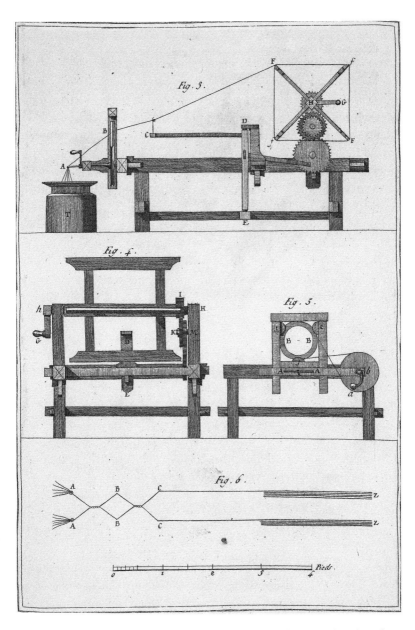

Fig. 6.9. Jacques Vaucanson's tour, as represented in the *Encyclopédie*. The image at the bottom represents the *croisade*, or the twisting of two silk threads around each other. From left to right, two threads are composed by bringing together several filaments. The two threads are twisted around each other twice, then separated to be reeled on two different wheels. The threads that will become warp will go through another twisting and reeling process, by means of water-operated mills (not shown). [Museo Galileo, Florence, Italy]

Vaucanson's discussion of the superiority of his machine echoed the philosophers' predilection for perfectly orderly mechanisms. He contrasted the subjective and improvisational nature of spinners' skills to the regular performance of his machine. Spinners, Vaucanson argued, worked with their fingers in very hot water (to dissolve the sericine holding the fibers together), and the heat weakened their sense of touch, with the result that the number of times they twisted the threads changed irregularly. De facto, this irregularity was a distinctive feature of artisanal practice before standardized production. Spinners, just like other artisans, built their reputation on their ability to be in the moment and improvise. Vaucanson characterized this practice as an inconvenience, which his new machine avoided "by giving the spinner a quick and easy means of twisting silk threads as many times as will be prescribed to her, and all this without touching any thread."[60] His machine required a new bodily discipline that instructed spinners to operate automatically and so prevented even the possibility that individual abilities might be acknowledged. Unlike their Piedmontese counterparts, spinners working at Vaucanson's machines did not have to decide how many times to twist the thread, because the number would have been decided by the manufacturer. As they complied with the new instructions, spinners became deskilled and replaceable, and their actions predictable and repetitive.

Based on his vision of industrial production wherein workers served only as motive forces, Vaucanson's mechanical inventions relocated skill from the spinners' bodies to his machines.[61] Just like the anonymous artisan in the portrait of Jean Raux that Nollet included in his *Lectures*, Vaucanson's spinners were deprived of subjectivity and individuality. This anonymization of the artisanal world excluded the possibility that intelligence or talent—in other words, *esprit*—could be found among manual workers. It also highlighted, by contrast, the unique features that distinguished Vaucanson as an inventor and a manager. With his machines and disciplined, not riotous workers, Vaucanson demonstrated to the state and to the French public that a mechanician such as himself was able to organize and manage production the way a military commander led his troops.

This was indeed how state administrators understood Vaucanson's work. In 1752, Vaucanson and the entrepreneur Henri Deydier established at Aubenas a Royal Manufacture of Silk under the auspices, and with the financial support, of the Bureau of Commerce. Vaucanson installed there one hundred of his spinning machines, entirely subsidized by the state, and implemented the organization of labor that such machines required. Even the architecture of the manufacture was conceived so as to optimize efficiency: the building received plenty of natural light so that spinners could work longer hours; a brick canal brought water from a nearby river to each spinning machine; the disposition of doors and windows facilitated the circulation of air (necessary for keeping the fire burning in the ovens and for fast drying silk) and prevented spinners from being impeded by smoke and vapors. Spinners and other workers were trained exclusively onsite.[62] The regimented organization of labor implemented at Aubenas elicited praise from government officials. During his routine visit, the local inspector commended the manufacture's "order, cleanliness, and extreme subordination." He saw two hundred spinners working on fifty spinning machines, and they all appeared, he remarked, as "disciplined as troops."[63]

As workers became interchangeable parts in the mechanism of industrial production, Vaucanson's machines began to be seen as artificial *virtuosi*, embodiments of their maker's *esprit* and of his conception of mechanized labor. An enthusiastic reporter for the *Mercure de France* observed that Vaucanson's newest invention, a weaving loom, was "a machine with which a horse, an axe, or an ass make fabrics much more beautiful and much more perfect than the most skilled workers."[64] The *Mercure de France*'s rhapsodic description of the loom recounted that one could see "the fabric weaving itself without any human help: the warp threads raise, the shuttle passes the weft, the reed presses the yarns; the fabric rolls up as it makes itself."[65] This image of the fabric that made itself was an implicit reference to César Du Marsais's definition of the philosopher as a self-winding clock, but whereas the self-winding clock eliminated the possibility of an external God, the self-weaving fabric took shape within a noisy and fast-moving loom whose presence was highly perceptible. Indeed, it was the loom, and not the self-weaving fabric, that the *Mercure de France* intended to celebrate. By exalting the virtues of the machine, the reviewer also highlighted the "genius" of Vaucanson, his dedication to the French state, and his pursuit of useful knowledge. The image of the self-weaving fabric that the *Mercure de France* presented to its readers was uncannily similar to the visual representations of artful machinery effortlessly operated by putti that circulated in the early eighteenth century. By concealing the presence of artisans and the fatiguing nature of artisanal labor, these visual representations highlighted the disembodied, graceful ingenuity that pertained to the world of the arts. It was this very same disembodied ingenuity that the self-weaving fabric evoked. By erasing the presence of manual workers and calling attention to the ingenuity that brought the loom into existence, the fabric that wove itself demonstrated the intelligence of its inventor and his dedication to the public good. If the metaphor of the self-winding clock served to eliminate the possibility of an external God and to affirm the philosopher's commitment to the common good, the image of the self-weaving fabric erased manual labor and revealed *l'esprit* in the machine. It also consolidated among the French public the idea that machines that performed better than humans were creations of superior minds who, because of their manual skills, did not fit any existing classification of knowledge.

Indeed, at the death of Vaucanson in 1784, the marquis de Condorcet struggled to find a proper definition to characterize him. Condorcet remarked that Vaucanson was neither a geometer, "who develops the theory of movement and the order of phenomena," nor a physicist, who, joining experience, observation, and calculation, applied his knowledge to "the construction of machines or to the practice of the arts." He was not even an *artiste*, who "owes his talents and successes to the practice of the arts," or an inventor who "had never made or operated a single machine."[66] Condorcet applied here Diderot's double standard in the evaluation of *artistes*. Diderot stated that the word *artiste* was always a compliment to an artisan who excelled in practical work, but it was "a sort of reprimand" to somebody who worked on a science that was both practical and theoretical, because it indicated that that person had privileged practice over theory.[67] Condorcet declared that Vaucanson was a mechanician, a maker of machines that "perform operations that we were obliged, before him, to entrust to the intelligence of men."[68] Originally indicating an *artiste* who built machines, the term *méchanicien* became widely associated with Vaucanson starting with the homonymous article that was published in the *Supplement to the Encyclopédie* in 1777. Condorcet's definition was meant to elevate Vaucanson above men of practice and men of theory, the goal that *artistes* had collectively pursued with the establishment of the Société des Arts. Paradoxically, however, the success of Vaucanson's individual trajectory went hand in hand with the downgrading of the *artiste* in the hierarchy of knowledge that academicians such as Condorcet sketched.

THE POLITICS OF MACHINES

The trajectories of Vaucanson and Nollet illuminate how the Société des Arts captured the *artistes'* ambition to play an official role in the decision-making processes of the French state in matters concerning the arts and manufactures. Although their interactions with the Société were short-lived, both Nollet and Vaucanson shared with other members the ambition to serve the Bureau of Commerce and to direct the attention of the elites toward the mechanical arts. Nollet rode the wave of Parisians' enthusiasm for natural philosophy by designing a course of lectures that was grounded in the principles of the mechanical arts. By metamorphosing artisanal gestures, tools, and operations into fashionable experimental demonstrations, he spread to the French public not only knowledge about the mechanical arts but above all the *artiste*'s way of reasoning. As with other former members of the Société des Arts, Nollet hoped that his expertise in the mechanical arts would earn him a position at the Bureau of Commerce, an aspiration that seemed more realistic when he started his collaboration with Réaumur, who actively supported interactions between the Académie and the bureau. Although in 1740, upon Dufay's death, the bureau preferred his fellow academician Jean Hellot for the position of inspector of the dyeing manufacture, the Bureau of Commerce did capitalize on Nollet's technical expertise. Not only was he requested to gather technical intelligence on silk manufacturing in Piedmont and other Italian countries, even after his return to Paris, the Bureau of Commerce repeatedly resorted to his expertise on the silk industry, treating him as a secret consultant. More importantly, through his teaching at court, at the Collège de Navarre, at the École du Génie, and at the École Royale d'Artillerie at La Fère, Nollet educated generations of aristocrats to appreciate and understand the mechanical arts.[69]

Vaucanson, for his part, used his position as inspector of the silk manufacture for the Bureau of Commerce to advance a complete program of improvement. If the education of the elites in the mechanical arts was essential for their financial and strategic choices, for Vaucanson it was equally, or even more, important to reconfigure the production processes of the silk industry. In his vision, the state should focus on perfecting machinery, not workers. Relocating skill out of the body of the artisan and into the machine, Vaucanson's automatic looms and spinning devices constituted powerful material arguments for the higher status of their maker, who elevated himself, simultaneously, above men of practice and men of theory. Vaucanson was eager for the Bureau of Commerce to understand this point. In 1765, after learning that leading members of the bureau were considering a proposal made by one of his rivals, Vaucanson explained to the bureau's director, Daniel Trudaine, what distinguished him from other inventors:

> Many people today have a fury to pass for creators. They delight in saying: I have invented this machine, I have improved this art, I have made this discovery.... They publish big books containing numerous discoveries useful to agriculture, the arts, manufactures, and among all these important innovations there is none that is of use: it is so that they present themselves as useful and industrious men.... He who invented the spinning wheel would not be regarded by the academicians of today more than as an *artiste*, and he would be despised as a maker of machines. These gentlemen, however, would easily

be humiliated if they reflected that this mechanician, by himself, has done much more good to mankind than all geometers and physicists that ever existed in their Company.[70]

Vaucanson's scathing remarks epitomized the *artistes'* attempts to equate useful knowledge with their own practice and expertise, and to dismiss other approaches to the mechanical arts as ill-informed or wrong. Vaucanson, as Henry Sully and other former members of the Société des Arts had done in the previous decades, stressed that investors and the state alike should rely on inventors who were able to demonstrate that their mechanical projects actually worked when built. This was not obvious at a time when the institution that was officially entrusted with the task of evaluating mechanical devices, the Académie des Sciences, was still more concerned with the theoretical underpinnings of the project than with the actual performance of the built machine.[71]

Vaucanson, who wrote after losing the promotion to pensionnaire at the Académie des Sciences to Jean d'Alembert, was troubled also by the rampant success of the physiocrats, whose economic vision and understanding of useful knowledge clashed with his own projects. With misplaced humanitarian fervor, he remarked, the so-called *économistes* opposed the establishment of state-subsidized manufactures as "larceny against the people of the countryside" and supported the deregulation of silk production. Invoking the public good, the physiocrats claimed that anyone interested in producing silk should be educated in the art and provided with the best machines available. Vaucanson's response illuminates his alternative conception of useful knowledge. The well-intentioned physiocrats, Vaucanson maintained, fundamentally misunderstood the notion of the public good because of their ignorance of the manufacture of silk and their naive conception of the nature of workers. They ignored, in particular, that spinners were negligent and naturally inclined to fraud. Moved by greed, Vaucanson explained, workers had no interest in self-improvement, or in learning the most advanced processes. The only viable means to perfect the manufacture of silk was to make it independent of them.[72]

For Vaucanson, improvement, and with it enlightenment, could only be achieved through rational machines. As the Académie des Sciences's report on his artificial flute player acknowledged, machines demonstrated their maker's intelligence. This *esprit* was even more powerfully revealed when the machine's internal workings were made transparent, an epistemological inversion of black boxing that distinguished machines from trickeries.[73] *Artistes* were eager to display the internal mechanisms of their creations. Julien Le Roy conceived of a special arrangement of the movement of the clock he made for Louis XV so that the king could see for himself the gears at work. Similarly, Vaucanson exposed to view the inner workings of his digesting duck because he wanted "to demonstrate, not simply to display, a machine."[74] He also allowed his audiences to see the mechanism of his flute player in order to dispel any suspicion of fraud. Nollet, for his part, published a three-volume text, *The Art of Experiments*, in which he detailed the material composition of the machines he had designed for his courses of experimental physics.

Opening the body of a machine, displaying its components, and demonstrating its inner mechanism were dissecting processes that, just like anatomical dissections, revealed intelligent design. Naturalists' repeated encounters with the order of nature, which was revealed through dissections, microscopic observations, or studies of the social life of insects, seemingly offered the most compelling argument for the existence of an intelligent creator.[75] Similarly, the exhibition of the inner workings of sophisticated machines pointed to the intelligence of their maker. If the *philosophes* mobilized the abstract notion of a self-winding clock to articulate their points of view on freedom of thought, religion, and materialism, *artistes* unveiled the mechanism of real machines to highlight their own intelligence, or *esprit*, and to claim a position for themselves in the decision-making processes related to the mechanical arts.

Epilogue

In 1751, the publication of the first volumes of Denis Diderot and Jean Le Rond d'Alembert's *Encyclopédie* brought to life the latest in a series of encyclopedic projects that had been articulated, revised, and interrupted since the times of Jean-Baptiste Colbert. With its broader scope and often irreverent tone, the *Encyclopédie* significantly departed from earlier attempts to offer a comprehensive work on the arts and crafts, yet it also displayed substantial continuity with the several projects that this book has discussed. The notion of the *enchaînement* of the branches of knowledge, so central to the projects developed within the Académie des Sciences, was also pivotal in Diderot and D'Alembert's work. As the "Preliminary Discourse" elucidated, and the title made clear, the *Encyclopédie; ou, Dictionnaire raisonné des sciences, arts, et métiers* had a double nature: it was an encyclopedia and a dictionary. As an encyclopedia, it was organized according to the editors' view of human knowledge as an interconnected landscape, where each science, art, and trade was linked to another. This *enchaînement*, visually represented by the *Encyclopédie*'s tree of knowledge, was verbally described in the text as a world map, which guided editors as they identified relevant articles and possible contributors. As a "reasoned dictionary" of the sciences, arts, and trades, the *Encyclopédie* offered small-scale maps that enabled readers to find their way through specific subjects. If, as Francis Bacon stated, knowledge was a labyrinth in which it was easy to get lost, the *Encyclopédie*, like Ariadne's thread, offered carefully demarcated paths to reach intellectual destinations. The alphabetical order that governed each volume allowed easy access to information, without forgoing the overarching *enchaînement* that was still traceable through a system of cross-references. It was thanks to this double organizing system that the volumes of the *Encyclopédie* could be published individually as they were completed, unlike the encyclopedic project of the Académie des Sciences. The *Encyclopédie*'s system of knowledge ensured completeness, its alphabetical order accessibility. The map analogy also demonstrated the arbitrary nature of the *enchaînement* presented by the *philosophes*. Just as maps of the same territory varied tremendously depending on the projections used to create them, so, too, by D'Alembert's own admission, there could have been multifarious ways of interconnecting the various branches of knowledge. But, he explained, and as the history of the many unaccomplished encyclopedic projects demonstrates, the search for an organizing principle could take longer than the actual completion of the work. So, the editors adopted a philosophical point of view, presenting a system in which philosophy reigned above all other branches of human learning, a hierarchy vividly illustrated in the *Encyclopédie*'s celebrated frontispiece.[1]

In the *Encyclopédie*'s reconfiguration of knowledge, the mechanical arts gained prominence and relevance. In his article "Art," Diderot exalted Colbert's encouragement of the arts, crafts, and industry, while he condemned the savants' characterization of the mechanical arts as a degrading activity that consisted in "laborious research, ignoble thoughts, difficult exposition, endless details, and minimum value."[2] With references to Bacon's *Novum organum*, Diderot denounced the imbalance between the advantages that society derived from the mechanical arts and the lack of prestige that its practitioners enjoyed. He further proclaimed that the man "who stole the stocking machine from the English, velvet from the Genoese, glass from the Venetians, did no less for the state than those who defeated its enemies and took their strongholds."[3] As Colbert's commitment to the world of the arts and manufactures demonstrated, the mechanical arts promised to improve the material conditions of human life and to secure the military and economic primacy of the French state. It was therefore urgent to include the mechanical arts in the world of learning, finally accomplishing Bacon's vision of a complete treatise of natural history. This compelling endeavor, Diderot claimed, was hampered by the paucity of relevant books:

> Too much has been written on the sciences, not enough has been well written on the liberal arts, almost nothing has been written on the mechanical arts. For, what is the scanty information available from the various authors, compared to the extent and richness of the subject? Among those who have treated it, one was not sufficiently informed, and succeeded more in demonstrating the necessity for a better work than in fulfilling the demands of his subject; another has only skimmed over the top of the material, treating it as

a grammarian and a man of letters rather than as an artisan; a third has more substance, to be sure, and is more workmanlike, but at the same time he is so brief that the operations of the artisans and the description of their machines, a matter in itself capable of filling substantial volumes, occupies only a very small part of his.[4]

In the absence of written sources and reliable authors, the *philosophes* directly addressed the most capable workers in Paris and France. Replicating a technique already tested by André Félibien, abbé Pluche, and other savants, they visited artisans' shops and interviewed various practitioners in the same or similar arts. By comparing the information thus obtained from observation and conversation, they interpreted the artisans' often incomplete sentences and corrected inexact details. Operating as "midwives" to the artisans' minds, they compensated for the imperfect language of the mechanical arts. Artisans, Diderot explained, understood each other "much more by the repetition of contingent actions than by the use of words." "In the workshop," he concluded, "it is the moment that speaks, and not the *artiste*."[5]

Artistes, however, did speak. As the previous chapters have demonstrated, *artistes* eagerly sought to make their voices heard within and beyond their workshops. They authored books, published articles, wrote memorandums, and designed machines that bore their names. Diderot's silence on the culture that *artistes* had produced before he wrote the *Encyclopédie*'s prospectus is even more striking after accounting for the relationship between the *Encyclopédie* and the Société des Arts.

The abbé Jean Paul de Gua de Malves, one of the Société's most active members, served as the original editor of the *Encyclopédie*.[6] In 1746, he signed a contract with the printers Le Breton, Durand, and David regarding the publication of an *Encyclopédie; ou, Dictionnaire universel des arts et des sciences*, with Diderot and D'Alembert as witnesses. In 1748, likely due to his financial mishandling of the endeavor, Gua de Malves withdrew from the project, and Diderot and D'Alembert replaced him as editors of the *Encyclopédie*.[7] The work was originally intended to be a translation of Ephraim Chambers's *Cyclopædia; or, Universal Dictionary of Arts and Sciences* (1728), but Gua de Malves expanded the project to include original new work addressing the mechanical arts in particular.[8] Gua de Malves's vision for the *Encyclopédie* was intertwined with his role for and activities he coordinated within the Société des Arts. He was one of the founding members and also one of the most active: he served on several committees and presented various essays throughout the life of the Société. In 1732, when the director, Daniel Medalon, revived an encyclopedic project on the arts, Gua de Malves held the role of *indicateur*, or indexer, for the Société. His task consisted of the compilation of annual catalogues of newly published European books, journals, articles, or any other work that concerned the arts or the sciences. Selecting, excerpting, and abridging were fundamental processes in the compilation of encyclopedic works, and it is clear that Gua de Malves's vision for the expanded translation of Chambers's dictionary was indebted both to his indexing activity and to his work on the encyclopedic project of the Société des Arts.

Gua de Malves envisioned the translation of Chambers's *Cyclopædia* as a collaborative endeavor. In his plan, the actual translation marked only the starting point for a much more ambitious project in which contributors would interpret, add, correct, and adapt the notions contained in the translation of Chambers's original text. Unlike Chambers or other authors of encyclopedic dictionaries up to that time, but in line with the Société des Arts's encyclopedic project, Gua de Malves intended to create a network of committed collaborators by sending invitations to well-connected individuals and academic institutions. As the *Encyclopédie*'s editor, his role would be to give stylistic uniformity to the work, while each contributor would be listed in the preface. In 1747, Gua de Malves prepared a memorandum for future collaborators in which he provided detailed directions on what they should accomplish. He specified that information that concerned only England, and that was of no interest to French readers, should be erased, and any religious view that contradicted France's Catholic doctrine should be accompanied by corrective remarks. Similarly, he encouraged the discussions of philosophical systems that Chambers obscured, such as those by Christian Wolff and Nicolas Malebranche, and he supported the addition of illustrations that could replace long textual descriptions. Contributors should read, or "at least scan with a quill in their hand," the five or six best books on their subject, focusing in particular on the most recent—just as the Société's indexer did.[9]

Gua de Malves planned a special team for the articles that concerned the arts, which mirrored the Société des Arts's approach. Unlike the savant-directed projects of the Académie des Sciences, the Société had organized smaller groups of contributors to its "History and Description" of the arts; these smaller groups collaborated with a member (savant or *artiste*) with proven writing skills. Similarly, Gua de Malves's team for the arts would consist of eight or ten *artistes* coordinated by an *artiste* who wrote well, or by a savant.[10] Team members would avail themselves of dictionaries, journals, encyclopedias, and other published or unpublished materials provided by Gua de Malves or the publishers.

Underscoring the Société des Arts's belief in the efficacy of the patronage system to encourage the arts, Gua de Malves requested that the team highlight the role of princes, aristocrats, or other patrons who had favored the advancement of the arts. The relationship with François Quesnay, formed through the Société des Arts, facilitated his connections with the Académie in Lyon, where Gua De Malves sent his letter of invitation to future collaborators.[11]

The *Encyclopédie* has been described as a "war machine" that aggressively spread Enlightenment ideals often at odds with those promoted by Gua de Malves.[12] Gua de Malves's commitment to the Catholic faith, in particular, clashed with Diderot's freethinking sympathies. While Diderot certainly used the *Encyclopédie* as a vehicle for spreading his unorthodox thoughts on a variety of sensitive subjects, the heterogeneity of collaborators makes it difficult to conclude that the *Encyclopédie* was conceived of as a war machine. This understanding of a belligerent *Encyclopédie* stemmed from the hostilities and censorship it faced over the years. While the comparison between Gua de Malves's memorandum and Diderot's prospectus highlights several crucial differences, it also demonstrates that Diderot and D'Alembert inherited from Gua de Malves a work in progress in which the legacy of the Société des Arts was conspicuous.

Gua de Malves's legacy, and the work's association with the Société des Arts, did not disappear when the editorship of the *Encyclopédie* shifted to Diderot and D'Alembert. If in his prospectus Diderot glossed over the role of Gua de Malves, possibly because of the abbé's dubious reputation at the time, the Preliminary Discourse acknowledged the material that Diderot and D'Alembert had inherited from their predecessor. They clearly did not want to lose the support of the subscribers that Gua de Malves had secured with his prospectus of 1745.[13] The Preliminary Discourse reassured the existing subscribers by explaining that the new project would not betray the initial spirit of the *Encyclopédie*; even without naming Gua de Malves, it acknowledged that his method of asking contributors to revise, adapt, and modify translated articles was very efficient. While Diderot and D'Alembert underscored the limitations of the previous project, they also admitted that the publishers had passed on to them the translation of Chambers's dictionary, and that they would accept the revised translations that several contributors had already produced.[14]

Diderot's article "Art" further demonstrates the legacy of Gua de Malves's plan. In his campaign for a rehabilitation of the mechanical arts in the Republic of Letters, Diderot emphasized the need for a "language of the arts," which would impose order on a world characterized by both the superabundance of synonyms and the scarcity of specific terms. He remarked that in different manufactures or workshops, the same tools were identified by different words, whereas in the various crafts the same word, such as "hammer," could identify very different objects. His proposal echoed the recommendation that Gua de Malves had addressed to his *artiste* collaborators, namely that they record all the words employed in their art and list the terms that different arts had in common.[15]

Numerous other sources document the *Encyclopédie*'s debt to the Société des Arts. As Jacques Proust has observed, Diderot was inspired by works produced by members of the Société, such as Germain Boffrand's book on the casting of equestrian statues and Jean-Gaffin Gallon's *Machines and Inventions Approved by the Academy of Sciences*.[16] But, even more significantly, a number of articles published in the *Encyclopédie* had been previously discussed at the Société des Arts. This was the case with the several articles authored by the wood engraver and member of the Société des Arts Jean-Michel Papillon. In 1734, Papillon read to the Société a series of essays on his art, which he intended to collect in a volume to be published under the auspices of the Société. He presented his texts to the Société over twelve meetings, received feedback from fellow members, and revised his essays accordingly. It was this revised material that appeared in the *Encyclopédie*.[17] Papillon produced vignettes for Gua de Malves's prospectus of 1745, and contributed vignettes and historiated capital letters for Diderot and D'Alembert's *Encyclopédie*. In addition to Papillon, several former members of the Société des Arts (Jean Paul Grandjean, Charles Marie de La Condamine, Jean-Baptiste de Puisieux, and François Quesnay) contributed to the *Encyclopédie*, together with the son of Julien Le Roy, who may have been recruited by Gua de Malves. More than forty members of the Société des Arts are mentioned in the *Encyclopédie*, several of them many times.[18] Furthermore, Diderot knew several former members of the Société des Arts personally. He attended the anatomy demonstrations of César Verdier at Saint-Côme, contributed to Antoine Deparcieux's mathematical treatise, and was friends with Jean Pigeon's daughter and her husband, Pierre Le Guay de Prémontval.[19] In 1752, when he was nominated for fellowship at the Royal Society of London, La Condamine, Alexis Clairaut, and Grandjean de Fouchy, three academicians who had all been members of the Société des Arts, supported his unsuccessful candidature.[20]

The political project of the *artiste*, however, was not Diderot's. The *Encyclopédie* was an Enlightenment project whose main goal was to educate savants, artisans, men of state, and people of leisure. It gathered and organized knowledge on behalf of its audience, because "a head stuffed with facts without order, without utility, and without connections" did not differ at all from the instinct of an artisan who carried out rote operations.[21] The arbiter of such an ordering project was the *philosophe*, whose bird's-eye view on the landscape of human knowledge structured the *Encyclopédie* itself. In demonstrating the efficacy of their way of organizing the work, Diderot and D'Alembert claimed for philosophy the highest position in this system of knowledge. They regarded the *artistes* who had contributed to the *Encyclopédie* as *gens de lettres*, or savants. Diderot and D'Alembert's intellectual upgrade, conflating *artistes* with men of theory, erased the political epistemology that *artistes* themselves had articulated in their written and material works, and that they had attempted to implement, most visibly, through the constitution of the Société des Arts.

Artistes did not see themselves as savants. They aimed to establish a state institution dedicated to the encouragement of the mechanical arts, which would bestow upon them official status and positions. The authority they claimed for themselves derived from their practical knowledge, and from their ability to make artifacts that could advance France's commercial, military, colonial, and imperial power. They were eager for the French public to understand that theory and practice could serve each other but were distinct ways of knowing that contributed differently to the public good. Diderot went so far as to advocate for the equal relevance of practice and theory when he declared that "a man who knows only theoretical geometry is usually not skillful, and an *artiste* who knows only practical geometry is very limited as a worker."[22] *Artistes*, however, maintained that theoretical knowledge had value only if it helped make progress in the practice of the arts. If savants, including Diderot, were fascinated by the thinking process that brought a machine into existence, *artistes* wanted to call the public's attention to the practical knowledge required to translate the conceptual design of a machine into a working, and economically efficient, device. They also wanted the French public and state administrators alike to realize that the expertise of the *artiste* was strategically crucial to France's plans for colonial and commercial expansion. Theory could offer insightful perspectives, but useful devices or improved techniques emerged only from the domain of practical knowledge, which *artistes* attempted to order and systematize, just as Diderot and D'Alembert would later do with human knowledge.

The ordering and systematization of the mechanical arts through the compilation of written works had important implications, and it did not concern *artistes* only. Since Colbert's era, the French state had regarded the mechanical arts as political matters. Colbert envisioned that a state-sponsored institution such as the Académie des Sciences would, as part of its mission, compile an encyclopedia of the arts. Colbert framed such an encyclopedia as a normative project that would guide future developments in the arts and, therefore, as strategic to France's economic, commercial, and colonial expansion. The editorship and authorship of this project impinged directly on the question whether practitioners or savants could most reliably write about the mechanical arts. While the Republic of Letters debated the question, the Académie des Sciences struggled to complete this task and the Société des Arts advanced an alternative model for the encyclopedic project. I have argued that the Société's project stemmed from the *artistes*' vision of the mechanical arts as a world in which improvement occurred frequently, a vision that radically differed from that of the Académie des Sciences. *Artistes* articulated their vision in a variety of publications in which they also characterized their differences from other practitioners of the arts.

Writing about the mechanical arts was a political act. *Artistes* wrote to educate the French public about the relevance of their practical knowledge and to claim for themselves an official role in the political economy of the state. Only an audience that was conversant in the mechanical arts would be able to appreciate the ingenuity embodied in complex mechanical devices such as clocks, watches, weaving looms, or in sophisticated surgical techniques. The same educated audience would also be able to acknowledge the *artistes*' unique contributions to the public good. Through writing, *artistes* defined this notion of the public good as connected to advancements in the state's commercial, military, and imperial power. Knowledge was useful when it contributed directly to the public good, precisely as *artistes* constantly did through their work. For this reason, in their writings, *artistes* did not merely describe how to make machines or other artifacts; rather, they also emphasized how their ways of knowing set them apart from savants and other practitioners of the arts. With their boundary-work separating practice from theory, and the sciences from the arts, *artistes* intended to curb the savants' claims to expertise in the mechanical arts and to present themselves as the only reliable experts on technical innovations, uniquely qualified to advise state administrators about how to promote improvement. Useful knowledge, in their view, could only come from the *artiste*.

The Société des Arts was the most ambitious moment of the *artistes*' boundary-work aimed at defining the mechanical arts as a domain of knowledge and practice that could benefit from the sciences but that was essentially distinct from them. Employing the rhetoric of collaboration that characterized contemporaneous business partnerships (*sociétés*), the *artistes* hoped that the Société des Arts would become a state institution that, by fostering collaboration among *artistes*, savants, entrepreneurs, and men of state, would determine the direction of improvement in France. Authoring texts about the mechanical arts was integral to this political vision. The Société des Arts's initial interest in an encyclopedia of the arts soon gave way to shorter and more specialized publications that presented the world of making as fast-moving and forward-looking. Writing was essential to educating audiences and promoting emulation; even more importantly, it was crucial to establishing the notion that improvement depended on the mechanical arts. This notion implied a new interpretation of the history of any art. As Diderot would later underscore in his *Encyclopédie* article "Art," this should not be a chronological account of successes and failures but rather a selection of episodes that created an impression of linear progress. This history was essentially distinct from the history of other branches of knowledge grounded in factual accuracy and contextualization.

The Société des Arts was not an isolated episode in the history of Old Regime France. It embodied ambitions of cultural, social, and economic reform that a variety of historical actors continued to voice over the course of the century. Soon after the dissolution of the Société des Arts in the early 1740s, prominent clockmakers such as Julien Le Roy, Antoine Thiout, and Ferdinand Berthoud attempted to revive an association dedicated to horology, whose main goal would have been to boost the trade of French clocks and watches through the reform of the apprenticeship system and the education of the French public. Grounding their claims in the relevance of clocks and watches for navigation, and therefore for France's commercial and colonial endeavors, their association would fight the widespread ignorance in horological matters that hindered commerce and imperial expansion. In the 1760s, Berthoud, who, like Henry Sully, was working on the solution to finding longitude at sea, reiterated this argument for the constitution of what he called an "academy of horology," to no avail. As he remarked, horology lacked *amateurs*. These wealthy and well-educated patrons, who made the fortune of the beaux arts, knew how to evaluate the quality of artworks and supported those who excelled in the arts. The lack of horological knowledge, by contrast, left *artistes horlogers* without patrons, and exposed the French public to frauds and deceptions.[23]

The repeated failure to establish an institution dedicated to the mechanical arts reveals the distrust of French statesmen for people who worked with their hands. Although the French state intended to encourage the mechanical arts, it continued to draw expertise from academicians and not from artisans, however intelligent and full of *esprit* the latter might be. This dismissive attitude toward practitioners characterized associations dedicated to the encouragement of the mechanical arts that were established in France later in the century. The Société

Libre d'Emulation, founded in Paris in 1776 with the patronage of the comte d'Artois, shared with the Société des Arts the ambition to constitute itself as a state institution for the evaluation of technical innovations. As Liliane Hilaire-Pérez has shown, it pursued this program by recruiting from the social and intellectual elites. Its members included academicians such as the marquis de Condorcet, Antoine Lavoisier, the duke of Chaulnes, César-François Cassini de Thury, as well as the entourage of the Count of Artois, who—unlike the Count of Clermont—was deeply involved in the Société. *Artistes* and other practitioners, however, amounted to just 5 percent of the Société's membership.[24]

Programs of social, economic, and cultural reform motivated several other societies dedicated to the encouragement of the arts that proliferated during the eighteenth and nineteenth centuries, in Europe and beyond.[25] With the ongoing industrialization of the Western world, these societies offered networking opportunities for investors, entrepreneurs, artisans, scientists, and government officials. They created common ground for entrepreneurs looking for promising investments and inventors eagerly seeking sponsors for their practical research. Not all of them looked at the state as the ultimate source of social authority and financial support. The Society of Arts, founded in London in 1754, is the best known case of an alternative approach, where artisanal participation was encouraged and state support was not pursued. The Society of Arts, whose origins are connected to the early history of the Paris Société des Arts, attracted landed aristocrats, inventors, natural philosophers, Fellows of the Royal Society, and government officials, none of whom balked at the prospect of working with artisans.[26] In Britain, the landed aristocracy did not have to confront the *dérogeance*—the loss of nobility deriving from engaging in trade—which haunted the French equivalent. Aristocrats joined practitioners and Newtonian natural philosophers alike in the simultaneous pursuit of public utility and self-interest.[27] Unlike their counterparts in France, the members of the Society of Arts were unconcerned with its status as a private association. They could find within the Society of Arts the sources for funding their entrepreneurial projects.

While associations dedicated to the mechanical arts continued to form, the trajectory of the *artiste* branched out in new directions over the course of the eighteenth century. When *esprit* became genius, and the commitment to improvement, useful knowledge, and the mechanical arts gave way to the pursuit of beauty, refinement, taste, and the cult of artistic imagination, the *artiste* metamorphosed into the artist in the modern sense of the word. But the French state did not cease to consider technical matters as political affairs. On the contrary, it invested in new institutions from which it could not only draw but, above all, create and standardize technical expertise. As Antoine Picon and Bruno Belhoste have demonstrated, the establishment of the École des Ponts et Chaussées (School of Roads and Bridges) in 1747, followed by the École Royale du Génie (School of Engineering) at Mézières (1748), the École des Mines (School of Mining) in 1783, and the Polytechnique (1794), created a new professional figure that was responsible for the organization and implementation of the infrastructural policies of the French state: the state engineer. Knowledgeable in the mathematical and physical sciences and trained—often in the field—in the realization of practical projects, state engineers were relevant players in decision-making processes related to technical improvements. By creating bridges, roads, and canals, they contributed the technical expertise that the French state needed to redefine its territorial reality.[28] As a *corps d'état*, they realized the *artiste*'s project of a state institution formally dedicated to technical matters. But state engineers were no *artistes*. Being trained by the state, they were disconnected from the world of the arts and from the bodily interaction with natural materials that had constituted the foundations for the *artiste*'s self-understanding. The figure of the *mécanicien*, as defined by Condorcet in 1788 in reference to Jacques de Vaucanson, and the subsequent history of those artisans who continued to regard themselves as intelligent, discerning, polite, and witty, would be interesting topics for other books yet to be written.

By focusing on the *artiste*, this book has uncovered the heterogeneous interests that gravitated around the mechanical arts in the years between the foundation of the Académie des Sciences and the publication of the first volumes of the *Encyclopédie*. I have discussed the mechanical arts as a domain of practical knowledge that the French state regarded as strategic to its commercial and colonial expansion. Practitioners of the mechanical arts who regarded themselves as *artistes* were eager to inform the decision-making processes of the state in these matters. They published, articulated projects of educational or commercial reform, and formed associations together with savants, entrepreneurs, and potential patrons, all with the goal of overturning the intellectual priority of theoretical over practical knowledge. In doing so, they constructed a discourse on useful knowledge centered on the notion of the public good, one that prioritized the ability to build devices that could effectively bring about improvement—commercial, economical, and industrial. They also articulated a new goal for the history of the mechanical arts, one that forfeited accuracy and context in favor of a celebratory presentation of heroic inventions and self-evident progress. I have argued that this political project undermined the authority of existing institutions, such as academies and guilds, and expressed itself through publications, artifacts, and the constitution of the Société des Arts. Its failure reveals both the *artiste*'s contradictory relationship with Old Regime structures and the general distrust of the French ruling elite for practitioners of the arts.

By retrieving the *artiste*'s voice, my purpose has not been to provide a sympathetic interpretation of his discourse or political project. I have argued that practices of making gave rise to intellectual, economic, and political experimentation, which, in turn, deeply influenced the culture of the French Enlightenment. The manufacture of complex artifacts, such as clocks, watches, or maps, required the collaboration of a variety of people possessing different forms of capital: money, skills, social connections, and access to natural resources. These multifaceted collaborations offered a new model of social interactions, a societal vision that de facto—if often unintentionally—challenged the social and cultural status quo. This vision impinged on the understanding of the mechanical arts as a field of human activity that was essentially different from any other for the contributions it made to improve the human condition. One important aspect of this vision was the articulation of the history of the mechanical arts as a linear account of successes, which stood as a normative example of how improvement should be achieved.

I have also argued that *artistes* provided material foundations for key Enlightenment ideals by publishing, developing educational programs, and making machines. Voltaire, to cite perhaps the most famous example, attributed his early optimistic views on human progress to the luxuries and comforts that the arts had brought to humanity. In *The Wordling* (1736), he defied the widespread conception of the prelapsarian state of nature as a golden age. Adam and Eve, he irreverently observed, were naked, slept on the ground, and had dirty fingernails. Unaware of industry or comforts, their happiness was only due to their ignorance. By contrast, the arts had guided humanity toward an "iron age" of luxury and wealth, which had turned Paris into the real Eden. Several *artistes* of the Société des Arts had contributed to the material landscape within which Voltaire's appreciation of the present age grew: he enjoyed the music of Jean-Philippe Rameau, the jewels of Thomas Germain and François-Julien Barrier, the clocks of Julien Le Roy, the Gobelin tapestries designed by Jean-Baptiste Oudry, the instruments of Jean-Antoine Nollet, and the medicines of Jean-Baptiste Silva.[29]

Despite their substantial contributions to the cultural, material, and economic life of eighteenth-century France, *artistes* have disappeared from the histories of the Enlightenment and its legacies, of encyclopedism, of early modern science and technology, and of the Old Regime. By bringing them to center stage, my broader intention has been to highlight the practical, material, discursive, and artifactual contexts within which the categories of improvement, useful knowledge, and progress emerged. *Artistes* campaigned for their political relevance heralding their belief that the "bettering of humanity" could only result from technical and material advances. They strove to demonstrate this exclusive connection. The widespread, current association of improvement, useful knowledge, and progress with the STEM disciplines to the detriment of the humanities echoes the main themes of the artisanal enlightenment.

Appendix

MEMBERS OF THE SOCIÉTÉ DES ARTS
(1728–1740s)

This appendix lists all members of the Société des Arts, with their year of admission, their profession, their role in the Société, and, when available, their affiliation to royal academies or Masonic lodges with dates of admission. To facilitate identification, I have provided, when known, the members' years of birth and death. The professions noted here are what is recorded in the available lists of members; if a member's profession is known from other sources, I have enclosed it in parentheses. I have kept the original French for professions that do not have an obvious English equivalent; in these cases, a more detailed explanation is offered in the notes.[1]

According to the *Règlement de la Société des Arts*, published in 1730, the members held different roles according to their expertise, social status, provenance, and commitment to the Société. The core members, independent of their profession, resided in Paris and could be "assiduous" (attending each biweekly meeting) or "free" (not expected to attend each meeting). Those who did not reside in Paris were grouped as "foreign" associates, even if they lived in the French provinces. "Responding" associates were drawn exclusively from the world of the arts and crafts and were called on a case-by-case basis. "Honorary" members were people "distinguished by birth, dignity, and love for the arts," from France or elsewhere (arts. 3–6, 9). The Société elected five officers among its assiduous or free members: the director, the secretary, his assistant, the treasurer, and the administrator. Another special role was that of the *indicateur*, or indexer, who compiled annual catalogues of all new books, journals, theses, and other works on the arts published in Europe. The existing documents do not record systematically who held these positions, but I have noted a member's position as an officer when known.

The Société des Arts is often characterized as tied to Masonic circles. This notion seems to have some grounds in the fact that the Société's patron, the Count of Clermont, became the Grand Maître of the Grande Loge de France in 1743, and that Jean-Baptiste de Puisieux, who offered his premises as the first meeting place of the Société des Arts in 1728, became venerable of the Loge des Arts-Sainte Marguerite around the same time. The existing documents make it impossible to establish with confidence the Masonic affiliations of the many members of the Société des Arts. As the list below shows, there were a few members who were also affiliated with Masonic lodges, but they are by no means the majority.

ABBREVIATIONS

AdA   Académie d'Architecture
AdC   Académie de Chirurgie
AdPS   Académie de Peinture et Sculpture
AdS   Académie des Sciences
AF   Académie Française
ML   Masonic Affiliation

Name	Profession	Role and year of admission	Affiliations
Abeille, (?)	Engineer	Foreign associate, 1730	
Andelot, abbé d'	Mechanician	Unknown	
André, (?)	Unknown	Free associate, 1729	
Angaut [Angot], (?)	Tanner	Responding associate, 1732	
Arbuthnot, John (1667–1735)	Mathematician	Foreign associate, 1730	
Arconati Visconti, comte de	Prominent aristocrat from Milan	Honorary associate, 1728	
Aubert, Jean (1680–1741)	Architect	Free associate, 1730	AdA
Auvray, Louis (b. 1703)	Geometer, *conseiller au Châtelet*[2]	Free associate, 1728	
Barrier, François-Julien (1684–1746)	Goldsmith	Assiduous associate, 1728	
Bassuel, Pierre (1706–57)	Physicist (surgeon)	Assiduous associate, 1730	AdC (1731)
Bélidor, Bernard Forest de (1698–1761)	Engineer	Foreign associate, 1728	AdS
Bellet, Isaac	Physician	Foreign associate, 1730	
Béthune, Marie-Henri, chevalier de	(Royal Navy officer)	Honorary associate, 1731	
Blakey, William (ca. 1688–1748)	Mechanician (maker of watch springs)	Free associate, 1728	

Appendix

Name	Profession	Role and year of admission	Affiliations
Blanchard, Edme	Carpenter	Responding associate, 1730	
Boffrand, Germain (1667–1754)	Architect	Free associate, 1731	AdA (1709)
Bompart, Jean	Physician	Foreign associate, 1734	
Bonneau, Charles	Master carpenter, entrepreneur	Free associate, 1729	
Bonnier de la Mosson, Joseph (1702–44)	(Aristocrat and collector)	Unknown	ML (Montpellier)
Bottée, Claude	(Author on military strategy, secretary to the Dukes of Croÿ)	Assiduous associate, 1732 Director, 1733–35	
Bourdier [Bourdrie], (?)	Gunsmith	Responding associate, 1730	
Buache, Philippe (1700–1773)	Geographer	Assiduous associate, 1730	AdS (1730)
Cabot, (Claude?)	Gardener	Responding associate, 1730	
Cayeux, Philippe (1668–1769)	Sculptor of wood	Responding associate, 1730	
Celsius, Anders (1701–44)	Astronomer	Foreign associate, 1735	
Chappuis, Pierre-Joseph	*Arithméticien juré expert pour les vérifications des comptes et calculs*[3]	Responding associate, 1732	
Chevotet, Jean-Michel (1698–1772)	Architect	Free associate, 1729	AdA (1732)
Chiquet de Champrenard, Claude (1703–66)	*Avocat au conseil*[4]	Free associate, 1728	
Clairac, Louis-André de La Mamie, chevalier de (1690–1752)	Engineer	Foreign associate, 1730	

Name	Profession	Role and year of admission	Affiliations
Clairaut, Alexis (1713–65)	Geometer (mathematician)	Assiduous associate, 1728	AdS (1731)
Clairaut, Jean-Baptiste	Geometer (teacher)	Assiduous associate, 1728 Administrator, 1729	
Clairaut, le cadet	(Mathematician)	Assiduous associate, 1732	
Clermont, Louis de Bourbon, comte de (1709–71)	(Prince of the blood)	Patron of the Société des Arts	AF (1753), ML (Grand Orient, 1747)
Cochin, Charles Nicolas (1688–1754)	Engraver	Free associate, 1730	AdPS (1731)
Cramer, Gabriel (1704–52)	Professor of mathematics	Foreign associate, 1730	
Crestelet Duplessis, Nicolas	Anatomist (surgeon)	Assiduous associate, 1728	
Crestelet Duplessis, Pierre	Anatomist (surgeon)	Assiduous associate, 1728	
Croissant de Garengeot, René-Jacques (1688–1759)	Botanist (surgeon)	Assiduous associate, 1730 Treasurer, 1733, 1735	AdC (1731)
Cronstedt, Carl Johan, comte de (1709–79)	Intendant of buildings to Her Majesty of Sweden	Foreign associate, 1735	
Dandrieu, Jean-François (c. 1682–1738)	Organist	Free associate, 1728	
Deparcieux, Antoine (1703–68)	(Mathematician)	Unknown	AdS (1746)
De Perne, (?)	Hydrographer, ship captain	Free associate, 1728	
Descemet, Jean (d. 1762 or 1763)	Apothecary gardener	Responding associate, 1730	ML (Paris, 1732)
Deschisaux, Pierre (1687–1730)	Botanist	Foreign associate, 1730	

Appendix

Name	Profession	Role and year of admission	Affiliations
Deseuttre, (?)	*Expert ecrivain juré, verificateur pour les ecritures*[5]	Assiduous associate, 1728	
Desrosiers, F.	Engraver of maps	Free associate, 1728	
Dietrichstein, Leopold Philipp Maria von (1711–73)	Unknown	Honorary associate, 1728	
Dubois, Jean Baptiste	Physician	Free associate, 1730	
Du Pin [Dupin], (?)	Unknown	Free associate, 1730	
Dutertre, Jean Baptiste I (d. 1773)	Clockmaker	Foreign associate (supernumerary), 1730[6] Foreign associate (non-supernumerary), 1732 Administrator, 1732	
Dutertre, Jean Baptiste II	Clockmaker	Assiduous associate, 1735	
Dutot, Nicolas (1684–1741)	Mechanician (economist)	Free associate, 1728	
Du Treyves [du Tryves, Du Treves], (?)	Physicist	Free associate, 1731	
Du Vicquet [du Viquet], (?)	Marble worker	Responding associate, 1730	
Duvivier, Jean (1687–1761)	Engraver of medals	Assiduous associate, 1728	AdPS (1717)
Enderlin, Henri (d. 1753)	Clockmaker	Assiduous associate, 1728	
Faget, Jean (d. 1762)	Anatomist (surgeon)	Assiduous associate, 1730	AdC (1731)
Fournier, (?)	Watch spring maker	Responding associate, 1730	
Franchini, abbé	Envoy of the Grand Duke of Tuscany	Honorary associate, 1728	

Name	Profession	Role and year of admission	Affiliations
Gallon, Jean-Gaffin (1706–75)	Engineer	Foreign associate, 1730	AdS (1735)
Gaudron, Pierre (d. 1745)	Clockmaker	Free associate, 1729	
Germain, Thomas (1673–1748)	Goldsmith	Assiduous associate, 1728	
Girardin, (?)	Metal door maker	Foreign associate, 1731	
Girault, (?)	*Taillandier*[7]	Responding associate, 1730	
Gourdain, Nicolas (d. 1753)	Clockmaker	Assiduous associate, 1735	
Gourdon, (?)	Geometer, astronomer	Free associate, 1728	
Grammare, (?)	Mechanician	Foreign associate, 1730	
Grandjean de Fouchy, Jean-Paul, (1707–88)	Astronomer	Assiduous associate, 1730 Secretary, 1729, second semester 1732	AdS (1731). ML (L'amité à l'épreuve, 1748)
Grimberghen, Louis Joseph d'Albert, prince de [comte d'Albert] (1672–1758)		Honorary associate, 1731	
Grosse, Jean (d. 1744)	Chemist	Foreign associate, 1730	AdS (1731)
Gua de Malves, Jean Paul de (1710–85)	Geometer	Assiduous associate, 1728 Indexer, 1732	AdS (1741)
Guyot, Edme	(Instrument maker)	Foreign associate, 1733	
Habert, C.	Chemist	Assiduous associate, 1730	
Harnoncourt, Pierre Durey d' (b. 1682)	Tax farmer, mechanician	Assiduous associate, 1733	
Harrach, Ernest, comte de	Auditor of the Roman Rota	Honorary associate, 1728	
Harrach, Ferdinand, comte de,	(Austro-Germanic aristocrat)	Honorary associate, 1728	

Name	Profession	Role and year of admission	Affiliations
Hillerin de Boistissandeau, Jean Baptiste de (1704–79)	Mechanician	Foreign associate, 1731	
Hinderlink, (?)	(Clockmaker)	Unknown, 1729	
Horrebow, Peder (1679–1764)	Astronomer	Foreign associate, 1732	
Hussenot, Claude	*Avocat au conseil*	Free associate, 1728	
Hynault, (?)	Physicist, *avocat en Parlement*[8]	Assiduous associate, 1730 Secretary, 1730, 1731, second semester 1733, 1734–35	
Jolly, (?)	Ship captain	Foreign associate, 1730	
Jousse, Daniel (1704–81)	Astronomer	Free associate, 1728	
Jullien, Nicolas (d. 1765)	Painter, maker of enamel dials	Responding associate, 1730	
Kerkove, (?)	Dyer	Free associate, 1730	
Kolthof, Herman	Unknown	Foreign associate, 1731	
Kriegseissen, Mathias	Mechanician	Unknown	
Kurdwanowski, Jean-Etienne Ligenza (1680–1780)	Geometer	Foreign associate, 1730	
La Condamine, Charles Marie de, chevalier de (1701–74)	Geometer	Assiduous associate, 1730 Director, 1731	AdS (1730)
La Grive, Jean de, abbé de (1689–1757)	(Geographer)	Assiduous associate, 1730	
La Hire, (?) de	Engineer	Assiduous associate, 1733	
Languet de Gergy, Jean-Baptiste (1674–1750)	Parish priest	Honorary associate, 1728	

Name	Profession	Role and year of admission	Affiliations
Lassay, Armand de Madaillan de Lesparre, marquis de (1652–1738)	(French aristocrat and writer)	Honorary associate, 1732	
Laumonier, François-Charles de	Pewter smith	Responding associate, 1730	
Le Blanc, Jean-Bernard, abbé (1707–81)	Physicist (literary author)	Assiduous associate, 1730	
Le Dran, Henry François (1685–1770)	Physicist (surgeon)	Free associate, 1730	AdC (1731)
Le Faure, (?)	Metal caster	Responding associate, 1730	
Le Feuvre, (?)	Physicist	Free associate, 1730	
Le Maire, Jacques	Mathematical instrument maker	Assiduous associate, 1728	
Le Maire, Pierre	(Mathematical instrument maker)	Unknown	
Lemoyne, François, (1688–1737)	Painter	Assiduous associate, 1730	AdPS (1701)
Lemoyne, Jean-Baptiste (1704–78)	Sculptor	Assiduous associate, 1730	
Le Normand, (?)	(General treasurer of the Mint)	Unknown	
Le Rat, François	Director of fire pumps	Foreign associate, 1731	
Le Roy, Julien (1686–1759)	Clockmaker	Assiduous associate, 1728 Director, 1729, 1730 Treasurer, 1732	
Le Roy, Pierre II	Clockmaker	Assiduous associate, 1728	
Liébaux, Henri (d. 1752)	Geographer	Assiduous associate, 1728 Secretary, 1729, 1730, first semester 1733	

Name	Profession	Role and year of admission	Affiliations
Loppin, Jean Claude, seigneur de Gemeaux (1679–1737)	Geometer	Foreign associate, 1730	
Luce, (?)	Turner of flutes	Responding associate, 1730	
Marci, abbé	Physicist	Foreign associate, 1728	
Marie (fils), (?)	Optician	Responding associate, 1730	
Marne, Louis-Antoine de, (1673–1755)	Draftsman of the Société	Assiduous associate, 1731	
Martigni [Martini], abbé	Geometer	Foreign associate, 1728	
Martin, (?)	Boilermaker	Responding associate, 1730	
Masson de Guérigny, Jacques (1663–1741)	(Banker)	Free associate, 1732 Administrator, 1733	
Medalon, Daniel	Physicist, mechanician	Free associate, 1729 Director, 1732	
Melian, (?)	*Maître des requêtes*[9]	Honorary associate, 1728	
Menieur [Menissier], (?)	Stonecutter	Responding associate, 1730	
Molières, Joseph Privat de (1677–1742)	(Mathematician)	Assiduous associate, 1732	AdS (1721)
Montcrif, François-Paradis de (1687–1770)	Physicist (poet and writer)	Free associate, 1730	AF (1733)
Morville, Charles Jean-Baptiste Fleuriau, comte de (1686–1732)	(Ambassador to Holland, secretary of state for the navy, secretary of state for foreign affairs, art collector)	Honorary associate, 1731	

Name	Profession	Role and year of admission	Affiliations
Nicolai, Henri-Albert (b. 1701)	Professor of anatomy	Foreign associate, 1730	
Nollet, Jean-Antoine (1700–1770)	Mechanician	Free associate, 1728	AdS (1739)
Oudry, Jean-Baptiste (1686–1755)	Painter	Assiduous associate, 1733	AdPS (1717)
Pachta, comte de	(Bohemian nobleman)	Honorary associate, 1731	
Papillon, Jean-Michel (1698–1776)	Wood engraver	Assiduous associate, 1733	
Parsecknecht, (?)	Physician	Foreign associate, 1728	
Pelays, Jacques-Louis (d. ca. 1731)	Distiller (assayer at the Paris Mint)	Assiduous associate, 1728	
Penetti, abbé	Secretary to the Grand Duke of Tuscany, Mechanician	Free associate, 1728	
Petit, (?)	Carpenter, entrepreneur of roads and bridges	Responding associate, 1728	
[Peyrot ?]	Locksmith	Responding associate, 1730	
Pigeon d'Osangis, Jean (1654–c. 1739)	Mechanician (mathematical instrument maker)	Assiduous associate, 1729	
Plélo, Louis Robert Hipolite de Bréhan, comte de (1699–1734)	(French army officer and diplomat in Denmark)	Honorary associate, 1728	
Pochet, (?)	Candlestick maker	Responding associate, 1730	
Polinière, Pierre (1671–1734)	Physicist	Foreign associate, 1730	
Pontcarré de Viarmes, Jean Baptiste Élie Camus de (1702–75)	*Maître des requêtes,* intendant of Britanny	Honorary associate, 1728	

Name	Profession	Role and year of admission	Affiliations
Porlier, (?)	Embroiderer	Responding associate, 1730	
Procope-Couteaux, Michel (1684–1753)	Physician	Free associate, 1731 Assiduous associate, 1733	ML (Saint-Jean de la Discretion)
Puisieux, Jean-Baptiste de (c. 1679–1776)	(Architect)	Treasurer, 1729	ML (Les Arts-Sainte Marguerite, 1729)
Quesnay, François (1694–1774)	Surgeon	Foreign associate, 1730	AdC (1739), AdS (1751)
Quillau, Gabriel-François	Printer	Assiduous associate, 1730	
Raillard, Claude IV Labey	Clockmaker	Assiduous associate, 1732	
Rameau, Jean-Philippe (1683–1764)	Organist (composer)	Assiduous associate, 1728	ML
Raux, Jean	Enameller	Assiduous associate, 1730	
Remond de Sainte-Albine, Pierre (1699–1778)	Geometer (playwright)	Assiduous associate, 1730 Indexer, 1731 Secretary, first semester 1732	
Renard [le cadet]	(Mint officer)	Unknown	
Renard du Tasta, Mathieu (d. 1738)	Chemist, director of the mint	Free associate, 1728 Director, 1731	
Restout, Jean II (1692–1768)	Painter	Assiduous associate, 1728	AdPS (1717)
Riccardi, marquis de	(Florentine nobleman)	Honorary associate, 1728	
Rigaut, (?)	Worker at the foundry at Cosne	Foreign associate, 1730	
Röettiers, (Joseph-Charles?)	(Engraver of medals for the mint)	Unknown	
Romieu, abbé de	Physicist	Free associate, 1728	
Rosenkrantz, (Holger?)	Engineer	Foreign associate, 1730	

Name	Profession	Role and year of admission	Affiliations
Royllet, Sébastien (1699–1767)	Calligrapher	Responding associate, 1732	
Saint-Marcel, (?) de	Physicist	Free associate, 1731 Foreign associate, 1732	
Saussay, (?)	Director of the gardens at Chantilly	Foreign associate, 1730	
Silva, Jean-Baptiste (1682–1742)	Physician	Free associate, 1730	
Simonnet, (?)	Entrepreneur for the manufacture of tapestries	Assiduous associate, 1730	
Slodtz, René-Michel, called Michel Ange (1705–64)	Sculptor	Foreign associate, 1728	AdPS
Sully, Henry (1680–1729)	Clockmaker	Founder, 1728	
Taglini, Carlo	Professor in philosophy	Foreign associate, 1730	
Texier, (?)	Small-dots painter	Responding associate, 1730	
Thomas, (?)	Engineer	Assiduous associate, 1730	
Thomin, Marc Mitouflet (1708–53)	(Optical instrument maker)	Unknown	
Thurah, Laurids Lauridsen de (1706–59)	Engineer	Foreign associate, 1730	
[Touenard ?]	Copperplate printer	Responding associate, 1730	
Vallée de Pimodan, abbé de la	Physicist	Free associate, 1730	
Varennes, (?) des	Geometer	Free associate, 1730	
Vassé, François Antoine, (1681–1736)	Sculptor	Assiduous associate, 1730	AdA
Vaucanson, Jacques de (1709–82)	Instrument maker (Inspector of silk manufacture)	Assiduous associate, 1735	AdS (1746)

Name	Profession	Role and year of admission	Affiliations
Verdier, César (1685–1759)	Anatomist, royal demonstrator	Free associate, 1730	AdC (1731)
Vigneron, (?)	Cutler	Responding associate, 1730	
Vigny, Pierre Vigné de (1690–1772)	Architect	Assiduous associate, 1728	AdA (1723)

NOTES

1. The manuscript lists of the members of the Société des Arts are located in Mss J 1750, Staatsbibliothek, Berlin; Mss CC 3459, Nationalmuseum, Stockholm; and the private archive of the Duke of Croÿ, Dülmen, Germany. See also http://www.clairaut.com/sa.html. The Masonic affiliations are derived from the Ficher Bossu, BnF. I have indicated the year of initiation and lodge, if available.
2. The *conseiller au Châtelet* was a magistrate who worked at the Châtelet (the Paris tribunal). See Rosset, "Les Conseillers au Châtelet."
3. This was a master calligrapher, specialized in examining calculations that were contested in judicial processes.
4. A magistrate who could represent cases to the various councils of the king.
5. A master calligrapher who specialized in examining texts that were contested in judicial cases.
6. Dutertre was initially supernumerary because the Société had already covered all the available positions reserved to clockmakers. He became an ordinary member in 1732, when he replaced Hinderlink.
7. An artisan who specialized in making cutting instruments within a specific guild (cutlers, cabinetmakers, etc.).
8. A magistrate to the Parlement of Paris, the local court of appeals.
9. A high-ranking judicial officer that examines appeals presented to the Council of the State

Notes

All translations, unless otherwise indicated, are my own.

ABBREVIATIONS

AASP	Archives de l'Académie des Sciences, Paris
AN	Archives Nationales, Paris
BnF	Bibliothèque Nationale de France, Paris
CNAM	Archive du Conservatoire des Arts et Métiers, Paris
De Croÿ	Papiers Bottée, Mons 555, Private Archive of the Duc de Croÿ, Dülmen, Germany
Encyclopédie	Denis Diderot and Jean Le Rond d'Alembert, eds., *Encyclopédie; ou, Dictionnaire raisonné des sciences, des arts et des métiers, etc.*, 17 vols., ed. Robert Massey (University of Chicago: ARTFL Encyclopédie Project, spring 2013 edition), http://artflsrv02.uchicago.edu/cgi-bin/philologic/getobject.pl?c.0:3094.encyclopedie0513.8035790
HARS	*Histoire et mémoires de l'Académie Royale des Sciences*, 97 vols. (Paris, 1699–1797), Section: Histoire
MARS	*Histoire et mémoires de l'Académie Royale des Sciences*, 97 vols. (Paris, 1699–1797), Section: Mémoires
MIAS	*Machines et inventions approuvées par l'Académie Royale des Sciences*, 7 vols. (Paris, 1735)
"Préface des arts"	René Réaumur, "Préface des arts," MS Typ 432.1 (1), Houghton Library, Harvard University, Cambridge, Mass.
PV	Procès Verbaux de l'Académie des Sciences, Archives de l'Académie des Sciences, Paris

INTRODUCTION

1. The story of Hephaestus as discussed here is drawn from Cuomo, *Technology and Culture*, chapter 1, and from Dacier, ed., *L'Iliade d'Homère*, 3:149n.

2. Cuomo, *Technology and Culture*, chapter 1.
3. Minutes, 15 June 1732, De Croÿ.
4. Dacier, ed., *L'Iliade d'Homère*.
5. On the Republic of Letters in eighteenth-century France, the classic works are Goodman, *The Republic of Letters*; Brockliss, *Calvet's Web*; and Goldgar, *Impolite Learning*.
6. Sturdy, *Science and Social Status*; Bertrand, *L'Académie des Sciences*; Hahn, *The Anatomy of a Scientific Institution*; Gillispie, *Science and Polity in France*.
7. My interpretation differs substantially from earlier studies on the Société des Arts such as Hahn, "Science and the Arts in France"; Hahn, "The Applications of Science to Society"; and Salomon-Bayet, "Un préambule théorique à une Académie des Arts." Here I am also moving beyond the interpretation I offered in collaboration with Olivier Courcelle in 2015; see Bertucci and Courcelle, "Artisanal Knowledge, Expertise, and Patronage."
8. Furetière, *Dictionnaire universel*, vol. 1, s.v.
9. Richelet, *Nouveau dictionnaire françois*, s.v. On things and self-definition in eighteenth-century Paris, see Roche, *A History of Everyday Things*; Goodman and Norberg, eds., *Furnishing the Eighteenth Century*; and Goodman, *Becoming a Woman in the Age of Letters*.
10. *Encyclopédie*, 1:713–17.
11. On Xenophon, see Cuomo, *Technology and Culture*, chapter 1.
12. Shiner, *The Invention of Art*.
13. Heinich, *Du peintre à l'artiste*.
14. Barolsky, *A Brief History of the Artist from God to Picasso*.
15. It would be too long to list all the relevant literature here. See at least Rossi, *I filosofi e le macchine*; Rossi, *Philosophy, Technology, and the Arts in the Early Modern Era*; Long, *Openness, Secrecy, Authorship*; Long, *Artisan/Practitioners and the Rise of the New Sciences*; Smith, *The Body of the Artisan*; and Harkness, *The Jewel House*. A very useful overview of the recent literature on early modern science is in Smith, "Science on the Move."
16. Roberts, Schaffer, and Dear, eds., *The Mindful Hand*; see especially the introduction for a methodological call not to essentialize such dichotomies. In 1962, Paolo Rossi offered a nuanced discussion of the relationships among science and the arts, scholars and artisans, and theory and practice in the early modern period; see Rossi, *I filosofi e le macchine*. Larry Stewart has worked extensively on these themes in the context of eighteenth-century Britain; see at least his *The Rise of Public Science* and "A Meaning for Machines." Other relevant works that address these dichotomies are Long, *Artisans/Practitioners*, and Klein, "Artisanal-Scientific Experts."
17. Cassirer, *Die Philosophie Der Aufklärung*; Gay, *The Enlightenment*.
18. See, for example, Venturi, *The End of the Old Regime in Europe*; Porter, *The Creation of the Modern World*; Porter and Teich, eds., *The Enlightenment in National Context*; Jacob, *The Radical Enlightenment*; Jacob, *Living the Enlightenment*; Sorkin, *The Religious Enlightenment*; Lehner and Printy, eds., *A Companion to the Catholic Enlightenment*; Israel, *Democratic Enlightenment*; Israel, *Radical Enlightenment*; and Israel, *Enlightenment Contested*.

19. Compare Brockliss, *Calvet's Web*, and Robertson, *The Case for the Enlightenment*.
20. Israel, *Radical Enlightenment*; Israel, *Democratic Enlightenment*; Mokyr, *The Enlightened Economy*.
21. Clark, Golinski, and Schaffer, eds., *The Sciences in Enlightened Europe*; Spary, *Eating the Enlightenment*; Withers, *Placing the Enlightenment*; Takats, *The Expert Cook in Enlightenment France*; Stewart, *The Rise of Public Science*; Stewart, "A Meaning for Machines"; Stewart, "Assistants to Enlightenment"; Jacob and Stewart, *Practical Matter*; Hilaire-Pérez, "Technology as Public Culture in the Eighteenth Century," 135–53; Hilaire-Pérez, "Technology, Curiosity, and Utility"; Venturino, "L'historiographie révolutionnaire française et les lumières"; and Edelstein, *The Enlightenment*.
22. *Encyclopédie*, 1:29. Latin in the original text. Here I have used the translation of the Collaborative Translation Project of the University of Michigan, "Preliminary Discourse," http://quod.lib.umich.edu/d/did/did2222.0001.083/1:4/-preliminary-discourse?rgn=div1;view=fulltext.
23. The ARTFL *Encyclopédie* is available at http://encyclopedie.uchicago.edu.
24. *Encyclopédie*, 9:808.
25. *Encyclopédie*, 5:583.
26. *Encyclopédie*, 9:769.
27. *Encyclopédie*, 12:476.
28. *Encyclopédie*, 7:584.
29. *Encyclopédie*, 19:698.
30. *Encyclopédie*, 9:431.
31. The earliest discussions on the role of artisans in the Scientific Revolution are Zilsel, "The Sociological Roots of Science," 544–62, and Rossi, *I filosofi e le macchine*. For subsequent developments, see Smith, "Science on the Move."
32. Stewart, *The Rise of Public Science*; Stewart, "A Meaning for Machines"; Stewart, "Assistants to Enlightenment"; Hilaire-Pérez, *La pièce et le geste*; Smith, *The Body of the Artisan*; Fox, *The Arts of Industry in the Age of Enlightenment*; Harkness, *The Jewel House*; Long, *Openness, Secrecy, Authorship*.
33. Richelet, *Dictionnaire de la langue françoise*, 1:186.
34. Edelstein, *The Enlightenment*.
35. *Dictionnaire de l'Académie Françoise* (1694), 1:57–58.
36. *Dictionnaire de l'Académie Françoise* (1762), 1:107. The term "genius" (*génie*) used here was defined as "talent, inclination or natural disposition for something estimable, which belongs to the *esprit*" (1:814).
37. Heinich, *Du peintre à l'artiste*. See also Guichard, "Arts libéraux et arts libres," 54–68, and Baudez, *Architecture et tradition académique*.
38. Larmessin, *Costumes grotesques*.
39. Portalis and Béraldi, *Les graveurs du dix-huitième siècle*, 1:529–36. See also Milliot, *Les cris de Paris*.
40. On the Royal Society's History of Trades, see Houghton, "The History of Trades," 33–60. See also Fox, *Art of Industry*, chapter 1.
41. Félibien, *Des principes de l'architecture, de la sculpture, de la peinture*, preface.
42. Félibien, *Des principes de l'architecture, de la sculpture, de la peinture*, preface.

43. *Encyclopédie*, 1:39.
44. Dempsey, *Inventing the Renaissance Putto*; Korey, *Putti, Pleasure, and Pedagogy*. On putti in scientific illustration, see Shapin, "The Invisible Technician," and Shea, *Science and the Visual Image in the Enlightenment*.
45. Van Veen, *Amorum emblemata*, 82.
46. Castiglione, *The Book of the Courtier*, 66–68.
47. On the politicization of technology, see Kaplan, *Les ventres de Paris*, and Hilaire-Pérez, *L'invention technique au siècle des lumières*.
48. Heinich, *Du peintre à l'artiste*; Guichard, "Arts libéraux et arts libres"; Shiner, *The Invention of Art*.
49. Papiers Bignon, Ms Fr. 22225, ff. 7–10, BnF.
50. On the emergence of the beaux arts, see Shiner, *The Invention of Art*, chapter 5.
51. Quoted in Hilaire-Pérez, "Technology as Public Culture"; Voskhul, *Androids in the Enlightenment*; Stewart, "A Meaning for Machines"; Fox, *Art of Industry*.
52. *Le mercure galant*, August 1691, 161–66.
53. Perrault, *Cabinet des beaux arts*. The *Cabinet* was sold on the rue St. Jacques in the shop of the engraver and printer Gérard Endelick, who had contributed to the work.
54. Perrault, *Cabinet des beaux arts*, 1.
55. Perrault, *Cabinet des beaux arts*, 39.
56. Perrault, *Cabinet des beaux arts*, 2.
57. Norman, *The Shock of the Ancient*.
58. On early modern natural knowledge and its relationship to the humanistic tradition, see Pomata and Siraisi, eds., *Historia*, and Siraisi, *History, Medicine, and the Traditions of Renaissance Learning*.
59. Vérin, "La technologie," 134–43.
60. Ash, "Introduction." On the notion of boundary-work, see Gieryn, "Boundary-Work and the Demarcation of Science from Non-Science."
61. Mokyr, *The Gifts of Athena*, chapter 2; Mokyr, *The Enlightened Economy*. See also Jacob and Stewart, *Practical Matter*, and Jones, *Industrial Enlightenment*.
62. Hilaire-Pérez, "Technology as Public Culture."
63. Venturi, *Le origini dell'Enciclopedia*; Hahn, "Science and the Arts in France," 77–93; Kafker, "Gua de Malves and the *Encyclopédie*," 93–102.
64. These documents are among the papers of Bottée, director of the Société des Arts between 1733 and 1735, and are kept in the archives of the Duke de Croÿ at Dülmen in Germany. The folder "Mons 555" in this archive comprises 183 uncatalogued pieces that concern directly the Société des Arts. They cover mainly the periods between the beginning of 1732 and April 1734 and, less systematically, the end of 1736 and the beginning of 1737. For earlier interpretations of the Société des Arts see Bertrand, *L'académie des sciences*; Hahn, "The Applications of Science to Society," 829–36; Hahn, "Science and the Arts in France"; and Irène Passeron and Olivier Courcelle, "La Société des Arts, espace provisoire de reformulation des rapports entre théories scientifiques et pratiques instrumentales," in Demeulenaere-Douyère and Brian, eds., *Règlement, usages et science dans la France de l'absolutisme*, 109–32.
65. Olivier Courcelle, "Chronologie de la vie de Clairaut (1713–1765)," 2015, http://clairaut.com/sa.html.

66. On Chambers's and other scientific dictionaries of the eighteenth century, see Yeo, *Encyclopaedic Visions*.
67. On the debate, see Kessler, *A Revolution in Commerce*.
68. Long, *Openness, Secrecy, Authorship*; Dubourg Glatigny and Vérin, eds., *Réduire en art*; Vérin, "La technologie."
69. Kang, *Sublime Dreams of Living Machines*; Voskhul, *Androids in the Enlightenment*; Riskin, *The Restless Clock*.

1. LOST KNOWLEDGE AND THE HISTORY OF THE ARTS

1. *Encyclopédie*, 1:713–17.
2. Bacon, *The Advancement of Learning*, 2:3. There are various online editions of this book. I consulted http://ebooks.adelaide.edu.au/b/bacon/francis/b12a/chapter9.html. The literature on Francis Bacon's natural history is vast. Here I am not concerned with Bacon's own notion of natural history but rather with how natural history and associated practices informed the way the mechanical arts were described. For recent studies on Bacon's notion of natural history, see the special issue of *Early Science and Medicine* 17 (2012): "The Place of Natural History in Francis Bacon's Philosophy," edited by Sorana Corneanu, Guido Giglioni, and Dana Jalobeanu. Relevant literature, classic and recent, on Bacon's natural history can be found there. Readers interested in the various meanings of the early modern notion of "history" should consult Pomata and Siraisi, eds., *Historia*.
3. Bacon, *Preparative Toward a Natural and Experimental History*, 258, in Spedding, Ellis, and Heath, eds., *The Works*. On Bacon's "experimental history," see Klein and Lefèvre, *Materials in Eighteenth-Century Science*, 22–28, and Fox, *The Arts of Industry*, 17–19.
4. Long, *Openness, Secrecy, Authorship*; Smith, *The Body of the Artisan*; Rossi, *I filosofi e le macchine*.
5. *Encyclopédie*, 1:715.
6. *Descriptions des arts et métiers, faites ou approuvées par messieurs de l'Académie Royale des Sciences* (Paris, 1761–88). The collection comprises 113 volumes, plus 3 supplements.
7. Venturi, *Le origini dell'Enciclopedia*; Pinault, "Aux sources de l'*Encyclopédie*"; Madeleine Pinault-Sørensen, "La *Description des arts et métiers* et le rôle de Duhamel du Monceau," in Corvol, ed., *Duhamel du Monceau*, 133–55; Pinault, "Diderot et les illustrateurs de l'*Encyclopédie*"; Jaoul and Pinault, "La collection 'Description des Arts et Métiers'"; Cole and Watts, *The Handicrafts of France*; Huard, "Les planches de l'*Encyclopédie*"; Sheridan, "Recording Technology"; Hahn, "Science and the Arts in France"; Hahn, "The Applications of Science to Society"; Hahn, *The Anatomy of a Scientific Institution*; Bertrand, *L'Académie des Sciences*; Passeron and Courcelle, "La Société des Arts"; Jammes, *La réforme de la typographie*.
8. Proust, *Diderot et l'*Encyclopédie; Cole and Watts, *The Handicrafts of France*; Huard, "Les planches de l'*Encyclopédie*"; Watts, "The *Encyclopédie* and the *Description des Arts et Métiers*," 444–54; Daumas and Tresse, "La *Description des Arts et Métiers*"; Belhoste, *Paris Savant*, chapter 4. See also Pinault, "Aux sources de l'*Encyclopédie*";

Madeleine Pinault-Sørensen, "La *Description des arts et métiers* et le rôle de Duhamel du Monceau," in Corvol, ed., *Duhamel du Monceau*, 133–55; Pinault, "Diderot et les illustrateurs de l'*Encyclopédie*," Jaoul and Pinault, "La collection 'Description des Arts et Métiers'."

9. These "hybrid" figures are discussed, notably, in Roberts, Schaffer, and Dear, eds., *The Mindful Hand*; Long, *Artisans/Practitioners*; Klein, "Artisanal-Scientific Experts".

10. See, for example, Pierre Petit to Henry Oldenburg, 23 October 1660, in Hall and Hall, eds., *The Correspondence of Henry Oldenburg*, 1:395–99. Petit asked Oldenburg whether the latest publication of Bacon's posthumous works was available in folio: "As I have the rest in a large folio volume I should be very glad to have this last work in the same format if they have been printed that way" (398).

11. Fox, *Art of Industry*, chapter 1.

12. Clément, ed., *Lettres, instructions et mémoires de Colbert*, vol. 2, part 1, annexe, 263–72. Clément explains that the memorandum was likely read aloud to the king, who approved one by one all of Colbert's points. On the role of merchants in the government's decisions after 1664, see Smedley-Weill, "La gestion du commerce français au xviième siècle," 473–96. On Colbert, see Soll, *The Information Master*, and Dessert, *Argent, pouvoir et société au grand siècle*. For a general overview on the French state in the early modern period, see Collins, *The State in Early Modern France*.

13. Clément, ed., *Lettres, instructions et mémoires de Colbert*, vol. 2, part 1, annexe, 268, 263, 271.

14. Oldenburg to Boyle, 28 September 1665, in Hall and Hall, eds., *The Correspondence of Henry Oldenburg*, 2:533.

15. The "Projet" is published in Christiaan Huygens, *Oeuvres complètes*, 4:325–29.

16. On these groups, see Maury, *L'ancienne Académie des Sciences*; Bertrand, *L'Académie Royale des Sciences*; Maidron, *L'Académie des Sciences*; Brown, *Scientific Organizations in Seventeenth-Century France*; Taton, *Les origines de l'Académie Royale des Sciences*; Wellman, *Making Science Social*; Mouy, *Le développement de la physique cartésienne, 1646–1712*; Bigourdan, *Les premières sociétés savantes de Paris*; Simone Mazauric, "Des académies de l'âge baroque à l'Académie Royale des Sciences," in Demeulenaere-Douyère and Brian, eds., *Règlement, usages et science*, 13–24; and Belhoste, *Paris savant*.

17. On Thévenot, see Dew, "Reading Travels in the Culture of Curiosity"; Turner, "Melchisédech Thévenot, the Bubble Level, and the Artificial Horizon"; and McClaughlin,"Sur les rapports entre la Compagnie de Thévenot et l'Académie Royale des Sciences."

18. Thévenot's cabinet is mentioned in Spon, *Recherche des antiquités et curiosités de la ville de Lyon*, 217, and Baudelot de Dairval, *De l'utilité des voyages*, 2:685.

19. Clément, ed., *Lettres, instructions et mémoires de Colbert*, vol. 2, part 1, annexe, 268, 263, 271.

20. Huygens, *Oeuvres complètes*, 4:325–29. Auzout explained in his dedication to the king in *L'éphéméride du comète*, published in the same year as the memorandum, that the improvement of "all the sciences and all useful arts" was a task that could only be accomplished by a state-funded institution. Auzout, "Au roy," in *L'éphéméride du comète*.

21. Huygens, *Oeuvres complètes*, 4:326–27.
22. Guido Panciroli (1523–99)—whose name was variously spelled as Pancirolli, Pancirollus, Pancirolle, and so on—was a professor at the University of Padua. His Italian manuscript on remarkable things that had disappeared since antiquity was translated into Latin and published posthumously in two volumes with a commentary by Heinrich Salmuth, one of his students. Panciroli, *Rerum memorabilium*. On Panciroli, see Keller, "Accounting for Inventions," and Andreoli, *Alcuni studi intorno a Guido Panciroli*.
23. Keller, "Accounting for Inventions," 235.
24. Huygens, *Oeuvres complètes*, 4:325–29. On desiderata, see Keller, *Knowledge and the Public Interest*.
25. On early modern French travel literature, see Lestringant, *Mapping the Renaissance World*, and Lindsay, "Pierre Bergeron," 31–38. On the "Orients," see Dew, *Orientalism in Louis XIV's France*.
26. Thévenot, *Relations de divers voyages curieux*.
27. I have discussed the category of "intelligent travel" in Bertucci, "Enlightened Secrets."
28. Huygens, *Oeuvres complètes*, 4:325–29.
29. See Hahn, *The Anatomy of a Scientific Institution*; Sturdy, *Science and Social Status*; Stroup, *Royal Funding of the Parisian Académie Royale des Sciences*; and Stroup, *A Company of Scientists*. On inventions and the state in eighteenth-century France, see Gillispie, *Science and Polity*, and Hilaire-Pérez, *L'invention technique*.
30. Quoted in Jammes, *La réforme de la typographie*, 16.
31. See in particular Hahn, *The Anatomy of a Scientific Institution*.
32. Hahn, "Science and the Arts in France." On Bignon's reform of the academy, see Bléchet, "L'abbé Bignon." On Bignon in general, see Bléchet, "Un précurseur de l'*Encyclopédie*"; Bléchet, "Le rôle de l'abbé Bignon"; Bléchet, "Fontenelle et l'abbé Bignon"; and Clarke, "Abbé Jean-Paul Bignon."
33. The term "history of the arts" was sometimes replaced by the synonymous "description of the arts" (in the singular). It is interesting to note that the Académie opted for the plural "descriptions" when it started publishing the work in the 1760s.
34. "Sur la description des arts," HARS 1699, 117.
35. Robert Boyle to Henry Oldenburg, 6 October 1664, quoted in Fox, *The Art of Industry*, 35.
36. HARS 1699, 145. On Parisians' reading habits, see Roche, *The People of Paris*.
37. Fontenelle, *Entretiens sur la pluralité des mondes*.
38. On Rohault, see Sutton, *Science for a Polite Society*. On natural philosophy as public spectacle, the classic work is Schaffer, "Natural Philosophy and Public Spectacle." See also Stewart, *The Rise of Public Science*; Golinski, *Science as Public Culture*; and Golinski, "A Noble Spectacle."
39. Fontenelle, "Éloge de Gilles Filleau des Billettes," HARS 1720, 122–24.
40. HARS 1666–99, 2:110.
41. Jaugeon, *Le jeu du monde*.
42. Jaugeon, "Calendrier perpétuel pour les mois, les semaines, les jours," BnF. On Jaugeon, see Sturdy, *Science and Social Status*. On 17 August 1681, Jaugeon obtained a

six-year privilege, valid in the entire French kingdom, for his "game of the world." Registres du secrétariat de la maison du roi, O1–25, 251, AN.
43. On Grollier de Servière, see Turner, "Grollier de Servière."
44. Truchet made a mechanical table for the king and several models of machines employed in manufactures, ranging from goldbeating to coinage and textiles. Fontenelle, "Éloge du P. Sébastien Truchet, Carme," *HARS* 1729, 93–101.
45. Sturdy, *Science and Social Status*; Stroup, *Royal Funding of the Parisian Académie Royale des Sciences*.
46. There has been some confusion as to the relationship of this commission to the Académie des Sciences. As Geraldine Sheridan observed, Hahn and other scholars overlooked the fact that at its foundation, the commission constituted what Réaumur would later call a "small separate academy"; see "Préface des arts." However, the three members of the commission would later continue their work on the history of the arts within the Académie des Sciences. This is discussed, notably, in Jammes, *La réforme de la typographie*.
47. On the Quarrel, see Norman, *The Shock of the Ancient*; Fumaroli, *La querelle des anciens et des modernes*; Levine, *The Battle of the Books*; and DeJean, *Ancients against Moderns*.
48. Perrault, *Parallele des anciens et des modernes*, 1:54–57.
49. Mukerji, *Impossible Engineering*.
50. Dossier Des Billettes, AASP.
51. PV 1690–93, f. 154. The folios 149–68v are the minutes of the meetings of the group that the abbé Bignon presided over during the years 1695 and 1696. The folios 154–60 of the minutes include an undated draft of a "Préface des arts" by Des Billettes.
52. PV 1690–93, f. 154.
53. PV 1690–93, f. 158.
54. PV 1690–93, f. 157.
55. On the theaters of machines, see Dolza and Vérin, "Figurer la méchanique"; Dolza, "Reframing the Language of Inventions"; and Blair, *The Theater of Nature*. On the history of early modern patents and the relevance of the notion of "reduction to practice," see Biagioli, "Patent Republic," 1129–72.
56. [Picot], *Explication des modèles*. See also Birembaut, "L'exposition de modèles," 141–58.
57. [Picot], *Explication des modèles*, 2–4.
58. [Picot], *Explication des modèles*, preface.
59. The short life of the exhibition was due to Colbert's death in 1683, a few weeks after the opening. The circulation of the catalogue is documented in several sale catalogues of aristocratic libraries, as discussed in Birembaut, "L'exposition de modèles."
60. *Encyclopédie*, 2:98 (quotation). On Diderot's article "Bas," see Stalnaker, *The Unfinished Enlightenment*, chapter 3, 115 (quotation); Barthes, "Les planches de l'Encyclopédie," in Barthes, ed., *Le degré zéro de l'écriture*, 89–105; "La documentation technique de Diderot dans l'*Encyclopédie*," 346–48.
61. Stalnaker, *The Unfinished Enlightenment*, 115.
62. PV 1690–93, f. 154. On the notion of an early modern "encyclopedia," see Blair, *Too Much to Know*, and Yeo, *Encyclopaedic Visions*.

63. Devey, ed., *The Moral and Philosophical Works of Francis Bacon*, 237.
64. Dossier Des Billettes, AASP.
65. PV 1690–93, f. 156.
66. PV 1690–93, ff. 154v–55.
67. Martin, *Livre, pouvoirs et société à Paris au XVIIe siècle*.
68. Dossier Des Billettes, AASP.
69. PV 1690–93, f. 155v.
70. On natural history and the world of erudition, see Pomata and Siraisi, *Historia*.
71. Dossier Des Billettes, AASP.
72. Dossier Des Billettes, AASP.
73. Des Billettes is ambiguous as to whether this new academy was the renovated Académie des Sciences or an academy especially dedicated to the arts.
74. See Stroup, *Royal Funding*, chapter 6.
75. Sheridan, "Recording Technology in France." Compare with Hahn, *The Anatomy of a Scientific Institution*, and Salomon-Bayet, "Un préambule théorique à une Académie des Arts," 229–50.
76. Jammes, *La réforme de la typographie*.
77. On the 1699 reform of the Académie des Sciences, see Bléchet, "Le rôle de l'abbé Bignon."
78. The list of the articles on the arts they read to the Académie (never published) can be found in *Table alphabétique des matières contenues dans l'histoire et les mémoires de l'Académie Royale des Sciences. Tome II: Années, 1699–1710* (Paris: Compagnie des Libraires, 1729), and *Tome III: Années, 1711–1720* (Paris: Compagnie des Libraires, 1731), under their names.
79. Nouvelles acquisitions françaises, 5148, BnF.
80. Rowlands, *The Financial Decline of a Great Power*; Dessert, *Argent, pouvoir et société*.
81. Jammes, *La réforme de la typographie*, 36n22. On the overall cost of the project, see Stroup, *Royal Funding*.
82. Jammes, *La réforme de la typographie*, 16.

2. RÉAUMUR AND THE SCIENCE OF THE ARTS

1. Daston and Park, *Wonders and the Order of Nature*; Eva Schulz, "Notes on the History of Collecting and Museums," in Pearce, ed., *Interpreting Objects and Collections*, 175–87; Burke, *A Social History of Knowledge*, 106–9. An excellent article on order in early modern libraries is Garberson, "Libraries, Memory and the Space of Knowledge," 105–36.
2. There is surprisingly little, especially in English, on Réaumur. A good source of biographical information is Grandjean de Fouchy's "Éloge de M. de Réaumur," in *MARS* 1757, 201–17. See also Torlais, *Réaumur*. On Réaumur's natural history, see Terrall, *Catching Nature in the Act*.
3. It is unclear when exactly Réaumur took over the leadership of the project from Des Billettes. De Fouchy's obituary of Réaumur states that Réaumur was assigned to the description of the arts as soon as he joined the academy. This is an ambiguous

statement that has been taken by some scholars as an indication that he was entrusted with the directorship of the project in 1708. However, there is no evidence for this. I believe that he began to take the project in his hands when Des Billettes's and Jaugeon's involvement started to wane and particularly during the years of the Regent's Survey (1716–18), which I discuss below. Des Billettes's last essay on the arts dates to 1712; he became a veteran in 1718. Jaugeon's last essay on the arts dates to 1718. An unsigned report on the Académie des Sciences's activities on the description of the arts, dated 1786, states that Réaumur was entrusted with the organization of the project around 1720. Dossier Lavoisier, 502–9, AASP.

4. *HARS* 1711, 101; *HARS* 1712, 80; *HARS* 1713, 75; *HARS* 1714, 106; *HARS* 1716, 76. Père Sébastien Truchet stopped working on the encyclopedic project on the arts when he became an honorary member of the Académie des Sciences.
5. For a list of Des Billettes's and Jaugeon's essays on the arts, see *Table alphabétique des matières continues dans l'histoire et les mémoires de l'Académie Royale des Sciences. Tome II: Années, 1699–1710*, and *Tome III: Années, 1711–1720*, under their names. On Des Billettes's status as a veteran, see *Almanach Royal* (1716): 223.
6. "Préface des arts," f. 5.
7. Terrall, *Catching Nature in the Act*.
8. Lorraine Daston, "Attention and the Values of Nature in the Enlightenment," in Daston and Vidal, eds., *The Moral Authority of Nature*, 100–126.
9. Réaumur, "Expériences et réflexions sur la prodigieuse ductilité de diverses matières," *MARS* 1713, 201.
10. Smith, *The Body of the Artisan*; Pamela H. Smith, "Vermillion, Mercury, Blood, and Lizards," in Klein and Spary, eds., *Materials and Expertise in Early Modern Europe*, 25–49.
11. The Académie's public campaign against alchemy was launched in 1725 by the physician and chemist Étienne François Geoffroy, who presented an essay on the topic at a public meeting of the Académie. The essay was summarized in various popular magazines, such as *Mélanges d'histoire et de littérature* 3 (1725): 314. Recently, Larry Principe has argued that this was only a public move, with members of the Académie still pursuing alchemical work. Principe, "The End of Alchemy?" 96–116. Réaumur was clearly on Geoffroy's side, as evidenced in a letter by his correspondent Jean-Pierre de Crousaz, who complained about having had to abandon any work on metals due to his own parents' disbelief that "one can have a laboratory without searching for the Philosopher's Stone and without breaking the bank [*se ruiner*]." Crousaz to Réaumur, 5 December 1721, Dossiers Réaumur, AASP. For the traditional view on the decline of alchemy within the Académie des Sciences, see Debus, *The French Paracelsians*.
12. Palissy, *Discours admirables*.
13. *HARS* 1720, 6.
14. Quoted from Crousaz to Réaumur, 5 December 1721, Dossiers Réaumur, AASP.
15. *HARS* 1722, 45.
16. "Préface des arts."
17. See Vérin, "La technologie," and Dubourg Glatigny and Vérin, eds., *Réduire en art*.

18. It should be evident from this discussion that the conception of Réaumur as a scientist who applied his theories to industry, which is commonplace in older scholarship, results from a simplified reading of his work. See, for example, Torlais, *Réaumur*, or the biographical sketch on Réaumur in the English translation of his *L'art de convertir le fer forgé en acier*: Réaumur, *Memoirs on Steel and Iron*, i–xxxiv.
19. "Préface des arts," f. 4.
20. Réaumur, "Expériences et réflexions sur la prodigieuse ductilité de diverses matières," MARS 1723, 201–22. For a recent discussion on the relationship between materials and the making of natural knowledge, see the authors' introduction in Klein and Spary, eds., *Materials and Expertise in Early Modern Europe*, 1–23.
21. Réaumur's conclusions were presented in the *Journal des savants*, July 1717, 7–11.
22. "Préface des arts," f. 23.
23. The draft was written on preused paper, which helps with dating the work. On the back of various pages there are crossed out notes that are or can be dated: one carries the date of 1711; another is a partial draft of Réaumur's essay "Observations sur le mouvement progressif de quelques coquillages de mer," read to the Académie in July 1712: HARS 1712, 128–30; the last is a draft in calligraphy of Réaumur's *L'art de convertir le fer forgé en acier* (Paris, 1722). Réaumur was certainly directing the project of the history of the arts by 1722, as he stated in the preface to the book.
24. Réaumur, "Observations sur les mines de turquoises du Royaume," MARS 1715, 179.
25. The witnesses to the duke's successful transmutation were the comte de Nocé, the chevalier deBéthune, and Jean Grosse (who worked as a laboratory assistant for the duke). Jean Hellot Papers, Ms in-4°171, vol. 5, ff. 252–53, Bibliothèque Municipal, Caen.
26. Petitfils, *Le régent*, 96–99, 188–92; Geoffroy, "Experiments upon Metals, Made with the Burning-Glass of the Duke of Orleans," 374–86.
27. Demeulenaere-Douyère and Sturdy, eds., *L'enquête du régent, 1716–1718*, 9–59.
28. Minard, *La fortune du colbertisme*.
29. The vast majority of the documents from the survey that are currently kept in the archives of the Paris Académie des Sciences have been published in Demeulenaere-Douyère and Sturdy, eds., *L'enquête du régent*. The term "personnes intelligentes" is recurrent in the survey's documents.
30. Demeulenaere-Douyère and Sturdy, eds., *L'enquête du régent*, 78.
31. Demeulenaere-Douyère and Sturdy, eds., *L'enquête du régent*, 78–81.
32. There is virtually no information on Foujean. See Demeulenaere-Douyère and Sturdy, eds., *L'enquête du régent*, passim.
33. "Idée générale des différentes manières de faire la porcelaine," *Mercure français*, June 1717, 1428.
34. Réaumur's works on anchors remained unpublished until the academician Henri-Louis Duhamel du Monceau found and published them in 1761: Réaumur and Duhamel du Monceau, *Fabrique des ancres*; Réaumur and Duhamel du Monceau, *Nouvel art*. Réaumur's work on cast iron and steel was published in 1722: Réaumur, *L'art de convertir le fer forgé en acier*.
35. Murphy, *John Law*.

36. Réaumur, *Memoirs on Steel and Iron*, 8.
37. Réaumur, *Memoirs on Steel and Iron*, 8–9.
38. This would be published by Duhamel du Monceau in 1761: Réaumur and Duhamel du Monceau, *Nouvel art*.
39. Réaumur, *Memoirs on Steel and Iron*, 120.
40. Réaumur, *Memoirs on Steel and Iron*, 120–21.
41. Réaumur, *Memoirs on Steel and Iron*, 11.
42. Réaumur, *Memoirs on Steel and Iron*, 5.
43. *Réflexions sur l'utilité dont l'académie pourrait être au royaume si le royaume luy donnoit le secours dont elle a besoin*, in Dossier Réaumur, AASP. This is published in Maindron, *L'Académie des Sciences*, 103–10.
44. The theme of the "decadence" of the Académie is explicit in Réaumur's *Réflexions*, in Dossier Réaumur, AASP.
45. *Réflexions*, 107, in Dossier Réaumur, AASP.
46. *Réflexions*, 107, in Dossier Réaumur, AASP.
47. "Préface des arts," f. 22.
48. The officers of the mint are listed in the individual issues of *Almanach Royal*. On the Paris Mint, see Abot de Bazinghen, *Traité des monnoies*.
49. Dutot, *Histoire du système de John Law*.
50. Principe, "The End of Alchemy?"
51. Quoted in Principe, "The End of Alchemy?" 106.
52. Marion, *Les bibliothèques privées à Paris*.
53. Inventory at death of Mathieu Renard Du Tasta (1738): MC/ET/CXV/508, AN.
54. Lorgues, "L'ancien hôtel de la monnaie de Paris et ses problèmes," 138–74.
55. On Réaumur and Du Tasta, see Jean Hellot Papers, Ms in-4°171, vol. 3, ff. 33–34, Bibliothèque Municipal, Caen. On Dufay's experiments at the mint, see PV 1727, f. 113v. On Dufay, see Bycroft, *Physics and Natural History*.
56. [Gayot], *Mémoire pour la dame Renard*; [De Beze de Lys], *Second mémoire pour la dame veuve Renard*; [De Gas de Fremainville], *Troisième mémoire signifié, pour la dame veuve Renard du Tasta*.
57. Haudrère, *La Compagnie Française des Indes au XVIIIème siècle*.
58. In addition to Renard Du Tasta the investors were Renard de Roussiac (Renard du Tasta's brother), Pierre de la Tour (the controller of the mint), Jean Aviat (the mint's controlleur au change), Octavian Souchet (the receveur au change), a M. Lambert (who in 1713 was president of the Cour des Monnoyes), Pierre Gaultier (a member of the Cour des Monnoyes in Lyon), Roettiers (one of the engravers at the Mint), Pierre Jarosson (a member of Parliament and Réaumur's close friend), Cyprien Bénezet (a member of the Royal Council), Nicholas Thibault (a military intendant related to Jarosson), and the Chevalier de Béthune (who headed the group and was an officer of the Royal Navy, close to the duke of Orléans). The information on the investors can be found in *Almanach Royal* (1713–31). I have been unable to find biographical details on the remaining two investors, Jean Rousselot and a Mr. Demon.
59. Réaumur and Duhamel du Monceau, *Nouvel art*, 42.

60. Demeulenaere-Douyère and Sturdy, eds., *L'enquête du régent*, 246; Réaumur and Duhamel du Monceau, *Nouvel art*; Réaumur and Duhamel du Monceau, *Fabrique des ancres*.
61. Jarosson to Réaumur, 1 December 1723, Dossier Réaumur, AASP.
62. PV 1725, f. 139.
63. Réaumur to Angelo Mazzoleni, 27 May 1724, Mss Ashburnoth 1522, Biblioteca Laurenziana, Florence, printed in Bontemps and Prade, "Un magasin parisien," 215–61.
64. Birembaut, "Une source inédite," 360.
65. PV 1725, ff. 139–40.
66. PV 1725, ff. 139–40.
67. PV 1725, f. 140.
68. Savary des Brûlons, *Dictionnaire universel de commerce*, 17.
69. *Mercure de France*, December 1726, 2730. The Cardinal de Fleury visited the Hôtel d'Uzès in March 1727 after chairing a public meeting at the Académie des Sciences. *La clef du cabinet des princes de l'Europe* 47 (July 1727): 11.
70. On Wunderkammern and alchemical laboratories, see Nummedal, *Alchemy and Authority in the Holy Roman Empire*.
71. *Visite d'expertise de l'hôtel d'Uzès sis rue Saint-Thomas du Louvre à Paris, procédé le 1er Juillet 1726*, Z1j 581, AN.
72. Savary des Brûlons, *Dictionnaire universel de commerce*, 17.
73. *Mercure de France*, December 1726, 2731.
74. *Etat des ouvrages de fer et acier fondus et adoucis, qui se trouvent dans le magasin de la manufacture royale d'Orléans, établie à Paris, rue S. Thomas du Louvre, à l'hôtel d'Uzès*, Bibliothèque Mazarin, Paris, printed in Bontemps and Prade, "Un magasin parisien."
75. Bignon to Réaumur, 27 October 1727, Papiers Bignon, Ms Fr. 22234, BNF.
76. On Réaumur's later work at the Hôtel d'Uzès, see Terrall, *Catching Nature in the Act*.
77. "Préface des arts."
78. "Préface des arts."
79. Baudouin, *De calceo antiquo*. The work was republished in 1667.
80. "Préface des arts."
81. "Préface des arts."
82. "Préface des arts."
83. Jordan remarked also on the details represented in the plates and on the title of the work, *Histoire des arts et crafts*. Jordan, *Histoire d'un voyage littéraire*, 98–99.
84. This is implied or explicitly stated in virtually all accounts on the *Descriptions des arts et métiers*.
85. *Mercure de France*, June 1727, 1422–30.
86. MARS 1761, 150–51.
87. Duhamel du Monceau, *Art du charbonnier*, iii.
88. On Duhamel du Monceau, see Corvol, ed., *Duhamel du Monceau*. On the published *Descriptions*, see Gillispie, *Science and Polity at the End of the Old Regime*.

89. "Préface des arts," f. 6.
90. Demeulenaere-Douyère and Sturdy, eds., *L'enquête du régent*, 9–59.
91. Terrall, *Catching Nature in the Act*, 58; see chapter 3 for an extended discussion of the relationship between Dumoustier and Réaumur.
92. Terrall, *Catching Nature in the Act*, 57.
93. Carton 60, undated, Dossier Réaumur, AASP.

3. THEORY, PRACTICE, IMPROVEMENT

1. On the London Society of Arts, see Allan, "The Society of Arts and Government," 434–52, and Allan, *William Shipley*.
2. Kersey's *New World of Words*, published in 1706, defines "artisan" as "an artificer, or tradesman," and "artist" as "a master of any art, an ingenious workman," s.v. In Boyer's *The Royal Dictionary, French and English, and English and French*, the English "artist" is translated as "artiste, artisan avec esprit." In Dyche and Pardon's *A New General English Dictionary*, published in 1740, there is only one entry for "artisan or artist," defined as "one skill'd in any art, a curious workman," s.v.
3. Given the focus of this book on the French context, I continue to use the French *artiste*. An analysis of the English term "artist" in the early modern period is beyond the scope of this work, but those interested in this question should start with Hale, *The Primitive Origination of Mankind*.
4. *Three Letters Concerning the Forming of a Society*, 4–5. A copy of this rare pamphlet is in the Archives of the Royal Society of Arts in London.
5. Minard, *La fortune du colbertisme*; Pincus, "Rethinking Mercantilism."
6. *Three Letters Concerning the Forming of a Society*, 8.
7. On the disinterested pursuit of truth, see Shapin, *A Social History of Truth*.
8. Liliane Hilaire-Pérez, "Les sociabilités industrielles."
9. *Three Letters Concerning the Forming of a Society*, 9.
10. *Three Letters Concerning the Forming of a Society*, 8–9.
11. This was, for example, the perspective that guided Roger Hahn in his analysis of the Société des Arts; see Hahn, "Science and the Arts in France," 77–93. For an overview on the topic, with useful references, see Bud, ed., "Focus."
12. Roberts, Schaffer, and Dear, eds., *The Mindful Hand*; Rossi, *I filosofi e le macchine*; Stewart, *The Rise of Public Science*; Stewart, "A Meaning for Machines," 259–94; Smith, *The Body of the Artisan*; Long, *Openness, Secrecy, Authorship*. On expertise in the early modern world, see Ash, "Expertise and the Early Modern State"; Rabier, *Fields of Expertise*; and Klein, "Artisanal-Scientific Experts."
13. Vérin, "La technologie."
14. For examples of this approach, see Hilaire-Pérez, *L'invention technique*, and Hilaire-Pérez and Garçon, eds., *Les chemins de la nouveauté*.
15. Bonnassieux and Lelong, *Conseil de commerce*.
16. For biographical information on Sully, see Sully and Le Roy, *Règle artificielle*, 407. On Sully as a Jacobite, see Harris, *Industrial Espionage and Technology Transfer*, chapter 1, and Hilaire-Pérez, "Transferts technologiques, droit et territoire," 547–79.

Hilaire-Pérez mentions that the Huguenot clockmaker Thomas Grignion, a member of the London Society of Arts, owned *Three Letters Concerning the Forming of a Society* and donated the document to the archives of the Royal Society of Arts. It is quite likely that Sully circulated the regulations of the Société des Arts among his fellow watchmakers when he was in London. See Hilaire-Pérez, "Transferts technologiques," 554n36.

17. Sully and Le Roy, *Règle artificielle*.
18. Leibniz to Rémond, 11 February 1715, in Leibniz, *Recueil de diverses pièces*, 170. For Sully's work on the art of clock- and watchmaking, see Sully, *Description abregée d'une horloge*, avertissement.
19. *HARS* 1716, 77–78.
20. Sully and Le Roy, *Règle artificielle*, 389.
21. Landes, *Revolution in Time*.
22. Dunn and Higgitt, *Finding Longitude*, 36.
23. Dunn and Higgitt, *Finding Longitude*; McClellan and Regourd, *The Colonial Machine*.
24. Sully and Le Roy, *Règle artificielle*, 382.
25. Blakey, *L'art de faire les ressorts des montres*.
26. Sully, *Description abregée d'une horloge*, avertissement.
27. On John Law, see Murphy, *John Law*; Faure, *La banqueroute de Law*; and Kaiser, "Money, Despotism, and Public Opinion."
28. Law's plan for French manufactures is published in John Law, "Restablissement du commerce" (September 1715), in Law, *Oeuvres complètes*, 2:67–259.
29. Murphy, *John Law*.
30. Documents about the operation coordinated by Sully are in SP 35/24, 25, 26, National Archives, London.
31. Harris, *Industrial Espionage and Technology Transfer*, chapter 1.
32. Sully read an essay on watchmaking to the Société des Arts in 1719. Sully and Le Roy, *Règle artificielle*, 252.
33. Harris, *Industrial Espionage and Technology Transfer*, chapter 1.
34. Papiers Bignon, Ms Fr. 22230, f. 372, BnF, lists the twenty-five members of the Société des Arts that was founded by Sully in 1728. The document explains that the new Société was the "renaissance" of the one Bignon directed years earlier. Julien Le Roy stated that a number of the Société's members had been members of the Société founded by Sully ten years earlier, though he did not indicate names or professions. Sully and Le Roy, *Règle artificielle*, 407. The tentative list offered here results from screening the twenty-five members by year of birth and presence in Paris in 1718. I was unable to find decisive data on other individuals listed in the 1728 documents who, based on age, might have been members of the earlier Société: the military engineer Bernard Forest de Belidor, the surgeons Crestelet du Plessis (father and son), the composers Jean-François Dandrieu and Jean-Philippe Rameau, the engraver and *écrivain juré* Deseuttre, and the director of the mint Mathieu Renard du Tasta. Of course, age and presence in Paris do not offer conclusive evidence about membership in the earlier Société des Arts. However, it is worth noting that all members listed here would

have shared an interest in metallurgy, which was an area of technical and natural knowledge particularly cherished by the regent. I am inclined to exclude Rameau and Dandrieu from the list of possible members of the early Société des Arts, given that music entered within the horizons of the Société des Arts only in its later manifestation. Indeed, metallurgy might well have been the broader research area pursued by the Société des Arts, with clock- and watchmaking as some of its specializations.

35. Sully, *Description abregée d'une horloge*, 82.
36. The painter was the Venetian Giovanni Antonio Pellegrini. The now lost ceiling of Law's bank is discussed in Garas, "Le plafond de la Banque Royale de Giovanni Antonio Pellegrini." See also Scott, *The Rococo Interior*.
37. Blakey, *L'art de faire les ressorts des montres*.
38. *Mercure de France*, January 1719, 141–45. On technology and the cultures of openness in a long-term perspective, see Long, *Openness, Secrecy, Authorship*, and Epstein, "Craft Guilds, Apprenticeship and Technical Change." More specifically on the eighteenth century, see Hilaire-Pérez, "Inventing in a World of Guilds."
39. Sully and Le Roy, *Règle artificielle*, 391.
40. *Mercure de France*, January 1719, 141–45.
41. Berthoud, *Histoire de la mesure du temps par les horloges*, 2:261; Fréron, *L'année littéraire*, 97–110. The standard work on the role of emulation in stimulating innovation is Hilaire-Pérez, *L'invention technique*.
42. Dequidt, *Horlogers des lumières*; Augarde, *Les ouvriers du temps*; Wilson, ed., *European Clocks in the J. Paul Getty Museum*.
43. On embodied skill and artisanal knowledge, see Smith, *Body of the Artisan*. More specifically on the French case, see Hilaire-Pérez, "Transferts technologiques."
44. Murphy, *John Law*; Faure, *La banqueroute de Law*.
45. Narbonne, *Journal des règnes de Louis XIV et Louis XV*, 54. See also Le Roi, *Histoire des rues de Versailles*.
46. Sully and Le Roy, *Règle artificielle*, 381–413.
47. This is clear in a letter that Bignon wrote to Grandjean in 1730, discussed below.
48. The existence of this earlier Société des Arts is revealed in a letter by Henri Liébaux to the abbé Bignon in which the author discusses the reestablishment of a Société des Arts. Papiers Bignon, Ms Fr. 22225, ff. 1–6, BnF.
49. Law's election is recorded in PV 1719, f. 294v (6 December 1719). Law had been informed of the election by Bignon a few days earlier. Law wrote Bignon a letter of thanks, which is kept in Law to Bignon, 3 December 1719, Papiers Bignon, Ms Fr. 22231, f. 30, BnF. Law attended only one meeting, on 13 December 1719. PV 1719, f. 298.
50. SP 35/24–26, National Archives, London.
51. *Three Letters Concerning the Forming of a Society*, 11.
52. Sully and Le Roy, *Règle artificielle*, Epître Dedicatoire.
53. On the debate, see Greenfeld, *The Spirit of Capitalism*; Mackrell, *The Attack on Feudalism in Eighteenth-Century France*, chapter 4; and Hecht, "Un problème de population."
54. Sully, *Description abregée d'une horloge*, 49–62.

55. Sully, *Description abregée d'une horloge*, 47.
56. Sully to Bignon, 1726, Ms Fr. 22233, ff. 210–33, BnF.
57. Sully, *Description abregée d'une horloge*, 200–203.
58. Sully and Le Roy, *Règle artificielle*, 402.
59. Angélique Delisle to her brother Joseph-Nicolas, 24 January 1728, 13 February 1729; undated draft by Delisle, Letters 15 and 19, Ms 1508, f. 43, Chambre des Députés, Paris.
60. Sully, *Description abregée d'une horloge*, 112.
61. Sully, *Description abregée d'une horloge*, 53.
62. Sully, *Description abregée d'une horloge*, 46.
63. Sully, *Description abregée d'une horloge*, 106.
64. Sully, *Description abregée d'une horloge*, 88.
65. Sully, *Description abregée d'une horloge*, 22.
66. Sully, *Description abregée d'une horloge*, 22.
67. On "heroization of invention," see MacLeod, *Heroes of Invention*. More specifically on the French context, see Hilaire-Pérez, "Diderot's Views on Artists' and Inventors' Rights," and Hilaire-Pérez, *L'invention technique au siècle des lumières*. See also chapter 5 for an extended discussion of how *artistes* understood invention.
68. Sully, *Description abregée d'une horloge*, 23.
69. Sully, *Description abregée d'une horloge*, 273.
70. Sully and Le Roy, *Règle artificielle*, 402–3.
71. Sully, "Description de la ligne meridienne de l'église de S. Sulpice," 1592.
72. Sully's royal pension, 1725, Ms Fr. 22233, f. 233, BnF.
73. Hamel, *Histoire de l'église Saint-Sulpice*; Chevreul, "Une belle figure bourguignonne."
74. Le Roy, "Mémoire pour servir à l'histoire de l'horlogerie, depuis 1715 jusqu'en 1729," in Sully and Le Roy, *Règle artificielle*, 381–413; *Suite de la clef*, November 1730, 321–23.
75. The full list is in Papiers Bignon, Ms Fr. 22230, f. 373, BnF. See also http://www.clairaut.com/n8novembre1728po1pf.html.
76. Minutes, 1731, De Croÿ. There is very scattered and fragmentary evidence of at least some members' interest in alchemy in the materials on the Société des Arts kept in Germany.
77. "Idée de l'établissement de la société académique des beaux arts," Ms Fr. 22225, f. 9, BnF.
78. Sully and Le Roy, *Règle artificielle*, 408.
79. Papiers Bignon, Ms Fr. 22230, f. 372, BnF.
80. Bignon to Grandjean, 13 October 1730, Ms Fr. 22235, f. 47, BnF.
81. Hamel, *Histoire de l'église Saint-Sulpice*, 178.
82. On Clermont, see Cousin, *Le comte de Clermont*, and Sainte-Beuve, *Le comte de Clermont et sa cour*. Clermont became Grand Maître in 1743.
83. *Mercure de France*, June 1730, 1185.
84. Schlüter and Hellot, *De la fonte, des mines, des fonderies*, 66.
85. Ms. Ashburnham 1804, Biblioteca Medicea Laurenziana, Florence; René Réaumur, "Second mémoire sur la porcelaine," *MARS* 1729, 336–38.

86. Inventory at death of the Count of Clermont: MC, LXXIII, 929, June 25, 1771, f. 40v, AN.
87. Languet's speech was published in *Mercure de France*, December 1728, 2893–95, 2894 (quotation).
88. On scientific patronage and the early modern court, see Moran, ed., *Patronage and Institutions*. On the Accademia del Cimento, see Beretta, Clericuzio, and Principe, eds., *The Accademia del Cimento and Its European Context*, and McClellan, *Science Reorganized*. On a Neapolitan attempt to emulate the patronage of Leopoldo de Medici, see Bertucci, "The Architecture of Knowledge."
89. *Mercure de France*, December 1728, 2894.
90. Ms J 1750, Staatsbibliothek, Berlin; Monteil, *Traité de matériaux manuscrits de divers genres d'histoire*, 1:41–43. The title "société académique" indicated that the king had granted permission to the group to gather periodically—such occasional gatherings were otherwise forbidden—and that he was well disposed toward the possibility of bestowing the academic status to the Société, provided the Société successfully demonstrated its public utility. This was the case with the Academy of Surgery; see chapter 4.
91. *Règlement de la Société des Arts*, art. 1.
92. Campbell, *Power and Politics in Old Regime France*.
93. McClellan and Regourd, *The Colonial Machine*.
94. Pelays was also entrusted with the secret mission of checking on the governor of Senegal, a delegate of the Compagnie, whose recent actions in Bambouk suggested he was not working for the Compagnie's benefit. Pelays sent lengthy accounts to the directors of the India Company and the minister of finance in 1730 and 1731. Following his research, he was atrociously murdered in Senegal, with the complicity of the governor. His story was recounted by the naturalist Michel Adanson in 1763. See Colonies, C6 15, AN. On Pelays's experiments in Africa, see Golbéry and Mudford, *Travels in Africa*. Pelays's alchemical work was known to French aristocrats such as the Duke of La Vallière, who owned *Thrésor chimique*, an alchemical compendium prepared by Pelays in 1729, when he was already a member of the Société des Arts. See Debure, *Catalogue des livres de la bibliothèque de feu M. le duc de La Vallière*, 509.
95. On his death, the surgeon to the king, François Gigot de la Peyronie, left part of his stocks of the Compagnie des Indies to Société des Arts member Daniel Medalon.
96. Sully, *Description abregée d'une horloge*, 46.
97. "Idée de l'etablissement de la société académique des beaux arts," MS Fr. 22225, f. 7v, BnF.

4. SOCIETY OF ARTS

1. Pluche, *Spectacle de la nature*, 7:229.
2. Moreri, *Dictionnaire historique*.
3. *Curiosités de Paris*; Martinière, *Le grand dictionnaire géographique et critique*.
4. Rousset de Missy, *Le ceremonial diplomatique des cours de l' Europe*.

5. King and Millburn, *Geared to the Stars*.
6. On the history of clocks, see Landes, *Revolution in Time*; Augarde, *Les ouvriers du temps*; Cipolla, *Clocks and Culture*; and Dequidt, *Horlogers des lumières*.
7. Pigeon, *Description d'une sphère mouvante*, 129–31.
8. Since the English makers had protected their secret by locking the clock case, it proved particularly challenging for the *horloger du roi* to regulate them. See *Anecdotes litteraires*, 389.
9. Sully, *Méthode pour régler les montres et les pendules*.
10. On immanent providence in the Newtonian universe and, more generally, on the relationship between Newton's theory of universal gravitation and his theological beliefs, see the essays in Force and Hutton, eds., *Newton and Newtonianism*, and Feingold, *The Newtonian Moment*.
11. Julien Le Roy, "Avis contenant les moyens de régler les montres, tant simples qu'à répétition," in Toussaint, *Observations périodiques sur la physique*, 395–401, 395 (quotation).
12. See Desauguliers, *The Newtonian System of the World*. On the politics of Newton's natural philosophy the classic work is Jacob, *The Newtonians and the English Revolution*. See also Force and Hutton, eds., *Newton and Newtonianism*, and Feingold, *The Newtonian Moment*.
13. *Encyclopédie*, 1:715.
14. La Peyronie, "Histoire de l'Académie Royale de Chirurgie," 3.
15. Meyssonnier, *La balance et l'horloge*; Minard, *La fortune du colbertisme*; Pincus, "Rethinking Mercantilism."
16. Savary, *Dictionnaire universel de commerce*, 2:1242.
17. Bonnassieux and Lelong, *Conseil de commerce*; Kammerling Smith, "Structuring Politics in Early Eighteenth-Century France."
18. Bonnassieux and Lelong, *Conseil de commerce*.
19. Minard, *La fortune du colbertisme*.
20. Bonnassieux and Lelong, *Conseil de commerce*.
21. *Réflexions*, in Dossier Réaumur, AASP.
22. Minard, *La fortune du colbertisme*, 166–68.
23. Bonnassieux and Lelong, *Conseil de commerce*, xxvii–xxviii.
24. Kammerling Smith, "Structuring Politics in Early Eighteenth-Century France."
25. Quoted in Augarde, *Les ouvriers du temps*, 52.
26. Gelfand, *Professionalizing Modern Medicine*, 26 (quotation), 218n91.
27. Julien Le Roy, "Mémoire contenant les moyens d'augmenter le commerce et la perfection des ouvrages d'horlogerie," Ms 165, CNAM.
28. Minutes, 15 June 1732, De Croÿ.
29. On subcontracting in eighteenth-century France, see Garçon, "Les dessous des métiers," 378–91.
30. On Le Roy's workshop, see Wilson, ed., *European Clocks in the J. Paul Getty Museum*, esp. 185–190. On subcontracting in clockmakers' workshops, see Augarde, *Les ouvriers du temps*, chapter 3.

31. Blakey, *L'art de faire les ressorts des montres*.
32. Blakey, *L'art de faire les ressorts des montres*. On the Blakey dynasty, see Hilaire-Pérez, "Steel and Toy Trade between England and France."
33. Jullien's membership in the Société des Arts is recorded in Lists of members, 22 October 1730, De Croÿ. On his collaboration with Le Roy, see Augarde, *Les ouvriers du temps*, 79.
34. Germain was one of the earliest members of the Société des Arts. Lists of members, 31 October 1728, De Croÿ. On the collaboration among Le Roy, Germain, and Caffieri, see Wilson, ed., *European Clocks in the J. Paul Getty Museum*.
35. *Encyclopédie*, 8:307–8.
36. Louis Berthoud, *Journal de caisse*, Ms 8° 187, CNAM.
37. Here and in the discussion below I refer to clocks that were used for civic purposes, not to astronomical clocks, unless otherwise stated.
38. Sully and Le Roy, *Règle artificielle*, 68–118.
39. Dufay, "Description d'une machine pour connoître l'heure vraye du soleil."
40. Minutes, 13 August 1733, De Croÿ.
41. Sully and Le Roy, *Règle artificielle*.
42. Augarde, *Les ouvriers du temps*, 311; Gersaint, *Catalogue raisonné d'une collection considerable*, 135.
43. Papiers Bignon, MS Fr. 22230, BnF.
44. On the Newton wars, see Shank, *The Newton Wars*.
45. Maupertuis, *La figure de la Terre*. On Maupertuis's expeditions, see Terrall, *The Man Who Flattened the Earth*.
46. The literature is too vast to be fully cited here. See, at least, Moran, ed., *Patronage and Institutions*; Biagioli, *Galileo, Courtier*; Findlen, *Possessing Nature*; Galluzzi et al., eds., *Scienziati a corte*; and Beretta et al., *The Accademia del Cimento and Its European Context*.
47. *Règlement de la Société des Arts*, art. 3.
48. Rousset de Missy, *Le ceremonial diplomatique des cours de l' Europe*, 424–29.
49. Sully and Le Roy, *Règle artificielle*, 370–81.
50. Julien Le Roy, "Mémoire contenant les moyens d'augmenter le commerce et la perfection des ouvrages d'horlogerie," Ms 165, CNAM.
51. Guichard, "Arts libéraux et arts libres," 54–68.
52. The king granted free lodgings at the galleries of the Louvre to a select number of *artistes*. Since they could work without the interferences of the guilds, they often kept their workshops there. See Merson, "Les logements d'artistes," 276–88.
53. They were the architects Jean Aubert, Germain Boffrand, and Pierre Vigné de Vigny; the painters François Lemoyne, Jean II Restout, and Jean-Baptiste Oudry; and the sculptor Jean-Baptiste Lemoyne. The sculptor François-Antoine Vassé was adjunct (*agréé*) to the Académie de Peinture et Sculpture in 1723, though he never became an academician. The engraver of medals Jean Duvivier was also already a member of the Académie de Peinture et Sculpture when he joined the Société des Arts in 1728.
54. Pedley, *The Commerce of Cartography*; Petto, *When France Was King of Cartography*; Petto, *Mapping and Charting in Early Modern England and France*.

55. Dawson and Vincent, *L'atelier Delisle*.
56. Humbert, *Institutions et gens de finance en Franche-Comté*, 53–60.
57. Lists of members, 1728–35, De Croÿ. On Grandjean de Fouchy, see Courcelle, "Grandjean de Fouchy."
58. In both cases the former apprentices had been admitted to the Société before their masters. It is unclear if Nollet introduced Raux, but Germain seems to have joined the Société through his connections with other members, not Chevotet (see the discussion of the Manufacture of Rolled Lead below).
59. Lists of members, 23 April 1733; Minutes, 23 April 1733; Oudry to Papillon, 23 April 1733, De Croÿ.
60. Carl Johan Cronstedt to his mother, May 1735, Ms 3447, National Archives, Stockholm.
61. Jonas De Meldercreutz to the Neapolitan Celestino Galiani after his visit to Paris, September 1736, XXXI.B1, f. 271v, Biblioteca di Storia Patria, Naples. See also the letter by the French ambassador in Denmark, the Count of Plélo, in MAR, 2JJ/62, 37, AN, and Carl Johan Cronstedt to his mother, May 1735, Ms 3447, National Archives, Stockholm (cited in Strandberg, "La reception," 95). Clermont's patronage of the Société des Arts was praised also by Voltaire in his *Temple du goût*, reprinted in Besterman, ed., *The Complete Works of Voltaire*, 9:180.
62. On the Republic of Letters as an ideal polity, see Goodman, *The Republic of Letters*, and Goldgar, *Impolite Learning*.
63. This is not to say that there were no women in the artisanal world. See, for example, Sheridan, *Louder than Words*. However, there was no equivalent of a *salonnière* in the world of the arts.
64. Sonenscher, *Work and Wages*; Kessler, *A Revolution in Commerce*.
65. *Règlement de la Société des Arts*, arts. 3–6.
66. The *arts du goût*, which included arts that pleased the senses, such as music, dance, and engraving, were part of the Société des Arts's interests in so far as they could be improved by means of other sciences or arts. The *Règlement* indicated that the Société would comprise a hundred members: twelve geometers, fifteen mechanicians, two astronomers, thirteen physicists, two physicians, three surgeon-anatomists, ten chemists, two botanists, five engineers, three architects, two shipbuilders, two pilots, two geographers, two hydrographers, four clockmakers, two mathematical instrument-makers, two painters, two sculptors, two goldsmiths, two engravers, one musician, one glassmaker, two spectacle makers, one enameller, and six entrepreneurs from different manufactures.
67. *Règlement de la Société des Arts*, art. 9.
68. *Règlement de la Société des Arts*, art. 22. The "officers" were the director, the treasurer, the administrator, the secretary, and his assistant.
69. Verdier was one of the five "royal demonstrators" who taught anatomy at Saint-Côme.
70. *Règlement de la Société des Arts*, arts. 9, 11.
71. Papiers Bignon, Ms Fr. 22225, ff. 1–6, BnF. For a comparison of the *Règlement* with the draft initially written by the Société's members, see Passeron and Courcelle, "La Société des Arts."

72. Augarde, *Les ouvriers du temps*, chapter 2.
73. *Mémoires de l'Académie Royale de Chirurgie*, 4 (1819): 3.
74. On surgeons' education, see Gelfand, *Professionalizing Modern Medicine*, esp. 56. Three of the surgeons in the Société des Arts (César Verdier, René-Jacques Croissant de Garengeot, and François Quesnay) took master of arts degrees. I have been unable to locate the inventories of two clockmakers' libraries (Pierre Gaudron's and Claude Raillard's), which Augarde discusses. He states that they read popular magazines, such as the *Journal des sçavans* and the *Mémoires de Trévoux*; travel and history books; classic Greek and Latin literary texts; Montaigne; Boileau; Molière; Racine; and religious books, especially on heresy and the Reformation. They did not seem to love contemporary authors (particularly Voltaire) as much. See Augarde, *Les ouvriers du temps*, chapter 2.
75. Papiers Bignon, Ms Fr. 22228, f 10, BnF. Only Clairaut would then be elected to the Académie. PV 1729, f. 210r, and PV 1731, f. 169r. On geometers, see Irène Passeron, "L'invention d'une nouvelle compétence," in Garçon and Hilaire-Pérez, eds., *Les chemins de la nouveauté*, 139–52.
76. Minutes, 1732–36, De Croÿ. A list of essays presented at the Société des Arts in 1730 is in Trew, *Commercium litterarium ad rei medicinae et scientiae naturalis*, 169–72.
77. Grandjean reported on a device for spinning silk and on a bridge of his own invention, while Clairaut spoke on fortifications, mathematical instruments, and teaching tools. De Gua commented on geometrical methods for calculating shadows, an important topic for several arts, including clockmaking. Minutes, 26 June 1732, 19 July 1733, 19 November 1733, De Croÿ.
78. Minutes, 23, 27 October 1729, De Croÿ. De Gua translated the recipe for making vinegar from the *Philosophical Transactions of the Royal Society*, even though it was a French recipe, commonly regarded as secret. Minutes, 3 April 1732, De Croÿ.
79. Minutes, 17 July 1732, De Croÿ.
80. Minutes, 16 August 1733, De Croÿ.
81. Minutes, 27 December 1733, De Croÿ.
82. Jacques Vaucanson to Daniel Trudaine, 5 October 1765, published in Doyon and Liaigre, *Jacques Vaucanson*, 440.
83. HARS 1731, 89.
84. Henri Enderlin, "Des irrégularités des pendules," in Thiout, *Traité de l'horlogerie méchanique et pratique*, 1:117–27.
85. Sainte-Albine, *Mémoire sur le laminage du plomb*, 45.
86. Sainte-Albine, *Mémoire sur le laminage du plomb*.
87. Minutes, 22 April 1731, De Croÿ; Sainte-Albine, *Mémoire sur le laminage du plomb*, 36–38.
88. The committee's members were the architect Jean Aubert, a member of the Royal Academy of Architecture; the director of the mint, Mathieu Renard Du Tasta; the chemist Habert; and the geometer Jean Paul de Gua de Malves.
89. Lists of members, 18 January 1731, De Croÿ.
90. *Règlement de la Société des Arts*, arts. 3, 5, 10. Each new candidate was first discussed during a meeting. His candidature was subjected to secret ballot, and, if he obtained

the majority of the votes and the approval of the Count of Clermont, his membership was finally established. If the new member was an "assiduous," he would then present a work of art or an essay to the Société. Exception was made for honorary members, who were often appointed directly by the count himself (the king or his representative in the case of the Académie des Sciences).

91. Minutes, 1732, De Croÿ. On the culture of academic prizes, see Caradonna, *The Enlightenment in Practice.*
92. HARS 1724, 94 (watch by Sully); MIAS 1716, 77 (watch by Sully); HARS 1717, 85 (clock by Le Roy); HARS 1728, 110–11 (clocks by Pierre and Julien Le Roy); HARS 1754, 14 (watch designed by Dutertre in 1724); MIAS 1727, 142 (mathematical instrument by Clairaut the father); HARS 1726, 45 (Alexis Clairaut on geometrical curves); HARS 1720, 140 (sawing machine by Guyot); HARS 1724, 37 (surgical instrument by Guyot); HARS 1726, 71 (machine by Le Maire); MIAS 1735, 6:61–62 (reflecting telescope by Le Maire); HARS 1726, 24; HARS 1728, 21–22 (anatomical observations by Garengeot); Joseph Saurin, "Remarques sur les horloges à pendule," MARS 1720, 208–30.
93. For the essay by Jacques Le Maire in Bignon's papers, see Ms Fr. 22230, ff. 28, 340–47, BnF.
94. In 1735, Jean Hellot asked for a collaboration with François-Julien Barrier, engraver of precious stones to the king, during his experiments on false gems. See Hellot, "Analyse chimique du zinc," MARS 1735: 221–42. The surgeon Pierre Bassuel presented an essay to the academy in 1731. HARS 1731, 26–27. The engineer Jean-Gaffin Gallon presented a method for launching sea vessels in 1731. HARS 1731, 90–91. The academy discussed the work on harmony by the composer Jean Philippe Rameau in 1737 and 1750. HARS 1737, 14; HARS 1750, 160. It discussed an essay by Pierre Deschisaux in 1738. PV 1738, f. 30. In 1738, Julien Le Roy presented a pendulum clock whose length did not vary with changes in temperature and that he built based on his knowledge of Musschenbroeks's experiments on the effects of heat on copper and iron. HARS 1741, 147–49. In 1742, Jean Claude Adrien Helvétius, an academician and physician to the queen, discussed the new bandage invented by the *ingénieur du roi* Abeille. HARS 1742, 153. The clockmakers Jean Baptiste II Dutertre and Nicolas Gourdain presented their inventions to the Académie in 1742 and 1743; see HARS 1742, 21–22, 158, and HARS 1743, 172. Jean Baptiste Hillerin de Boistissandeau presented his odometer to the Académie in 1744. HARS 1744, 61.
95. Ms. Ashburnham 1804, Biblioteca Medicea Laurenziana, Florence.
96. Réaumur, "Second mémoire sur la porcelaine."
97. The position went to Buache, who was married to Guillaume Delisle's daughter and who took over Delisle's workshop upon his death.
98. MIAS, vol. 1, i.
99. Papiers Bignon, Ms Fr. 22234, f. 178 (2 May 1728), BnF.
100. *Journal de Trévoux,* December 1733, 2196–98.
101. Lists of members, 6 August 1730, De Croÿ.
102. PV 1730 (24, 26 May, 7, 14 June). For a list of the Académie's members, see Institut de France, *Index biographique de l'Académie des Sciences.*

103. *Réflexion*, in Dossier Réaumur, AASP.
104. Bertrand, *L'Académie des Sciences*, 95–97; Hahn, "Science and the Arts in France"; Hahn, *The Anatomy of a Scientific Institution*.
105. La Condamine to Grandjean de Fouchy (undated), L I a 685, f. 659r, Öffentliche Bibliothek der Universität Basel, Basel.
106. In the archives of CNAM, there is an essay by Julien Le Roy, dated 1741, which he read to the Société des Arts: Ms 165 (15). This seems to be the latest surviving document on the Société des Arts.
107. Louis, *Éloges de Messieurs Bassuel, Malaval et Verdier*, 9–10.
108. MAR, 2JJ/62, 37, AN.
109. Minutes, 1731–33, De Croÿ. We do not have enough evidence to draw conclusions about the other members of the Société who became members of the Académie des Sciences.
110. Minutes, 1732, De Croÿ. On Privat de Molières, see Shank, *The Newton Wars*, 38–48.
111. La Condamine to Bottée, November 1733, De Croÿ.
112. Réaumur, "Second mémoire sur la porcelaine."
113. Bignon to Réaumur, 10 March, 21 July 1730, Ms Fr. 22230, ff. 83, 88, BnF; Lists of members, 19 March 1730, De Croÿ.
114. On 18 December 1731, the king granted to a group of sixty-eight surgeons permission to constitute a Société Académique de Chirurgie. Although the date is often referred to as the first meeting of the Royal Academy of Surgery, it was in fact—similar to the case of the Société des Arts—the beginning of a probationary status (evidenced by the term *académique*). During this time, the Société's members could meet (group meetings were otherwise forbidden by law) and demonstrate, through their collective work, that their association deserved to become an academy. It was only in 1748 that the Académie de Chirurgie was officially constituted. However, its officers retrospectively dated its constitution to 1731. Here I follow their example.
115. *Mémoires de l'Académie Royale de Chirurgie* 1 (1748): 48–54.
116. *Mémoires de l'Académie Royale de Chirurgie* 4 (1768): 39.
117. *Règlement de la Société des Arts*, arts. 13, 14.
118. "Discours de la première sèance de l'Académie de Chirurgie," Ms 61, n. 103, Archives de l'Académie de Chirurgie, Paris. See also *Règlement de la Société des Arts*, art. 23. The article stated that the author of an essay could ask to first read it without interruptions, then read it again during the next meeting, taking comments from his colleagues.
119. Gelfand, *Professionalizing Modern Medicine*, chapter 3. For a more recent interpretation, see Rabier, "La disparition du barbier chirurgien," 679–711.
120. *Mémoires de l'Académie Royale de Chirurgie* 4 (1768): 34.
121. *Mémoires de l'Académie Royale de Chirurgie* 4 (1768): 45.
122. Minutes, 20 July 1732, 31 May 1733, De Croÿ.
123. Daniel Medalon, "Sur la différence des tumeurs à extirper ou à ouvrir simplement," in *Recueil des pièces qui ont concouru pour le Prix de l'Académie Royale de Chirurgie*,

3–65. The award stirred a controversy within the Académie de Chirurgie about whether the competition should be reserved to surgeons. Guardia, "Feuilleton," 323n–328n.
124. Ms 3505, Bibliothèque de l'Arsenal, Paris.
125. *Règlement de la Société des Arts*, art. 1; Voltaire, *The Complete Works*, 9:180.
126. Minutes, 27 March 1732, De Croÿ.
127. *Journal de Trévoux*, February 1733, 357–59; *Mercure de France*, May 1734, 937–38.
128. Minutes, 25 January 1733, De Croÿ.
129. Cousin, *Le comte de Clermont*; Sainte-Beuve, *Le comte de Clermont et sa cour*.
130. Lists of members, 22 April 1731, De Croÿ. On the debate, see Greenfeld, *The Spirit of Capitalism*; Mackrell, *The Attack on Feudalism in Eighteenth-Century France*, chapter 4; and Hecht, "Un problème de population."
131. In the French language the word *expert* was not only an adjective—as in early modern English—but also a noun. See Richelet, *Nouveau dictionnaire françois*; Furetière, *Dictionnaire universel*, vol. 2; and Savary des Brûlons, *Dictionnaire universel de commerce*, vol. 2.
132. Minutes, 19 December 1733, De Croÿ.
133. Minutes, 19 December 1733, De Croÿ. Changes to this initial project were made on 24 December 1733 and on 11 and 18 February 1734. The Société added a report by Clairaut Sr. and replaced a report by Julien Le Roy (which would be later published as "Description et usage d'un nouveau cadran horizontal, universel, et propre à tracer les méridiennes," *Mercure de France*, September 1735, 1898–1905, and Sully and Le Roy, *Règle artificielle*, 293–304) with another (which would also be published: "Description d'un nouveau cadran universel, portatif et à boussole," *Mercure de France*, September 1735, 1906–15, and Sully and Le Roy, *Règle artificielle*, 318–23).
134. Count of Clermont to the Société des Arts, 23 December 1733, De Croÿ.
135. Bottée to Hynault, 23 October 1733, De Croÿ.
136. Minutes, 17 February 1737, De Croÿ.
137. In addition to Le Roy's 1741 essay, Ms 165 (15), CNAM, there is a volume of papers presented to the Société des Arts, dated 1740, at the Bibliothèque de l'Arsenal in Paris. The volume belonged to Joseph Bonnier de la Mosson. See Ms 6130.
138. Angélique Delisle to her brother Joseph-Nicolas, 14 February 1730, Ms 1508, f. 64r, Bibliothèque de l'Assemblée Nationale, Paris.
139. Minutes, 12 June 1732, De Croÿ.
140. "Mémoire pour la perfection de la Société des Arts," undated, De Croÿ.
141. Minutes, 19 December 1733, De Croÿ.
142. Goldgar, *Impolite Learning*.
143. Dutot, *Histoire du système de John Law*. On Dutot, see Velde, "The Life and Times of Nicolas Dutot," 67–107.
144. Bottée to Moncrif, 14 January 1734, De Croÿ.
145. Jean-Antoine-Nicolas de Caritat, marquis de Condorcet, "Éloge de M. de Fouchy," *HARS* 1788, 37–49.
146. D'Alembert, *Œuvres philosophiques, historiques et littéraires*, 11:414–15. On D'Alembert's attitude toward artisans, see Schaffer, "Enlightened Automata."

147. *Encyclopédie*, 11:414–15.
148. Berthoud, *Essai sur l'horlogerie*; Louis, *Éloges*, 8.
149. Royllet to Bottée, 2 September 1733, De Croÿ; Royllet, *Nouveaux principes de l'art d'écrire*.
150. Maillet, *Telliamed*. Liébaux is mentioned on page 8.
151. On Quesnay's career, see Vardi, *The Physiocrats and the World of the Enlightenment*, and Gelfand, *Professionalizing Modern Medicine*, 71–76. On Nollet, see Bertucci, "Enlightened Secrets," 820–52. On Liébaux, see Pernety, *Dictionnaire portative*, 85n.
152. Royllet to Bottée, 2 September 1733, De Croÿ.

5. THE POLITICS OF WRITING ABOUT MAKING

1. Long, *Openness, Secrecy, Authorship*; Long, *Artisan/Practitioners*; Dubourg Glatigny and Vérin, eds., *Réduire en art*; Vérin, "La technologie"; Smith, "In the Workshop of History," 22, 25 (quotations); Smith, "Why Write a Book?"; Blair, *Too Much to Know*; Yeo, *Encyclopaedic Visions*.
2. Intersemiotic translation is discussed in Eco, *Experiences in Translation*.
3. Dubourg Glatigny and Vérin, eds., *Réduire en art*; Vérin, "La technologie."
4. Quesnay, *Lettres sur les disputes*, 142. On the changing status of surgery in this period, see Rabier, "La disparition du barbier chirurgien," 679–711; Gelfand, *Professionalizing Modern Medicine*; and Brockliss and Jones, *The Medical World of Early Modern France*.
5. Dornau and Hurtaut, *L'art de peter*; *L'art de plumer la poulle sans crier*.
6. *Journal de Trévoux*, February 1742, 301–3.
7. Folard, *Nouvelles découvertes sur la guerre*, 1–18.
8. De Pas, *Mémoires sur la guerre*, 12.
9. De Pas, *Mémoires sur la guerre*, 5.
10. *Journal de Trévoux*, April 1732, 608.
11. Sully, *Regle artificielle*, unpaginated preface.
12. *Journal de Trévoux*, February 1742, 301, 303, 306–7.
13. *Règlement de la Société des Arts*, art. 13.
14. Compare Ms Fr. 22225, ff. 1–6, BnF, with *Règlement de la Société des Arts*.
15. Minutes, 1733–34, De Croÿ.
16. Sully, *Description abregée d'une horloge*, avertissement, 43.
17. Sully envisioned six volumes. The first would present an updated version of his *Artificial Regulation of Time* as a necessary introduction to key horological concepts and methods for understanding clocks and watches. The second would acquaint readers with great inventors and their contributions to horology. The third would offer textual and visual details on each component of watches and clocks, so as to instruct them on what distinguished a well-made piece. The fourth would discuss the elements of geometry, mathematics, physics, and mechanics that underpinned the making of watches and clocks. The fifth would provide directions for other makers on how to build each piece, and the sixth culminated with the descriptions of Sully's marine

chronometer, the pinnacle of current knowledge. See Sully, *Description abregée d'une horloge*.

18. May, "Histoire et sources de l'*Encyclopédie*," 21.
19. The chevalier de Jaucourt, the most prolific contributor to the *Encyclopédie*, received twelve thousand livres. The abbé Mallet, who contributed on theology, ecclesiastical history, commerce, and minting of coins, received nine hundred livres, while Monsieur Toussaint, who wrote on law, received three hundred livres. May, "Histoire et sources de l'*Encyclopédie*," 23–29.
20. Proust, *Diderot et l'*Encyclopédie, 47–58; Darnton, *The Business of Enlightenment*.
21. Demeulenaere-Douyère and Sturdy, eds., *L'enquête du régent*.
22. Sully and Le Roy, *Règle artificielle*, 195–96.
23. Lepaute, *Traité d'horlogerie*, 13.
24. Quesnay, *Mémoires*, preface, quoted in Gelfand, *Professionalizing Modern Medicine*, 74.
25. Quesnay, *Recherches*, 23.
26. Quesnay, *Recherches*, 24.
27. Berthoud, *Essai sur l'horlogerie*, 34.
28. Berthoud, *Essai sur l'horlogerie*, 37.
29. Berthoud, *Essai sur l'horlogerie*, 42. The various workmen are listed on pp. 38–41.
30. Le Dran, *Parallèle des différentes manières de tirer la pierre hors de la vessie*; Garengeot, *Nouveau traité des instrumens de chirurgie*.
31. Berthoud, *Essai sur l'horlogerie*, 44.
32. This is consistent with discussions of innovations occurring within, and not in spite of, the guild system. See Epstein and Prak, *Guilds, Innovation and the European Economy*.
33. Berthoud, *Essai sur l'horlogerie*, 44. On the Académie de Saint-Luc, see Guichard, "Arts libéraux et arts libres."
34. Ferrand, *L'art du feu*, 186.
35. Félibien, *Des principes de l'architecture, de la sculpture, de la peinture*, preface.
36. Shapin, *A Social History of Truth*.
37. Long, *Openness, Secrecy, Authorship*.
38. Eamon, *Science and the Secrets of Nature*; Kavey, *Books of Secrets*.
39. Nollet, *L'art des expériences*, 1:444.
40. Le Dran, *Observations de chirurgie*, xv.
41. Gelfand, *Professionalizing Modern Medicine*, 27.
42. Le Dran, *Observations de chirurgie*, xv.
43. Le Dran, *Observations de chirurgie*, xv–xvi.
44. Sully and Le Roy, *Règle artificielle*, 380.
45. Sully and Le Roy, *Règle artificielle*, 371–72.
46. Sully and Le Roy, *Règle artificielle*, 414–23.
47. On the spiritual and religious understandings of technology through time, see Noble, *The Religion of Technology*, and Merchant, *Reinventing Eden*. For discussions of alchemy and redemption, see Principe, *The Secrets of Alchemy*; Smith, *The Business of*

Alchemy; and Smith, *The Body of the Artisan*. The absence of archival sources makes it difficult to offer well-grounded interpretations of how the *artistes'* religious beliefs informed these narratives of the inventive process. We do know that at least Gaudron's father was interested in the history of heresy and of the reform; see Augarde, *Les ouvriers du temps*, 72.

48. Le Roy, "Lettre écrite aux auteurs des mémoires pour l'histoire des sciences et beaux arts," *Journal de Trévoux*, March 1733, 544.
49. *Mercure de France*, June 1735, 135.
50. Ferrand, *L'art du feu*, 218.
51. Nollet, *Programme*, 38.
52. Ferrand, *L'art du feu*, 186.
53. Berthoud, *Essai sur l'horlogerie*, 36, 43, 46.
54. Quesnay, *Lettres sur les disputes*, 140.
55. Thomin, *Traité d'optique mécanique*, 257–58.
56. Smith, *The Body of the Artisan*; Roberts, "The Death of the Sensuous Chemist," 503–29; Vila, ed., *A Cultural History of the Senses*.
57. Ferrand, *L'art du feu*, 204–5.
58. For an insightful discussion of the practice of writing while making, see Smith, "In the Workshop of History."
59. Lepaute, *Traité d'horlogerie*, 22.
60. Augarde, *Les ouvriers du temps*, 60.
61. Nollet, *Programme*.
62. Dubourg Glatigny and Vérin, eds., *Réduire en art*.
63. Furetière, *Dictionnaire Universel*, vol. 2, s.v. "Invention."
64. Sully, *Description abregée d'une horloge*, 266.
65. Voltaire, *Lettres philosophiques*, letter 12.
66. Le Roy, "Nouvelle maniere de construire de grosses horloges," *Mercure de France*, June 1735, 1312–13.
67. Sully, *Description abregée d'une horloge*, 265.
68. Sully, *Description abregée d'une horloge*, 86. For the role of analogy, see Vérin, "La technologie."
69. Sully and Le Roy, *Règle artificielle*, 414–23, 418 (quotation).
70. Sully, *Description abregée d'une horloge*, 265.
71. Alder, "Making Things the Same," 499–545. See also Alder, *Engineering the Revolution*.
72. Stalnaker, *The Unfinished Enlightenment*, 106.
73. Bender and Marrinan, *The Culture of Diagram*; Pannabecker, "Representing Mechanical Arts in Diderot's *Encyclopédie*," 33–73; Pinault-Sørensen, "A propos des planches de l'*Encyclopédie*," 143–52; Koepp, "The Alphabetical Order"; Sewell, "Visions of Labor"; Picon, "Gestes, ouvriers, opération et processus technique," 131–47.
74. Sheridan, "Recording Technology in France," 329–54. On technical illustrations, see also Lefèvre, ed., *Picturing Machines*.
75. On the negotiations among authors, printers, draftsmen, and engravers, see Kusukawa, *Picturing the Book of Nature*.

76. Papillon, *Traité historique et practique de la gravure en bois,* vols. 2–3, 386–88.
77. Plumier, *L'art de tourner,* 2.
78. Plumier slightly misquotes from Ecclesiastes, which reads, "res difficiles non potest eas homo explicare sermone" (Eccl. 1:8). The verse had been quoted by Albertus Magnus, whose works had been published in Lyon in 1651. Magnus, *Albertus Magnus, opera, Edid. Petrus Iammy,* 55; Plumier, *L'art de tourner,* 24.
79. Sainte-Albine, *Mémoire sur le laminage du plomb,* 10.
80. Nollet, *Art des experiences,* 1:174–77.
81. Garengeot, *Nouveau traité des instrumens de chirurgie,* 145, 197.
82. Garengeot, *Nouveau traité des instrumens de chirurgie,* 218.
83. Sully and Le Roy, *Règle artificielle,* 383. According to Le Roy, Sully could read Dutch and German in addition to English and French.
84. Sully and Le Roy, *Règle artificielle,* 284.
85. Papillon, *Traité historique et practique de la gravure en bois,* x; Gallon's preface in *MIAS,* vol. 1.
86. *Notice des livres et des objets principaux.*
87. 8° KA 13, CNAM.
88. Sully and Le Roy, *Règle artificielle,* 44–45; Sully, *Règle artificielle.*
89. Garengeot, *Nouveau traité des instrumens de chirurgie,* 414–15.
90. Sully, *Description abregée d'une horloge,* 36.
91. Sully, *Description abregée d'une horloge,* 37.
92. Sully, *Description abregée d'une horloge,* 10.
93. Sully, *Description abregée d'une horloge,* 43.
94. "Discours préliminaire sur l'étude de l'astronomie," in Dossier Fouchy, AASP.
95. *The Encyclopedia of Diderot and d'Alembert Collaborative Translation Project,* translated by Nelly S. Hoyt and Thomas Cassirer (Ann Arbor: Michigan Publishing, University of Michigan Library, 2003), http://hdl.handle.net/2027/spo.did2222.0000.139. Translation of "Art," in *Encyclopédie; ou, Dictionnaire raisonné des sciences, des arts et des métiers,* vol. 1 (Paris, 1751).
96. Edelstein, *The Enlightenment.*

6. L'ESPRIT IN THE MACHINE

1. Du Marsais, *Nouvelles libertés de penser,* 173–204, 174–75 (quotation). For an analysis of the *Encyclopédie* article "Philosophe" and its relationship with Du Marsais's text, see Dieckmann, *Le Philosophe.*
2. Du Marsais, *Nouvelles libertés de penser,* 187–200.
3. On the public culture of technology in eighteenth-century France, see Hilaire-Pérez, "Technology, Curiosity, and Utility."
4. Kang, *Sublime Dreams of Living Machines*; Voskhul, *Androids in the Enlightenment*; Riskin, *The Restless Clock.*
5. Winner, "Do Artifacts Have Politics?"; Pinch and Bijker, "The Social Construction of Facts and Artefacts"; Bijker, *Of Bicycles, Bakelites, and Bulbs.*

6. The *bureau typographique* is described in Dumas, *La bibliotheque des enfans*. On the *bureau typographique* and children's education, see Grandière, "Louis Dumas et le système typographique." On rational recreations, see Stafford, *Artful Science*, and Riskin, "Amusing Physics." The printer's desk was used also for the education of girls; see Goodman, *Becoming a Woman*.
7. Minutes, 1732, De Croÿ.
8. The report by the Société des Arts was published in *Mercure de France*, September 1731, 2199–2206.
9. Grandière, "Louis Dumas et le système typographique."
10. *Enfants prodiges* were often girls. See Mazzotti, *The World of Maria Gaetana Agnesi*, and Bertucci, "The In/visible Woman."
11. Dumas, *Bibliothèque des enfans*, 78.
12. *Mercure de France*, June 1732, 1102.
13. Rousseau, *Discours*; Voltaire, *Le mondain*. On the history of education in early modern France, see Chartier et al., *L'éducation en France*, and Mormiche, *Devenir prince*.
14. On Nollet, see Pyenson and Gauvin, eds., *The Art of Teaching Physics*.
15. Trew, *Commercium litterarium ad rei medicinae et scientiae naturalis*, 169–72. On Nollet's globes, see Ronfort, "Two Acquisitions."
16. Bycroft, *Physics and Natural History*; Bycroft, "What Difference Does a Translation Make?"
17. Terrall, *Catching Nature in the Act*; Gauvin, "The Instrument That Never Was."
18. De Fouchy, "Eloge de Nollet." The reference to Nollet as the author of the volume on the art of enameling, glassmaking, and glazing is in PV 1758, ff. 597–99. There is no evidence on whether Nollet ever started this work, though it would be worth studying the connections with the equivalent article in the *Encyclopédie*, attributed to Diderot.
19. Nollet, *Programme*, xi.
20. Voltaire to Thieriot, 27 October 1738, in Besterman, *Voltaire: Correspondance*, 1:1181.
21. Nollet, *Programme*, 113.
22. Nollet, *Programme*, 114.
23. Nollet, *Oratio*; Nollet, *Arts des expériences*.
24. Nollet, *Cours*.
25. Stafford, *Artful Science*; Sutton, *Science for a Polite Society*; Bensaude-Vincent and Blondel, *Science and Spectacle in the European Enlightenment*; Stewart, *The Rise of Public Science*; Schaffer, "Natural Philosophy and Public Spectacle in the Eighteenth Century"; Schaffer, "Machine Philosophy."
26. Sutton, *Science for a Polite Society*.
27. Nollet, *Leçons*, 1:xxi. On the "Newton wars," see Shank, *Newton Wars*.
28. Nollet, *Leçons*, 1:xxix.
29. Nollet, *Leçons*, 1:xxii.
30. Nollet, *Leçons*, 1:xxv.
31. Nollet, *Leçons*, 1:xxiv.
32. Nollet, *Programme*, vi, xxviii.

33. Nollet, *Programme*, xxiii–xxiv. On the abbé Pluche, see Koepp, "Advocating for Artisans."
34. Nollet, *Programme*.
35. Nollet, *Programme*, xxxi, xxxii, xxxvi.
36. Algarotti, *Opere*, 16:16.
37. Bertucci, *Viaggio nel paese delle meraviglie*, chapter 2.
38. *Gentlemen's Magazine* 15 (1745): 193–97. On eighteenth-century electricity, see Bertucci, "Sparks in the Dark."
39. Nollet, *Oratio*, 41–42.
40. Nollet, *Oratio*, 29, 9, 43.
41. On French salons, see Goodman, *The Republic of Letters*, and Lilti, *The World of the Salons*.
42. Quesnay, *Observations sur la saignée*. On Vaucanson, see Doyon and Liaigre, *Jacques Vaucanson*, and Riskin, *The Restless Clock*.
43. Ms 266, XIII, ff. 187–88, Ms 189, ff. 87–92, Académie Royale des Sciences, Belles-Lettres et Arts, Lyon, quoted in Doyon and Liaigre, *Jacques Vaucanson*, 148n6.
44. Doyon and Liaigre, *Jacques Vaucanson*, 147–60. On the king's support, see Condorcet, "Eloge de Vaucanson."
45. Doyon and Liaigre, *Jacques Vaucanson*, 151.
46. Schaffer, "Enlightened Automata"; Riskin, "Defecating Duck"; Riskin, "Eighteenth-Century Wetware." On later androids, see Voskhul, *Androids in the Enlightenment*.
47. Vaucanson, *Mécanisme du fluteur automate*.
48. Report by Villars, 4 September 1777, F12/1453b, AN.
49. On silk in Lyon, see Doyon and Liaigre, *Jacques Vaucanson*; Ballot, *L'introduction du machinisme*; Poni, "Fashion as flexible production"; and Pérez, "Cultures techniques."
50. C 2272, dossier 30, Archive Départementale Hérault, printed in Doyon and Liaigre, *Jacques Vaucanson*, 456.
51. Vaucanson's compensation for his spinning machine is listed in F12/823, AN. The fund F12/821, AN, lists several other generous compensations that Vaucanson received from the bureau. On the failure of Vaucanson's "great design," see Doyon and Liaigre, *Jacques Vaucanson*.
52. Quoted in Doyon and Liaigre, *Jacques Vaucanson*, 231.
53. Gillispie, *Science and Polity*; *Mercure de France*, November 1745, 116–20.
54. Bertucci, "Enlightened Secrets."
55. MacKenzie, "How Do We Know the Properties of Artifacts?"
56. *Affiches de Paris*, 21 April 1749. On the public interest in technological innovations, see Stewart, "A Meaning for Machines," and Hilaire-Pérez, "Technology, Curiosity, and Utility."
57. Vaucanson, "Construction d'un nouveau tour," 142–54. On the political economy of the bureau, see Minard, *La fortune du colbertisme*.
58. Vaucanson, "Construction d'un nouveau tour," 142–54.
59. Vaucanson, "Construction d'un nouveau tour," 142–54, 145 (quotation).
60. Vaucanson, "Construction d'un nouveau tour," 150.

61. Riskin, "Defecating Duck"; Riskin, "Eighteenth-Century Wetware."
62. Doyon and Liaigre, *Jacques Vaucanson*, 294.
63. C 2291, dossier 16, pp. 17–26, Archive Départementale Hérault, quoted in Doyon and Liaigre, *Jacques Vaucanson*, 298.
64. *Mercure de France*, November 1745, 116–17.
65. *Mercure de France*, November 1745, 116–17.
66. Condorcet, *Eloge de Vaucanson*, 160.
67. *Encyclopédie*, 1:713–17.
68. Condorcet, "Eloge de Vaucanson," 160.
69. Bertucci, "Enlightened Secrets."
70. Jacques Vaucanson to Daniel Trudaine, 5 October 1765, published in Doyon and Liaigre, *Jacques Vaucanson*, 439–40.
71. See Biagioli, "Patent Republic."
72. Vaucanson, "Second mémoire," 443–53.
73. The classic work on "black boxes" is Latour, *Pandora's Hope*.
74. Vaucanson, *Le mécanisme du fluteur automate*, 20. This text was translated into English by Desaguliers, making Vaucanson's sentence even more explicit: Vaucanson's intention was to "demonstrate the manner of actions [rather] than to shew a machine. . . . I would not be thought to impose upon the spectators by any conceal'd or juggling contrivance." Jacques Vaucanson, *An Account of the Mechanism of an Automaton* (London, 1742), 23, quoted in Schaffer, "Enlightened Automata," 144.
75. Terrall, *Catching Nature in the Act*; Daston, "Attention and the Values of Nature in the Enlightenment."

EPILOGUE

1. *Encyclopédie*, 1:i–xlv.
2. *Encyclopédie*, 1:714.
3. *Encyclopédie*, 1:714.
4. *Encyclopédie*, 1:xxxix. This section was taken from Diderot's "Prospectus," published in November 1750, and available here: https://encyclopedie.uchicago.edu/node/88.
5. *Encyclopédie*, 1:xxxix. This section was taken from Diderot's "Prospectus," published in November 1750 and available at https://encyclopedie.uchicago.edu/node/88.
6. Kafker and Loveland, "La vie agitée de l'abbé De Gua de Malves."
7. Kafker and Loveland, "La vie agitée de l'abbé De Gua de Malves"; Favre and Durr, "Un texte inédit de l'abbé Gua de Malves concernant la naissance de l'*Encylopédie*," 51–68.
8. May, "Histoire et Sources de l'*Encyclopédie*," 18; Loïc and Théré, "Un nouvel élément."
9. Favre and Durr, "Un texte inédit de l'abbé Gua de Malves concernant la naissance de l'*Encylopédie*."
10. Favre and Durr, "Un texte inédit de l'abbé Gua de Malves concernant la naissance de l'*Encylopédie*," 55.

11. Favre and Durr, "Un texte inédit de l'abbé Gua de Malves concernant la naissance de l'*Encylopédie*," 55.
12. Kafker and Loveland, "La vie agitée de l'abbé De Gua de Malves." For a recent repropostion of the *Encyclopédie* as a war machine, see, for example, Le Ru, *Subversives lumières*.
13. A copy of Gua de Malves's 1745 prospectus is in Littérature et art, X 1246, BnF.
14. *Encyclopédie*, 1:xxxvi.
15. Gua de Malves, memorandum, printed in Favre and Durr, "Un texte inédit de l'abbé Gua de Malves concernant la naissance de l'*Encylopédie*," 7.
16. Proust, *Diderot et l'*Encyclopédie, 49.
17. Minutes, February–April 1734, De Croÿ; Papillon, *Traité historique et practique de la gravure en bois*, x.
18. This emerges from various keyword searches through the ARTFL edition of the *Encyclopédie*.
19. Ballstadt, *Diderot*, 176, 13. On Diderot's friendship with the Premontavals, see his *Jacques le fataliste*.
20. Strugnell, "La candidature de Diderot."
21. *Encyclopédie*, 1:xiii.
22. *Encyclopédie*, 1:716.
23. Berthoud, *Essai sur l'horlogerie*, 45. On the amateurs, see Guichard, *Les amateurs d'art*.
24. Hilaire-Pérez, "Les sociabilités industrielles."
25. Hahn, "The Applications of Science to Society"; Hilaire-Pérez, "Les sociabilités industrielles."
26. On the Society of Arts, see Hilaire-Pérez, "Les sociabilités industrielles," and Matthew Paskins, *Sentimental Industry*.
27. Stewart, *The Rise of Public Science*; Jacob and Stewart, *Practical Matter*.
28. Picon, *French Architects and Engineers*; Belhoste et al., eds., *Le Paris des polytechniciens*.
29. Voltaire, *Le mondain*.

Bibliography

Full references for manuscript sources and periodicals are in the notes. To avoid confusion, I have modernized the accents in the titles of early modern French books.

MANUSCRIPT COLLECTIONS

Archives de l'Académie de Chirurgie, Paris
Archives de l'Académie des Sciences, Paris
Archive Départementale Hérault, Montpellier, France
Archive du Conservatoire des Arts et Métiers, Paris
Archive du Duc de Croÿ, Dülmen, Germany
Archives Nationales, Paris
Biblioteca Medicea Laurenziana, Florence
Biblioteca Napoletana di Storia Patria, Naples
Bibliothèque de l'Arsenal, Paris
Bibliothèque de l'Assemblée Nationale, Paris
Bibliothèque Mazarin, Paris
Bibliothèque Municipale, Caen, France
Bibliothèque Nationale de France, Paris
Chambre des Députés, Paris
Houghton Library, Harvard University, Cambridge, Mass.
National Archives, London
National Archives, Stockholm
Nationalmuseum, Stockholm
Öffentliche Bibliothek der Universität Basel, Basel, Switzerland
Royal Society of Arts, Archives, London
Staatsbibliothek, Berlin

PRINTED WORKS

Abot de Bazinghen, François-André. *Traité des monnoies, et de la jurisdiction de la cour des monnoies.* 2 vols. Paris: Guillyn, 1764.

Alder, Ken. *Engineering the Revolution: Arms and Enlightenment in France, 1763–1815*. Princeton, N.J.: Princeton University Press, 1997.

———. "Making Things the Same: Representation, Tolerance and the End of the Ancien Regime in France." *Social Studies of Science* 28 (1998): 499–545.

Algarotti, Francesco. *Opere*. 17 vols. Venice: Palese, 1794.

Allan, D. G. C. "The Society of Arts and Government, 1754–1800: Encouragement of Arts, Manufactures, and Commerce in Eighteenth-Century England." *Eighteenth-Century Studies* 7 (1974): 434–52.

———. *William Shipley, Founder of the Royal Society of Arts: A Biography with Documents*. London: Scholar Press, 1979.

Andreoli, Aldo Bacchi. *Alcuni studi intorno a Guido Panciroli*. Reggio-Emilia, Italy: Calderini, 1903.

Ash, Eric. "Introduction: Expertise and the Early Modern State." *Osiris* 25 (2010): 1–24.

Augarde, Jean-Dominique. *Les ouvriers du temps: La pendule à Paris de Louis XIV à Napoléon Ier: Ornamental Clocks and Clockmakers in Eighteenth Century Paris*. [Geneva]: Antiquorum Editions, 1996.

Auzout, Adrien. *L'éphéméride du comète*. [Paris]: s. n., [1665].

Bacon, Francis. *The Two Books of Francis Bacon. Of the Proficience and Advancement of Learning*. London: Thomes, 1605.

———. *The Works of Francis Bacon*. Edited by James Spedding, Robert Leslie Ellis, and Douglas Denon Heath. 14 vols. London: Longmans, Green, 1868.

Ballot, Charles. *L'introduction du machinisme dans l'industrie française*. Lille: O. Marquant, 1923.

Ballstadt, Kurt. *Diderot: Natural Philosopher*. Oxford: Voltaire Foundation, 2008.

Barolsky, Paul. *A Brief History of the Artist from God to Picasso*. University Park: Pennsylvania State University Press, 2010.

Barthes, Roland, ed. *Le degré zéro de l'écriture*. Paris: Éditions du Seuil, 1972.

Baudelot de Dairval, Charles-César. *De l'utilité des voyages: Et de l'avantage que la recherche des antiquitez procure aux sçavans*. Paris: Chez Pierre Auboüin et Pierre Emery, 1686.

Baudez, Basile. *Architecture et tradition académique: Au temps des lumières*. Rennes: Presses universitaires de Rennes, 2013.

Baudouin, Benoît. *De calceo antiquo*. Amsterdam: Frisi, 1667.

Becq, Annie, ed. *L'encyclopédisme: Actes du colloque de Caen, 12–16 Janvier 1987*. Paris: Aux Amateurs de Livres, Diffusion Klincksieck, 1991.

Belhoste, Bruno. *Paris savant: Parcours et rencontres au temps des lumières*. Paris: Colin, 2011.

Belhoste, Bruno, Francine Masson, and Antoine Picon, eds. *Le Paris des polytechniciens: Des ingénieurs dans la ville, 1794–1994*. Paris: Délégation à l'Action Artistique de la Ville de Paris, 1994.

Bender, John, and Michael Marrinan. *The Culture of Diagram*. Stanford, Calif.: Stanford University Press, 2010.

Benguigui, Isaac. *Théories électriques du XVIIIe siècle: Correspondence entre l'abbé Nollet (1700–1770) et le physicien genevois Jean Jallabert (1712–1768)*. Geneva: Georg, 1984.

Benítez, Miguel. *La face cachée des lumières: Recherches sur les manuscrits philosophiques clandestins de l'âge classique*. Paris: Universitas, 1996.

Bensaude-Vincent, Bernadette, and Christine Blondel, eds. *Science and Spectacle in the European Enlightenment: Science, Technology and Culture, 1700–1945*. Aldershot, U.K.: Ashgate, 2008.

Beretta, Marco, Antonio Clericuzio, and Lawrence Principe. *The Accademia del Cimento and Its European Context*. Sagamore Beach, Mass.: Science History, 2009.

Berthoud, Ferdinand. *Essai sur l'horlogerie; dans lequel on traite de cet art relativement à l'usage civil, à l'astronomie et à la navigation, en etablissant des principes confirmés par l'expérience*. Paris: Jombert, 1763.

———. *Histoire de la mesure du temps par les horloges*. Paris: Imprimerie de la République, 1802.

Bertrand, Joseph. *L'Académie des Sciences et les académiciens de 1666 à 1793*. Amsterdam: B. M. Israël, 1969.

Bertucci, Paola. "The Architecture of Knowledge: Science, Collecting, and Display in Eighteenth-Century Naples." In *New Approaches to Naples c. 1500–c. 1800: The Power of Place*, edited by Melissa Calaresu and Helen Hills, 149–74. Aldershot, U.K.: Ashgate, 2013.

———. "Enlightened Secrets: Silk, Industrial Espionage, and Intelligent Travel in 18th-Century France." *Technology and Culture* 54 (2013): 820–52.

———. "The In/Visible Woman: Mariangela Ardinghelli and the Circulation of Knowledge Between Paris and Naples in the Eighteenth Century." *Isis* 104 (2013): 226–49.

———. "Sparks in the Dark: The Attraction of Electricity in the Eighteenth Century." *Endeavour* 31 (2007): 88–93.

———. *Viaggio nel paese delle meraviglie. Scienza e curiosità nell'Italia del Settecento*. Turin: Bollati Boringhieri, 2007.

Bertucci, Paola, and Olivier Courcelle. "Artisanal Knowledge, Expertise, and Patronage in Early Eighteenth-Century Paris: The Société des Arts (1728–36)." *Eighteenth-Century Studies* 48 (2015): 159–79.

Besterman, Theodore, ed. *The Complete Works of Voltaire*. 143 vols. Geneva: Institut et Musée Voltaire, 1968.

———. *Voltaire: Correspondance*. 2 vols. Paris: Gallimard, 1963.

Biagioli, Mario. *Galileo, Courtier: The Practice of Science in the Culture of Absolutism*. Chicago: University of Chicago Press, 1993.

———. "Patent Republic: Representing Inventions, Constructing Rights and Authors." *Social Research* 73 (2006): 1129–72.

Bigourdan, Guillaume. *Les premières sociétés savantes de Paris au XVIIe siècle et les origines de l'Académie des Sciences*. Paris: Gauthier-Villars, 1919.

Bijker, Wiebe E. *Of Bicycles, Bakelites, and Bulbs: Toward a Theory of Sociotechnical Change*. Boston: MIT Press, 1997.

Birembaut, Arthur. "L'exposition de modèles de machines à Paris en 1683." *Revue d'histoire des sciences et de leurs applications* 20 (1967): 141–58.

Blair, Ann. *The Theater of Nature: Jean Bodin and Renaissance Science*. Princeton, N.J.: Princeton University Press, 1997.

———. *Too Much to Know: Managing Scholarly Information Before the Modern Age*. New Haven, Conn.: Yale University Press, 2010.

Blakey, William, Jr. *L'art de faire les ressorts des montres*. Amsterdam: Rey, 1780.

Bléchet, Françoise. "L'abbé Bignon, président de l'Académie Royale des Sciences." In *Règlement, usages et science dans la France de l'absolutisme*, edited by Christiane Demeulenaere-Douyère and Eric Brian, 51–69. Paris: Lavoisier, 2002.

———. "Fontenelle et l'abbé Bignon, du president de l'Académie Royale des Sciences au secretaire perpétuel: Quelques lettres de l'abbé Bignon à Fontenelle." *Corpus, Revue de Philosophie* 13 (1990): 51–62.

———. "Le rôle de l'abbé Bignon dans l'activité des sociétés savantes au XVIIIe siècle." In *Actes du 100e Congrès National des Sociétés Savantes, section d'histoire moderne et contemporaine et commission d'histoire des sciences et des techniques*, 31–41. Paris: Bibliothèque Nationale, 1975.

———. "Un précurseur de l'*Encyclopédie* au service de l'état: L'abbé Bignon." In *L'encyclopédisme: Actes du colloque de Caen, 12–16 Janvier 1987*. Edited by Annie Blecq, 395–412. Paris: Aux Amateurs de Livres, Diffusion Klincksieck, 1991.

Bonnassieux, Pierre, and Eugène Lelong. *Conseil de commerce et Bureau de commerce (1700–1791)*. Geneva: Mégariotis Reprints, 1979.

Bontemps, Daniel, and Catherine Prade. "Un magasin parisien d'ouvrages en fonte de fer ornée au XVIIIe siècle." *Bulletin de la Société de l'Histoire de Paris et de l'Ile-de-France* 118 (1991): 215–61.

Boyer, Abel. *The Royal Dictionary, French and English, and English and French*. London: J. and J. Knapton, 1729.

Brockliss, Lawrence W. B. *Calvet's Web: Enlightenment and the Republic of Letters in Eighteenth-Century France*. Oxford: Oxford University Press, 2002.

Brockliss, Lawrence W. B., and Colin Jones. *The Medical World of Early Modern France*. Oxford: Clarendon Press, 1997.

Brown, Harcourt. *Scientific Organizations in Seventeenth Century France (1620–1680)*. Baltimore: Williams and Wilkins, 1934.

Bud, Robert, ed. "Focus: Applied Science." *Isis* 103 (2010): 515–63.

Burke, Peter. *A Social History of Knowledge: From Gutenberg to Diderot*. Cambridge: Polity Press, 2000.

Bycroft, Michael Trevor. "Physics and Natural History in the Eighteenth Century: The Case of Charles Dufay." Ph.D. diss., University of Cambridge, 2014.

———. "What Difference Does a Translation Make? The *Traité des vernis* (1723) in the Career of Charles Dufay." In *Translating Early Modern Science*, edited by Sietske Fransen and Niall Hodson. Boston: Brill, forthcoming.

Campbell, Peter. *Power and Politics in Old Regime France, 1720–1745*. London: Routledge, 2003.

Caradonna, Jeremy L. *The Enlightenment in Practice: Academic Prize Contests and Intellectual Culture in France, 1670–1794*. Ithaca, N.Y.: Cornell University Press, 2012.

Cassirer, Ernst. *Die Philosophie Der Aufklärung*. Tübingen: Mohr, 1932.

Castiglione, Baldassarre. *The Book of the Courtier*. Baltimore: Penguin, 1967.

Charles, Loïc, and Christine Théré. "Un nouvel élément pour l'histoire de l'*Encyclopédie*: Le 'Plan' inédit du premier éditeur, Gua de Malves." *Recherches sur Diderot et sur l'*Encyclopédie" 39 (2005): 105–22.

Chartier, Roger, Dominique Julia, and Marie-Madeleine Compère. *L'éducation en France du XVIe au XVIIIe siècle*. Paris: Société d'Édition d'Enseignement Supérieur, 1976.

Chaussinand-Nogaret, Guy. *La noblesse au XVIIIe siècle: De la féodalité aux lumières*. Brussels: Editions Complexe, 2000.

Chevreul, Raoul. "Une belle figure bourguignonne: L'abbé Jean-Baptiste Joseph Languet de Gergy (1675–1750)." In *37e Congrès de l'Association Bourguignonne des Sociétés Savantes*, 45–47. Dijon, 1966.

Cipolla, Carlo M. *Clocks and Culture, 1300–1700*. London: Collins, 1967.

Clark, William, Jan Golinski, and Simon Schaffer, eds. *The Sciences in Enlightened Europe*. Chicago: University of Chicago Press, 1999.

Clarke, Jack A. "Abbé Jean-Paul Bignon 'Moderator of the Academies' and Royal Librarian." *French Historical Studies* 8 (1973): 213–35.

Clément, Pierre, ed. *Lettres, instructions et mémoires de Colbert*. Vol. 2. Paris: Imprimerie Impériale, 1863.

Cole, Arthur Harrison, and George B. Watts. *The Handicrafts of France as Recorded in the Descriptions des Arts et Métiers, 1761–1788*. Boston: Baker Library, Harvard Graduate School of Business Administration, 1952.

Collins, James B. *The State in Early Modern France*. Cambridge: Cambridge University Press, 1995.

Condorcet, Jean-Antoine-Nicolas de Caritat, Marquis de. "Eloge de M. de Vaucanson." *Histoire et mémoires de l'Académie Royal des Sciences pour l'année 1782* (1785): 156–68.

Corvol, Andrée, ed. *Duhamel du Monceau, 1700–2000: Un européen du siècle des lumières*. Orléans: Académie d'Orléans, 2001.

Courcelle, Olivier. "Grandjean de Fouchy et la Société des Arts à Stockholm." *Revue d'histoire des sciences* 61 (2008): 203–4.

Cousin, Jules. *Le comte de Clermont, sa cour et ses maitresses: Lettres familières, recherches et documents inédits*. Paris: Académie des Bibliophiles, 1867.

Cuomo, Serafina. *Technology and Culture in Greek and Roman Antiquity*. Cambridge: Cambridge University Press, 2007.

Curiosités de Paris, de Versailles, Marly, Vincennes, Saint-Cloud, et des environs. Paris: Les Libraires Associés, 1771.

Dacier, Anne Le Fèvre, ed. *L'Iliade d'Homère, traduite en françois, avec de remarques par Madame Dacier*. Paris: Rigaus, 1719.

D'Alembert, Jean Le Rond. *Œuvres philosophiques, historiques et littéraires*. 18 vols. Paris: Jean-François Bastien, 1805.

Darnton, Robert. *The Business of Enlightenment: A Publishing History of the* Encyclopédie. Cambridge, Mass.: Belknap Press of Harvard University Press, 1979.

Daston, Lorraine, and Katharine Park. *Wonders and the Order of Nature, 1150–1750*. New York: Zone Books, 1998.

Daston, Lorraine, and Fernando Vidal, eds. *The Moral Authority of Nature*. Chicago: University of Chicago Press, 2004.

Daumas, Maurice, and Rene Tresse. "La *Description des arts et métiers* de l'Académie des Sciences et le sort de ses planches gravées en taille douce." *Revue d'histoire des sciences et de leurs applications* 7 (1954): 163–71.

Dawson, Nelson-Martin, and Charles Vincent. *L'atelier Delisle: L'Amerique du Nord sur la table a dessin*. Paris: Les Éditions du Septentrion, 2000.

[De Beze De Lys]. *Second mémoire pour la Dame veuve Renard, plaignante contre le Sieurs Renard, héritiers de leur frère*. Paris: Le Breton, [1739].

Debure, Guillaume-François. *Catalogue des livres de la bibliothèque de feu M. le duc de La Vallière*. N.p.: Guillaume de Bure Fils Aîné, 1783.

Debus, Allen G. *The French Paracelsians: The Chemical Challenge to Medical and Scientific Tradition in Early Modern France*. Cambridge: Cambridge University Press, 1991.

De Fouchy, Jean Paul Grandjean. "Eloge de M. l'abbé Nollet." *Histoire et mémoires de l'Académie Royale des Sciences pour l'année* 1770 (1773): 121–36.

[De Gas De Fremainville]. *Troisième mémoire signifié, pour la Dame veuve Renard du Tasta*. Paris: Le Breton, n.d.

DeJean, Joan E. *Ancients Against Moderns: Culture Wars and the Making of a Fin de Siècle*. Chicago: University of Chicago Press, 1997.

Demeulenaere-Douyère, Christiane, and E. Brian, eds. *Règlement, usages et science dans la France de l'absolutisme*. Paris: Lavoisier, 2002.

Demeulenaere-Douyère, Christiane, and David J. Sturdy, eds. *L'enquête du régent, 1716–1718: Sciences, techniques et politique dans la France pré-industrielle*. Turnhout, Belgium: Brepols, 2008.

Dempsey, Charles. *Inventing the Renaissance Putto*. Chapel Hill: University of North Carolina Press, 2001.

De Pas, Antoine, Marquis de Feuquières. *Mémoires sur la guerre*. Amsterdam: Changuion, 1731.

Dequidt, Marie-Agnès. *Horlogers des lumières: Temps et société à Paris au XVIIIe siècle*. Monts, France: Editions du Comité des Travaus Historiques et Scientifiques, 2014.

Desaguliers, John T. *The Newtonian System of the World, the Best Model of Government: An Allegorical Poem*. Westminster, England: Campbell, 1728.

Dessert, Daniel. *Argent, pouvoir et société au grand siècle*. Paris: Fayard, 1984.

Devey, Joseph, ed. *The Moral and Historical Works of Francis Bacon*. London: Bell and Sons, 1882.

Dew, Nicholas. *Orientalism in Louis XIV's France*. Oxford: Oxford University Press, 2009.

———. "Reading Travels in the Culture of Curiosity: Thévenot's Collection of Voyages." *Journal of Early Modern History* 10 (2006): 39–59.

Dictionnaire de l'Académie Françoise. Vol. 1. Paris: Coignard, 1694.

Dictionnaire de l'Académie Françoise. Vol. 1. Paris: Bernard Brunet, 1762.

Diderot, Denis. *Jacques le fataliste*. Paris: Garnier, 1796.

Diderot, Denis, and Jean d'Alembert, eds. *Encyclopédie; ou, Dictionnaire raisonné des sciences, des arts, et des métiers*. Edited by Robert Morrissey. University of Chicago,

ARTFL, Encyclopédie Project, winter 2008 edition. http://encyclopedie.uchicago.edu/.
Dieckmann, Herbert. *Le Philosophe: Texts and Interpretation.* St. Louis, 1948.
Dolza, Luisa. "Reframing the Language of Inventions." In *The Power of Images in Early Modern Science,* edited by Wolfgang Lefèvre, Jürgen Renn, and Urs Schoepflin, 89–104. Basel: Verlag, 2003.
Dolza, Luisa, and Hélène Vérin. "Figurer la mécanique: L'énigme des théâtres de machines de la Renaissance." *Revue d'histoire moderne et contemporaine* 51 (2004): 7–37.
Dornau, Caspar, and Pierre-Thomas-Nicolas Hurtaut. *L'art de peter: Essay théori-physique et méthodique.* Westphalie [Paris]: Florent-Q., 1751.
Doyon, André, and Lucien Liaigre. *Jacques Vaucanson: Mécanicien de génie.* Paris: Presses Universitaires de France, 1966.
Dubourg Glatigny, Pascal, and Hélène Vérin, eds. *Réduire en art: La technologie de la Renaissance aux lumières.* Paris: Editions de la Maison des Sciences de l'Homme, 2008.
Dufay, Charles François de Cisternay. "Description d'une machine pour connoître l'heure vraye du soleil tous les jours de l'année." *Histoire et mémoires de l'Académie Royale des Sciences pour l'année 1725* (1727): 67–78.
Duhamel du Monceau, Henri-Louis. *Art du charbonnier; ou, Maniere de faire le charbon de bois.* Paris: Desaint et Saillant, 1761.
Du Marsais, César Chesneau. *Nouvelles libertés de penser.* Amsterdam, 1743.
Dumas, Louis. *La bibliothèque des enfans.* Paris: Simon, 1732.
Dunn, Richard, and Rebekah Higgitt. *Finding Longitude.* Glasgow: Collins, 2014.
Dutot, [Nicolas]. *Histoire du système de John Law, 1716–1720: Publication intégrale du manuscrit inédit de Poitiers.* Paris: Institut National d'Études Démographiques, 2000.
Dyche, Thomas, and William Pardon. *A New General English Dictionary.* London: Richard Ware, 1740.
Eamon, William. *Science and the Secrets of Nature: Books of Secrets in Medieval and Early Modern Culture.* Princeton, N.J.: Princeton University Press, 1994.
Eco, Umberto. *Experiences in Translation.* Toronto: University of Toronto Press, 2001.
Edelstein, Daniel. *The Enlightenment: A Genealogy.* Chicago: University of Chicago Press, 2010.
Epstein, S. R. "Craft Guilds, Apprenticeship and Technical Change in Preindustrial Europe." *Journal of Economic History* 58 (1998): 684–713.
———. "Property Rights to Technical Knowledge in Premodern Europe, 1300–1800." *American Economic Review* 94 (2004): 382–87.
Epstein, S. R., and Maarten Prak, eds. *Guilds, Innovation and the European Economy, 1400–1800.* Cambridge: Cambridge University Press, 2008.
Faure, Edgar. *La banqueroute de Law, 17 juillet 1720.* Paris: Gallimard, 1977.
Favre, Robert, and Michel Durr. "Un texte inédit de l'abbé Gua de Malves concernant la naissance de *l'Encyclopédie.*" *Mémoires de l'Académie des Sciences, Belles-Lettres et Arts de Lyon* 55 (2000): 51–68.
Feingold, Mordechai. *The Newtonian Moment: Isaac Newton and the Making of Modern Culture.* Oxford: Oxford University Press, 2004.

Félibien, André. *Des principes de l'architecture, de la sculpture, de la peinture, et des autres arts qui en dependent.* Paris: Coignard, 1676.

Ferrand, Jacques-Philippes. *L'art du feu; ou, De peindre en email.* Paris: Collombat, 1721.

Findlen, Paula. *Possessing Nature: Museums, Collecting, and Scientific Culture in Early Modern Italy.* Berkeley: University of California Press, 1994.

Follard, Jean Charles, Chevalier de. *Nouvelles découvertes sur la guerre, dans une dissertation sur Polybe.* Brussels: Foppens, 1753.

Fontenelle, Bernard le Bovier de. *Entretiens sur la pluralité des mondes.* Paris: C. Blageart, 1686.

Force, James E., and Sarah Hutton, eds. *Newton and Newtonianism: New Studies.* Dordrecht: Kluwer Academic, 2004.

Fox, Celina. *The Arts of Industry in the Age of Enlightenment.* New Haven, Conn.: Yale University Press, 2009.

Fréron, Elie Catherine. *L'année littéraire; ou, Suite des lettres sur quelques écrits de ce temps.* Paris: Lambert, 1757.

Fumaroli, Mario. "Feu et glace: Le comédien de Rémond de Saint-Albine (1747), antithèse du paradoxe." *Revue d'histoire littéraire de la France* 93 (1993): 702–23.

———. *La querelle des anciens et des modernes.* Paris: Gallimard-Folio, 2001.

Furetière, Antoine. *Dictionnaire universel.* Vol. 1. The Hague and Rotterdam: 1690.

———. *Dictionnaire universel.* Vol. 2. Rotterdam: Leers, 1708.

Gallet, Michel. *Les architectes parisiens du XVIIIe siècle.* Paris: Mengès, 1995.

Galluzzi, Paolo, ed. *Scienziati a corte: L'arte della sperimentazione nell'Accademia Galileiana del Cimento, 1657–1667.* Livorno, Italy: Sillabe, 2001.

Garas, Clara. "Le plafond de La Banque Royale de Giovanni Antonio Pellegrini." *Bulletin du musée hongrois des beaux-arts* 21 (1962): 75–93.

Garberson, Eric. "Libraries, Memory and the Space of Knowledge." *Journal of the History of Collections* 18 (2006): 105–36.

Garçon, Anne-Françoise. "Les dessous des métiers: Secrets, rites et sous-traitance dans la France du XVIIIe siècle." *Early Science and Medicine* 10 (2005): 378–91.

Garengeot, René-Jacques Croissant de. *Nouveau traité des instrumens de chirurgie les plus utiles.* Paris: Cavelier, 1727.

Gauvin, Jean-François. "The Instrument That Never Was: Inventing, Manufacturing, and Branding Réaumur's Thermometer During the Enlightenment." *Annals of Science* 69 (2012): 515–49.

Gay, Peter. *The Enlightenment: An Interpretation: The Rise of Modern Paganism.* New York: Knopf, 1967.

[Gayot]. *Mémoire pour la Dame Renard, plaignante en faux principal.* Paris: Le Breton, [1742].

Gelfand, Toby. *Professionalizing Modern Medicine: Paris Surgeons and Medical Science and Institutions in the 18th Century.* Westport, Conn.: Greenwood Press, 1980.

Geoffroy, Monsieur. "Experiments upon Metals, Made with the Burning-Glass of the Duke of Orleans. By Monsieur Geoffroy, F. R. S." *Philosophical Transactions* 26 (1708): 374–86.

Gersaint, Edme-François. *Catalogue raisonné: D'une collection considerable de diverses curiosités en tous genres, contenuës dans les cabinets de feu Monsieur Bonnier de la Mosson.* Paris: Chez Jaques Barois et Pierre-Guillaume Simon, 1744.

Gieryn, Thomas. "Boundary-Work and the Demarcation of Science from Non-Science." *American Sociological Review* 48 (1983): 781–795.

Gillispie, Charles Coulston. *Science and Polity in France: The End of the Old Regime.* Princeton, N.J.: Princeton University Press, 2004.

Golbéry, Sylvain Meinrad Xavier de, and William Mudford. *Travels in Africa: Performed During the Years 1785, 1786, and 1787.* London: Jones, 1803.

Goldgar, Anne. *Impolite Learning: Conduct and Community in the Republic of Letters, 1680–1750.* New Haven, Conn.: Yale University Press, 1995.

Golinski, Jan. "A Noble Spectacle: Phosphorus and the Public Cultures of Science in the Early Royal Society." *Isis* 80 (1989): 11–39.

———. *Science as Public Culture: Chemistry and Enlightenment in Britain, 1760–1820.* Cambridge: Cambridge University Press, 1992.

Goodman, Dena. *Becoming a Woman in the Age of Letters.* Ithaca, N.Y.: Cornell University Press, 2009.

———. *The Republic of Letters: A Cultural History of the French Enlightenment.* Ithaca, N.Y.: Cornell University Press, 1994.

Goodman, Dena, and Kathryn Norberg, eds. *Furnishing the Eighteenth Century: What Furniture Can Tell Us About the European and American Past.* New York: Routledge, 2007.

Grandière, Marcel. "Louis Dumas et le système typographique, 1728–1744." *Histoire de l'éducation* 81 (1999): 35–62.

Greenfeld, Liah. *The Spirit of Capitalism: Nationalism and Economic Growth.* Cambridge, Mass.: Harvard University Press, 2009.

Guardia, J. M. "Feuilleton." *Gazette Médicale de Paris* 3 (1868): 323n–328n.

Guichard, Charlotte. "Arts libéraux et arts libres à Paris au XVIIIe siècle: Peintres et sculpteurs entre corporation et Académie Royale." *Revue d'histoire moderne et contemporaine* 49 (2002): 54–68.

———. *Les amateurs d'art à Paris au XVIIIe siècle.* Seyssel, France: Champ Vallon, 2008.

Hahn, Roger. *The Anatomy of a Scientific Institution: The Paris Academy of Sciences, 1666–1803.* Berkeley: University of California Press, 1971.

———. "The Applications of Science to Society: The Societies of Arts." *Studies on Voltaire and the Eighteenth Century* 25 (1963): 829–36.

———. "Science and the Arts in France: The Limitations of an Encyclopedic Ideology." *Studies in Eighteenth-Century Culture* 10 (1981): 77–93.

Hale, Matthew. *The Primitive Origination of Mankind, Considered and Examined According to the Light of Nature.* London: Godbid, 1677.

Hall, A. Rupert, and Marie Boas Hall, eds. *The Correspondence of Henry Oldenburg.* Madison: University of Wisconsin Press, 1965.

Hamel, Charles. *Histoire de l'église Saint-Sulpice.* Paris: Lecoffre, 1909.

Harkness, Deborah E. *The Jewel House: Elizabethan London and the Scientific Revolution.* New Haven, Conn.: Yale University Press, 2007.

Harris, John Raymond. *Industrial Espionage and Technology Transfer: Britain and France in the Eighteenth Century.* Aldershot, U.K.: Ashgate, 1998.
Haudrère, Philippe. *La Compagnie Française des Indes au XVIIIe siècle (1719–1795).* 4 vols. Paris: Librarie de l'Inde, 1989.
Hecht, Jacqueline. "Un problème de population active au XVIIIe siècle en France: La querelle de la noblesse commerçante." *Population* 19 (1964): 267–90.
Heinich, Nathalie. *Du peintre à l'artiste: Artisans et académiciens à l'age classique.* Paris: Editions de Minuit, 1993.
Hilaire-Pérez, Liliane. "Cultures techniques et pratiques de l'échange, entre Lyon et le Levant: Inventions et réseaux au XVIIIe siècle." *Revue d'histoire moderne et contemporaine* 49 (2002): 89–114.
——— . "Diderot's Views on Artists' and Inventors' Rights: Invention, Imitation and Reputation." *British Journal for the History of Science* 35 (2002): 129–50.
——— . "Inventing in a World of Guilds: Silk Fabrics in Eighteenth-Century Lyon." In *Guilds, Innovation and the European Economy, 1400–1800,* edited by S. R. Epstein and Maarten Prak, 232–63. Cambridge: Cambridge University Press, 2008.
——— . *La pièce et le geste: Artisans, marchands et savoir technique à Londres au XVIIIe siècle.* Paris: Michel, 2013.
——— . "Les sociabilités industrielles en France et en Angleterre au XVIIIe siècle." In *Encourager l'innovation en France et en Europe,* edited by Serge Benoit, Gérard Emptoz, and Denis Woronoff, 201–38. Paris: Editions du CTHS, 2006.
——— . *L'invention technique au siècle des lumières.* Paris: Albin Michel, 2000.
——— . "Steel and Toy Trade Between England and France: The Huntsmans' Correspondence." *History of Metallurgy* 42 (2008): 127–47.
——— . "Technology as Public Culture in the Eighteenth Century: The Artisan's Legacy." *History of Science* 45 (2007): 135–53.
——— . "Technology, Curiosity, and Utility in France and England in the Eighteenth Century." In *Science and Spectacle in the European Enlightenment,* edited by Bernadette Bensaud-Vincent and Christine Blondel, 25–42. Aldershot, U.K.: Ashgate, 2008.
——— . "Transferts technologiques, droit et territoire: Le cas franco-anglais au XVIIIe siècle." *Revue d'histoire moderne et contemporaine* 44 (1997): 547–79.
Hilaire-Pérez, Liliane, and Anne-Françoise Garçon, eds. *Les chemins de la nouveauté: Innover, inventer au regard de l'histoire.* Paris: Editions du CTHS, 2003.
Hilaire-Pérez, Liliane, and Marie Thébaud-Sorger. "Les techniques dans l'espace public: Publicité des inventions et littérature d'usage au XVIIIe siècle (France, Angleterre)." *Revue de synthèse* 5 (2006): 393–428.
Houghton, Walter E. "The History of Trades: Its Relation to Seventeenth-Century Thought: As Seen in Bacon, Petty, Evelyn, and Boyle." *Journal of the History of Ideas* 2 (1941): 33–60.
Horn, Jeff. *Economic Development in Early Modern France.* Cambridge: Cambridge University Press, 2015.
Huard, Georges. "Les planches de l'*Encyclopédie* et celles de la *Description des Arts et Métiers* de l'Académie des Sciences." *Revue d'histoire des sciences et de leurs applications* 4 (1951): 238–49.

Humbert, Roger. *Institutions et gens de finance en Franche-Comté, 1674–1790.* Paris: Diffusion Les Belles Lettres, 1996.
Huygens, Christiaan. *Oeuvres complètes.* Vol. 4. The Hague: M. Nijhoff, 1888.
Institut de France. *Index biographique de l'Académie des Sciences du 22 décembre 1666 au 1er octobre 1978.* Paris: Gauthier-Villars, 1979.
Israel, Jonathan I. *Democratic Enlightenment: Philosophy, Revolution, and Human Rights, 1750–1790.* Oxford: Oxford University Press, 2011.
———. *Enlightenment Contested: Philosophy, Modernity, and the Emancipation of Man, 1670–1752.* Oxford: Oxford University Press, 2006.
———. *Radical Enlightenment: Philosophy and the Making of Modernity, 1650–1750.* Oxford: Oxford University Press, 2001.
Jacob, Margaret C. *Living the Enlightenment: Freemasonry and Politics in Eighteenth-Century Europe.* New York: Oxford University Press, 1991.
———. *The Newtonians and the English Revolution, 1689–1720.* New York: Gordon and Breach, 1990.
———. *The Radical Enlightenment: Pantheists, Freemasons, and Republicans.* London: Allen and Unwin, 1981.
Jacob, Margaret C., and Larry Stewart. *Practical Matter: Newton's Science in the Service of Industry and Empire, 1687–1851.* Cambridge, Mass.: Harvard University Press, 2004.
Jammes, André. *La réforme de la typographie royale sous Louis XIV, Le Grandjean.* Paris: P. Jammes, 1961.
Jaoul, Martine, and Madeleine Pinault. "La collection 'Description des Arts et Métiers': Étude des sources inédites de la Houghton Library Université Harvard." *Ethnologie Française* 12 (1982): 335–60.
Jaugeon, Jacques. *Le jeu du monde; ou, L'intelligence des plus curieuses choses qui se trouvent dans tous les etats, les terres, et les mers du monde.* Paris: Chez Amable Auroy, 1684.
Jones, Peter. *Industrial Enlightenment: Science, Technology and Culture in Birmingham and the West Midlands, 1760–1820.* Manchester: Manchester University Press, 2008.
Jordan, Charles-Étienne. *Histoire d'un voyage littéraire fait en MDCCXXXIII en France, en Angleterre, et en Hollande.* The Hague: Chez Adrien Moetjens, 1735.
Kafker, Frank A. "Gua de Malves and the *Encyclopédie.*" *Diderot Studies* 19 (1978): 93–102.
Kafker, Frank A., and Jeff Loveland. "La vie agitée de l'abbé de Gua de Malves et sa direction de l'*Encyclopédie.*" *Recherches sur Diderot et sur l'*Encyclopédie 47 (2012): 187–205.
Kaiser, Thomas E. "Money, Despotism, and Public Opinion in Early Eighteenth-Century France: John Law and the Debate on Royal Credit." *Journal of Modern History* 63 (1991): 1–28.
Kammerling Smith, David. "Structuring Politics in Early Eighteenth-Century France: The Political Innovations of the French Council of Commerce." *Journal of Modern History* 74 (2002): 490–537.
Kang, Minsoo. *Sublime Dreams of Living Machines: The Automaton in the European Imagination.* Cambridge, Mass.: Harvard University Press, 2011.

Kaplan, Steven. "Les corporations, les 'faux ouvriers' et le faubourg Saint-Antoine au XVIIIe siècle." *Annales* 43 (1988): 353–78.

———. *Les ventres de Paris: Pouvoir et approvisionnement dans la France d'ancien régime*. Paris: Fayard, 1988.

Kavey, Allison. *Books of Secrets: Natural Philosophy in England, 1550–1600*. Urbana: University of Illinois Press, 2007.

Keller, Vera. "Accounting for Invention: Guido Pancirolli's Lost and Found Things and the Development of Desiderata." *Journal of the History of Ideas* 73 (2012): 223–45.

———. *Knowledge and the Public Interest, 1575–1725*. Cambridge: Cambridge University Press, 2015.

Kersey, John. *The New World of Words; or, Universal English Dictionary*. London: J. Phillips, 1706.

Kessler, Amalia D. *A Revolution in Commerce: The Parisian Merchant Court and the Rise of Commercial Society in Eighteenth-Century France*. New Haven, Conn.: Yale University Press, 2007.

King, Henry C., and John R. Millburn. *Geared to the Stars: The Evolution of Planetariums, Orreries, and Astronomical Clocks*. Toronto: University of Toronto Press, 1978.

Klein, Ursula, ed. "Artisanal-Scientific Experts in Eighteenth-Century France and Germany." *Annals of Science* 69 (2012): 303–446.

Klein, Ursula, and Wolfgang Lefèvre. *Materials in Eighteenth-Century Science: A Historical Ontology*. Cambridge, Mass.: MIT Press, 2007.

Klein, Ursula, and E. C. Spary, eds. *Materials and Expertise in Early Modern Europe: Between Market and Laboratory*. Chicago: University of Chicago Press, 2010.

Koepp, Cynthia J. "The Alphabetical Order: Work in Diderot's Encyclopédie." In *Work in France: Representations, Meaning, Organization, and Practice*, edited by Steven L. Kaplan and Cynthia J. Koepp, 229–57. Ithaca, N.Y.: Cornell University Press, 1986.

———. "Advocating for Artisans: The Abbé Pluche's *Spectacle de la Nature* (1732–51)." In *The Idea of Work in Europe from Antiquity to Modern Times*, edited by Josef Ehmer and Catharina Lis, 245–73. Aldershot, U.K.: Ashgate 2009.

Korey, Alexandra M. "Putti, Pleasure, and Pedagogy in Sixteenth-Century Italian Prints and Decorative Arts." Ph.D. diss., University of Chicago, 2007.

Kusukawa, Sachiko. *Picturing the Book of Nature: Image, Text, and Argument in Sixteenth-Century Human Anatomy and Medical Botany*. Chicago: University of Chicago Press, 2012.

Landes, David S. *Revolution in Time: Clocks and the Making of the Modern World*. Cambridge, Mass.: Belknap Press of Harvard University Press, 2000.

La Peyronie, François de. "Histoire de l'Académie Royale de Chirurgie dans son établissement jusqu'à 1743." *Mémoires de l'Académie Royale de Chirurgie* 4 (1768): 1–62.

Larmessin, Nicolas II de. *Costumes grotesques*. Paris, 1695.

L'art de plumer la poulle sans crier. Cologne: Robert Le Turc, 1710.

Latour, Bruno. *Pandora's Hope: Essays on the Reality of Science Studies*. Cambridge, Mass.: Harvard University Press, 1999.

Law, John. *Oeuvres complètes*. Edited by Paul Harsin. 3 vols. Paris: Librairie du Recueil Sirey, 1934.

Le Dran, Henry-François. *Observations de chirurgie*. Paris: Osmont, 1731.

———. *Parallèle des différentes manières de tirer la pierre hors de la vessie*. Paris: Osmont, 1730.

Lefèvre, Wolfgang, ed. *Picturing Machines, 1400–1700*. Cambridge, Mass.: MIT Press, 2004.

Lehner, Ulrich L., and Michael O'Neill Printy, eds. *A Companion to the Catholic Enlightenment in Europe*. Boston: Brill, 2010.

Leibniz, Gottfried Wilhem, Samuel Clarke, and Isaac Newton. *Recueil de diverses pièces*. Vol. 2. Amsterdam: Changuion, 1740,

Lepaute, Jean André. *Traité d'horlogerie*. Paris: Samson, 1767.

Le Roi, Joseph Adrien. *Histoire des rues de Versailles et de ses places et avenues, depuis l'origine de cette ville jusqu'à nos jours*. Versailles: A. Montalant, 1861.

Le Roy, Julien. "Lettre écrite aux auteurs des *Mémoires pour l'histoire des sciences et beaux arts*." *Journal de Trévoux*, March 1733, 541–47.

———. "Nouvelle manière de construire de grosses horloges." *Mercure de France*, June 1735, 1312–26.

Le Ru, Véronique. *Subversives lumières: L'Encyclopédie comme machine de guerre*. Paris: CNRS, 2007.

Lestringant, Frank. *Mapping the Renaissance World: The Geographical Imagination in the Age of Discovery*. Cambridge: Polity Press, 1994.

Levine, Joseph M. *The Battle of the Books: History and Literature in the Augustan Age*. Ithaca, N.Y.: Cornell University Press, 1991.

Lilti, Antoine. *The World of the Salons: Sociability and Worldliness in Eighteenth-Century Paris*. New York: Oxford University Press, 2015.

Lindsay, Robert O. "Pierre Bergeron: A Forgotten Editor of French Travel Literature." *Terrae Incognitae* 7 (1975): 31–38.

Long, Pamela O. *Artisan/Practitioners and the Rise of the New Sciences, 1400–1600*. Corvallis: Oregon State University Press, 2011.

———. *Openness, Secrecy, Authorship: Technical Arts and the Culture of Knowledge from Antiquity to the Renaissance*. Baltimore: Johns Hopkins University Press, 2001.

Lorgues, Christiane. "L'ancien Hôtel de la Monnaie de Paris et ses problèmes." *Revue numismatique* 6 (1968): 138–74.

Louis, Antoine. *Éloges de messieurs Bassuel, Malaval et Verdier, prononcés aux écoles de chirurgie*. Paris: P. Guillaume Cavelier, 1759.

MacKenzie, Donald. "How Do We Know the Properties of Artifacts? Applying the Sociology of Knowledge to Technology." In *Technological Change: Methods and Themes in the History of Technology*, edited by Robert Fox, 247–63. Amsterdam: Harwood Academic, 1998.

Mackrell, J. Q. C. *The Attack on Feudalism in Eighteenth-Century France*. London: Routledge, 2013.

MacLeod, Christine. *Heroes of Invention: Technology, Liberalism and British Identity, 1750–1914*. Cambridge: Cambridge University Press, 2010.

Maerker, Anna Katharina. *Model Experts: Wax Anatomies and Enlightenment in Florence and Vienna, 1775–1815*. Manchester: Manchester University Press, 2011.

Magnus, Albertus. *Albertus Magnus, Opera, Edid. Petrus Iammy.* Leiden: Prost, 1651.
Maillet, Benoît de. *Telliamed; ou, Entretiens d'un philosophe indien avec un missionnaire françois sur la diminution de la mer.* The Hague: Gosse, 1755.
Maindron, Ernest. *L'Académie des Sciences.* Paris: Félix Alcan, 1888.
Marion, Michel. *Les bibliothèques privées à Paris au milieu du XVIIIe siècle.* Paris: Bibliothèque Nationale, 1978.
Martin, Henri-Jean. *Livre, pouvoirs et société à Paris au XVIIe siècle: 1598–1701.* 2 vols. Geneva: Droz, 1969.
Martinière, Antoine Augustin Bruzen de la. *Le grand dictionnaire géographique et critique.* Venice: Pasquali, 1737.
Maupertuis, Pierre Louis Moreau de. *La figure de la terre, déterminée par les observations de messieurs de Maupertuis, Clairaut, Camus, le Monnier, de l'Académie Royale des Sciences.* Amsterdam: Catuffe, 1738.
Maury, Louis-Ferdinand-Alfred. *Les académies d'autrefois: L'ancienne Académie des Sciences.* Paris: Didier, 1864.
May, Louis-Philippe. "Histoire et sources de l'*Encyclopédie.*" *Revue de synthèse* 15 (1938): 7–30.
Mazzotti, Massimo. *The World of Maria Gaetana Agnesi, Mathematician of God.* Baltimore: Johns Hopkins University Press, 2007.
McClaughin, Trevor. "Sur les rapports entre la Compagnie de Thévenot et l'Académie Royale des Sciences." *Revue d'histoire des sciences* 28 (1975): 235–42.
McClellan, James E., III. *Science Reorganized: Scientific Societies in the Eighteenth Century.* New York: Columbia University Press, 1985.
McClellan, James E., III, and François Regourd. *The Colonial Machine: French Science and Overseas Expansion in the Old Regime.* Turnhout, Belgium: Brepols, 2011.
Medalon, Daniel. "Sur la différence des tumeurs à extirper ou à ouvrir simplement; et sur le choix du cautere ou de l'instrument tranchant dans ces différens cas." In *Recueil des pièces qui ont concouru pour le Prix de l'Académie Royale de Chirurgie,* 3–65. Paris: Delaguette, 1753.
Merchant, Carolyn. *Reinventing Eden: The Fate of Nature in Western Culture.* New York: Routledge, 2004.
Merson, Olivier. "Les logements d'artistes au Louvre à la fin du XVIIIe siècle." *Gazette des beaux-arts* 23 (1881): 264–70.
Meyssonnier, Simon. *La balance et l'horloge: La genèse de la pensée libérale en France au XVIIIe siècle.* Montreuil: Editions de la Passion, 1989.
Milliot, Vincent. *Les cris de Paris; ou, Le peuple travesti: Les représentations des petits métiers parisiens (XVIe–XVIIIe siècles).* Paris: Publications de la Sorbonne, 1995.
Minard, Philippe. *La fortune du colbertisme: État et industrie dans la France des lumières.* Paris: Fayard, 1998.
Mokyr, Joel. *The Enlightened Economy: An Economic History of Britain, 1700–1850.* New Haven, Conn.: Yale University Press, 2009.
———. *The Gifts of Athena: Historical Origins of the Knowledge Economy.* Princeton, N.J.: Princeton University Press, 2002.

Monteil, Amans Alexis. *Traité de matériaux manuscrits de divers genres d'histoire*. 2 vols. Paris: Duverger, 1835.
Moran, Bruce T., ed. *Patronage and Institutions: Science, Technology, and Medicine at the European Court, 1500–1750*. Rochester, N.Y.: Boydell Press, 1991.
Moreri, Louis. *Dictionnaire historique*. Paris: Vincent, 1732.
Mormiche, Pascale. *Devenir prince: L'école du pouvoir en France, XVIIe-XVIIIe siècles*. Paris: CNRS, 2009.
Mouy, Paul. *Le développement de la physique cartésienne, 1646–1712*. Paris: Vrin, 1934.
Mukerji, Chandra. *Impossible Engineering: Technology and Territoriality on the Canal du Midi*. Princeton, N.J.: Princeton University Press, 2009.
Murphy, Antoin E. *John Law: Economic Theorist and Policy-Maker*. Oxford: Oxford University Press, 1997.
Narbonne, Pierre. *Journal des règnes de Louis XIV et Louis XV: L'année 1701 à l'année 1744*. Paris: Durand et Pedone Lauriel, 1866.
Noble, David F. *The Religion of Technology: The Divinity of Man and the Spirit of Invention*. New York: Knopf Doubleday, 2013.
Nollet, Abbé (Jean-Antoine). *Cours de physique expérimentale*. Paris: Herault, 1735.
——. *L'art des expériences; ou, Avis aux amateurs de la physique*. Paris: Durand, 1770.
——. *Leçons de physique expérimentale*. Paris: Guerin, 1743–64.
——. *Oratio habita a Joanne-Antonio Nollet*. Paris: Thiboust, 1753.
——. *Programme; ou, Idée générale d'un cours de physique expérimentale*. Paris: Le Mercier, 1738.
Norman, Larry F. *The Shock of the Ancient: Literature and History in Early Modern France*. Chicago: University of Chicago Press, 2011.
Notice des livres et des objets principaux de physique et d'optique, appartenans à feu M. L'abbé Noel. [Paris]: Demonville, 1783.
Nummedal, Tara E. *Alchemy and Authority in the Holy Roman Empire*. Chicago: University of Chicago Press, 2007.
Palissy, Bernard. *Discours admirables*. Paris: M. le Jeune, 1580.
Panciroli, Guido. *Rerum memorabilium jam olim deperditarum et contrà recens atque ingeniose inventarum libri duo*. Amberg, Germany: Typis Fosterianis, 1599.
Pannabecker, John R. "Representing Mechanical Arts in Diderot's *Encyclopédie*." *Technology and Culture* 39 (1998): 33–73.
Papillon, Jean-Baptiste Michel. *Traité historique et pratique de la gravure en bois*. Paris: Simon, 1766.
Paskins, Matthew. "Sentimental Industry: The Society of Arts and the Encouragement of Public Useful Knowledge, 1754–1848." Ph.D. diss., University College London, 2014.
Passeron, Irène. "L'invention d'une nouvelle compétence: Géomètre au XVIIIème siècle." In *Les chemins de la nouveauté: Innover, inventer au regard de l'histoire*, edited by Françoise Garçon and Liliane Hilaire-Pérez, 139–52. Paris: Editions du CTHS, 2003.
Passeron, Irène, and Olivier Courcelle. "La Société des Arts, espace provisoire de reformulation des rapports entre théories scientifiques et pratiques instrumentales."

In *Règlement, usages et science dans la France de l'absolutisme*, edited by Christiane Demeulenaere-Douyère and Eric Brian, 109–32. Paris: Lavoisier, 2002.

Pearce, Susan M., ed. *Interpreting Objects and Collections*. London: Routledge, 1994.

Pedley, Mary Sponberg. *The Commerce of Cartography: Making and Marketing Maps in Eighteenth-Century France and England*. Chicago: University of Chicago Press, 2005.

Pernety, Dom Antoine-Joseph. *Dictionnaire portatif de peinture, sculpture et gravure*. Paris: Bauche, 1757.

Perrault, Charles. *Le cabinet des beaux arts*. Paris: G. Edelinck, 1690.

——. *Parallele des anciens et des modernes, en ce qui regarde les arts et les sciences*. 2 vols. Paris: J. B. Coignard, 1693.

Petitfils, Jean-Christian. *Le régent*. Paris: Fayard, 1986.

Petto, Christine Marie. *Mapping and Charting in Early Modern England and France: Power, Patronage, and Production*. Lanham, Md.: Lexington Books, 2015.

——. *When France Was King of Cartography: The Patronage and Production of Maps in Early Modern France*. Lanham, Md.: Lexington Books, 2007.

Picon, Antoine. *French Architects and Engineers in the Age of Enlightenment*. Cambridge: Cambridge University Press, 1992.

——. "Gestes, ouvriers, opération et processus technique: La vision de travail des encyclopédistes." *Recherches sur Diderot et sur l'*Encyclopédie 13 (1992): 131–47.

[Picot, Jean-Baptiste]. *Explication des modèles des machines et forces mouvantes*. Paris: Guillery, 1683.

Pigeon, Jean. *Description d'une sphère mouvante*. Paris: Quillau, 1714.

Pinault, Madeleine. "Aux sources de l'*Encyclopédie*: La description des arts et métiers." Ph.D. diss., École Pratique des Hautes-Études, Paris, 1984.

——. "Diderot et les illustrateurs de l'*Encyclopédie*." *Revue de l'art* 66 (1984): 17.

Pinault-Sørensen, Madeleine. "A propos des planches de l'*Encyclopédie*." *Recherches sur Diderot et sur l'*Encyclopédie 15 (1993): 143–52.

Pinch, Trevor J., and Wiebe E. Bijker. "The Social Construction of Facts and Artefacts; or, How the Sociology of Science and the Sociology of Technology Might Benefit Each Other." *Social Studies of Science* 14 (1984): 399–441.

Pincus, Steve. "Rethinking Mercantilism: Political Economy, the British Empire, and the Atlantic World in the Seventeenth and Eighteenth Century," *The William and Mary Quarterly* 69 (2012): 3–34.

Pluche, Noël Antoine. *Spectacle de La Nature . . . Tr. from the Original French*. London: L. Davis and C. Reymers, 1763.

Plumier, Charles. *L'art de tourner; ou, De faire en perfection toutes sortes d'ouvrages au tour*. Paris: Jombert, 1749.

Pomata, Gianna, and Nancy G. Siraisi, eds. *Historia: Empiricism and Erudition in Early Modern Europe*. Cambridge, Mass.: MIT Press, 2005.

Poni, Carlo. *Fashion as Flexible Production: The Strategies of the Lyons Silk Merchants in the Eighteenth Century*. Edited by Charles F. Sabel. Cambridge: Cambridge University Press, 1997.

Portalis, Roger, and Henri Béraldi. *Les graveurs du dix-huitième siècle*. 3 vols. Paris: Morgand et Fatout, 1880.

Porter, Roy. *The Creation of the Modern World: The Untold Story of the British Enlightenment.* New York: Norton, 2000.
Porter, Roy, and Mikuláš Teich, eds. *The Enlightenment in National Context.* Cambridge: Cambridge University Press, 1981.
Principe, Lawrence M. "The End of Alchemy? The Repudiation and Persistence of Chrysopoeia at the Académie Royale des Sciences in the Eighteenth Century." *Osiris* 29 (2014): 96–116.
———. *The Secrets of Alchemy.* Chicago: University of Chicago Press, 2012.
Proust, Jacques. *Diderot et l'*Encyclopédie. 3rd ed. Paris: A. Michel, 1995.
———. "La documentation technique de Diderot dans l'*Encyclopédie.*" *Revue d'histoire littéraire de la France* 57 (1957): 335–52.
Pyenson, Lewis, and Jean-François Gauvin, eds. *The Art of Teaching Physics: The Eighteenth-Century Demonstration Apparatus of Jean Antoine Nollet.* Sillery, Québec: Septentrion, 2002.
Quesnay, François. *Lettres sur les disputes.* N.p., 1738.
———. *Observations sur la saignée.* Paris: Osmont, 1730.
Rabier, Christelle, ed. *Fields of Expertise: A Comparative History of Expert Procedures in Paris and London, 1600 to Present.* Cambridge: Cambridge Scholars, 2007.
———. "La disparition du barbier chirurgien: Analyse d'une mutation professionnelle au XVIIIe siècle." *Annales: Histoire, Sciences Sociales* 65 (2010): 679–711.
[Raynal, Abbé]. *Anecdotes litteraires.* Paris: Durand, 1750.
Réaumur, René-Antoine Ferchault de. *L'art de convertir le fer forgé en acier et l'art d'adoucir le fer fondu.* Paris: Brunet, 1722.
———. *Memoirs on Steel and Iron: A Translation from the Original Printed in 1722.* Edited by Anneliese Grünhaldt Sisco. Chicago: University of Chicago Press, 1956.
Réaumur, René-Antoine Ferchault de, and Henri-Louis Duhamel du Monceau. *Fabrique des ancres: Lue à l'académie en juillet 1723.* [Paris], 1761.
———. *Nouvel art d'adoucir le fer fondu et de faire des ouvrages de fer fondu aussi finis que de fer forgé.* [Paris], 1761.
Règlement de la Société des Arts. Paris: Quillau, 1730.
Richelet, Pierre. *Dictionnaire de la langue françoise, ancienne et moderne.* Lyon: Duplain, 1728.
———. *Nouveau dictionnaire françois.* Amsterdam: Jean Elzevir, 1709.
Riskin, Jessica. "Amusing Physics." In *Science and Spectacle in the European Enlightenment,* edited by Bernadette Bensaud-Vincent and Christine Blondel, 43–63. Aldershot, U.K.: Ashgate, 2008.
———. "The Defecating Duck; or, The Ambiguous Origins of Artificial Life." *Critical Inquiry* 29 (2003): 599–633.
———. "Eighteenth-Century Wetware." *Representations* 83 (2003): 97–125.
———. *The Restless Clock: A History of the Centuries-Long Argument over What Makes Living Things Tick.* Chicago: University of Chicago Press, 2016.
Roberts, Lissa. "The Death of the Sensuous Chemist: The 'New' Chemistry and the Transformation of Sensuous Technology." *Studies in History and Philosophy of Science* 26 (1995): 503–29.

Roberts, Lissa, Simon Schaffer, and Peter Dear, eds. *The Mindful Hand: Inquiry and Invention from the Late Renaissance to Early Industrialisation*. Amsterdam: Koninkliijke Nederlandse Akademie van Wetenschappen, 2007.

Robertson, John. *The Case for the Enlightenment: Scotland and Naples, 1680–1760*. Cambridge: Cambridge University Press, 2005.

Roche, Daniel. *A History of Everyday Things: The Birth of Consumption in France, 1600–1800*. Cambridge: Cambridge University Press, 2000.

———. *The People of Paris: An Essay in Popular Culture in the 18th Century*. Berkeley: University of California Press, 1987.

Ronfort, Jean-Nérée. "Science and Luxury: Two Acquisitions by the J. Paul Getty Museum." *J. Paul Getty Museum Journal* 17 (1989): 47–82.

Rosset, Philippe. "Les conseillers au Châtelet de Paris à la fin du XVIIe siècle (1661–1700)." *Bibliothèque de l'école des chartes* 143 (1985): 117–52.

Rossi, Paolo. *I filosofi e le macchine (1400–1700)*. Milan: Feltrinelli, 1962.

———. *Philosophy, Technology, and the Arts in the Early Modern Era*. Edited by Benjamin Nelson. Translated by Salvator Attanasio. New York: Harper and Row, 1970.

Rousseau, Jean-Jacques. *Discours qui a remporté le prix à l'académie de Dijon, en l'année 1750: Sur cette question proposée par la même académie: Si le rétablissement des sciences et des arts a contribué à épurer les moeurs*. Geneva: Barillot et Fils, 1750.

Rousset de Missy, Jean. *Le ceremonial diplomatique des cours de l' Europe*. Amsterdam: Wetstein et al., 1739.

Rowlands, Guy. *The Financial Decline of a Great Power: War, Influence, and Money in Louis XIV's France*. Oxford: Oxford University Press, 2012.

Royllet, Honoré Sébastien. *Les nouveaux principes de l'art d'écrire*. Paris: Mesnier, 1731.

Sainte-Albine, Rémond de. *Mémoire sur le laminage du plomb*. 3rd ed. Paris: Guerin, 1746.

Sainte-Beuve, Charles Augustin. *Le comte de Clermont et sa cour: Étude historique et critique*. Paris: Académie des Bibliophiles, 1868.

Salomon-Bayet, Claire. "Un préambule théorique à une Académie des Arts." *Revue d'histoire des sciences et de leurs applications* 23 (1970): 229–50.

Savary des Brûlons, Jacques. *Dictionnaire universel de commerce*. 6th ed. 4 vols. Paris: La Veuve Estienne, 1750.

Savary des Bruslons, Jacques, and Philémon-Louis Savary. *Dictionnaire universel de commerce*. Amsterdam: Janson, 1726.

Schaffer, Simon. "Enlightened Automata." In *The Sciences in Enlightened Europe*, edited by William Clark, Jan Golinski, and Simon Schaffer, 126–65. Chicago: University of Chicago Press, 1999.

———. "Machine Philosophy: Demonstration Devices in Georgian Mechanics." *Osiris* 9 (1994): 157–82.

———. "Natural Philosophy and Public Spectacle in the Eighteenth Century." *History of Science* 21 (1983): 1–43.

Schlüter, Christophe-André, and Jean Hellot. *De la fonte, des mines, des fonderies, &c: Qui traite des assais des mines et métaux, de l'affinage et raffinage de l'argent, du départ de l'or, &c*. Paris: Pissot, 1750.

Scott, Katie. *The Rococo Interior: Decoration and Social Spaces in Early Eighteenth-Century Paris.* New Haven, Conn.: Yale University Press, 1995.
Sewell, William H. "Visions of Labor: Illustrations of the Mechanical Arts Before, In, and After Diderot's *Encyclopédie.*" In *Work in France: Representations, Meaning, Organization, and Practice,* edited by Steven L. Kaplan and Cynthia J. Koepp, 258–86. Ithaca, N.Y.: Cornell University Press, 1986.
Shank, John Bennett. *The Newton Wars and the Beginning of the French Enlightenment.* Chicago: University of Chicago Press, 2008.
Shapin, Steven. "The Invisible Technician." *American Scientist* 77 (1989): 554–63.
——. *A Social History of Truth: Civility and Science in Seventeenth-Century England.* Chicago: University of Chicago Press, 1994.
Shea, William R., ed. *Science and the Visual Image in the Enlightenment.* Canton, Mass.: Science History, 2000.
Sheridan, Geraldine. *Louder Than Words: Ways of Seeing Women Workers in Eighteenth-Century France.* Lubbock: Texas Tech University Press, 2009.
——. "Recording Technology in France: The Descriptions des Arts, Methodological Opportunities at the Turn of the Eighteenth Century." *Cultural and Social History* 5 (2008): 329–54.
Shiner, Larry E. *The Invention of Art: A Cultural History.* Chicago: University of Chicago Press, 2001.
Siraisi, Nancy G. *History, Medicine, and the Traditions of Renaissance Learning.* Ann Arbor: University of Michigan Press, 2007.
Smedley-Weill, Anette. "La gestion du commerce français au xviiième siècle: Impulsions gouvernementales et besoins des échanges." *Histoire, économie et société* 12 (1993): 473–86.
Smith, Pamela H. *The Body of the Artisan: Art and Experience in the Scientific Revolution.* Chicago: University of Chicago Press, 2004.
——. *The Business of Alchemy: Science and Culture in the Holy Roman Empire.* Princeton, N.J.: Princeton University Press, 1994.
——. "In the Workshop of History: Making, Writing, and Meaning." *West 86th* 19 (2012): 4–31.
——. "Science on the Move: Recent Trends in the History of Early Modern Science." *Renaissance Quarterly* 62 (2009): 345–75.
——. "Why Write a Book? From Lived Experience to the Written Word in Early Modern Europe." *Bulletin of the German Historical Institute* 47 (2010): 25–50.
Soll, Jacob. *The Information Master: Jean-Baptiste Colbert's Secret State Intelligence System.* Ann Arbor: University of Michigan Press, 2009.
Sonenscher, Michael. *Work and Wages: Natural Law, Politics, and the Eighteenth-Century French Trades.* Cambridge: Cambridge University Press, 1989.
Sorkin, David Jan. *The Religious Enlightenment: Protestants, Jews, and Catholics from London to Vienna.* Princeton, N.J.: Princeton University Press, 2008.
Spary, Emma C. *Eating the Enlightenment: Food and the Sciences in Paris.* Chicago: University of Chicago Press, 2012.
Spon, Jacob. *Recherche des antiquités et curiosités de la ville de Lyon.* Lyon: Antoine Cellier Fils, 1675.

Stafford, Barbara Maria. *Artful Science: Enlightenment, Entertainment, and the Eclipse of Visual Education.* Cambridge, Mass.: MIT Press, 1994.

Stalnaker, Joanna. *The Unfinished Enlightenment: Description in the Age of the Encyclopedia.* Ithaca, N.Y.: Cornell University Press, 2010.

Stewart, Larry. "Assistants to Enlightenment: William Lewis, Alexander Chisholm and Invisible Technicians in the Industrial Revolution." *Notes and Records of the Royal Society* 62 (2008): 17–29.

———. "A Meaning for Machines: Modernity, Utility, and the Eighteenth-Century British Public." *Journal of Modern History* 70 (1998): 259–94.

———. *The Rise of Public Science: Rhetoric, Technology, and Natural Philosophy in Newtonian Britain, 1660–1750.* Cambridge: Cambridge University Press, 1992.

Strandberg, Runar. "La réception de Carl Johan Cronstedt à la Société des Arts et des Sciences à Paris." *Konsthistorisk tidskrift* 31 (1962): 95–104.

Stroup, Alice. *A Company of Scientists: Botany, Patronage, and Community at the Seventeenth-Century Parisian Royal Academy of Sciences.* Berkeley: University of California Press, 1990.

———. *Royal Funding of the Parisian Académie Royale des Sciences During the 1690s.* Philadelphia: American Philosophical Society, 1987.

Strugnell, Anthony R. "La candidature de Diderot à la Société Royale de Londres." *Recherches sur Diderot et sur l'*Encyclopédie 4 (1988): 37–41.

Sturdy, David J. *Science and Social Status: The Members of the Academie des Sciences, 1666–1750.* Woodbridge, U.K.: Boydell and Brewer, 1995.

Sully, Henry. *Description abregée d'une horloge d'une nouvelle invention.* Paris: Briasson, 1726.

———. "Description de la ligne méridienne de l'église de Saint Sulpice, avec l'explication de ses usages." *Mercure de France,* July 1728, 1591–1610.

———. *Méthode pour régler les montres et les pendules.* Paris: G. Dupuis, 1728.

———. *Règle artificielle du temps.* Vienna: Heyinger, 1714.

Sully, Henry, and Julien Le Roy. *Règle artificielle du temps.* Paris: Dupuis, 1737.

Sutton, Geoffrey V. *Science for a Polite Society: Gender, Culture, and the Demonstration of Enlightenment.* Boulder, Col.: Westview Press, 1995.

Takats, Sean. *The Expert Cook in Enlightenment France.* Baltimore: Johns Hopkins University Press, 2011.

Taton, René. *Les origines de l'Académie Royale des Sciences.* Paris: Palais de la découverte, 1966.

Terrall, Mary. *Catching Nature in the Act: Réaumur and the Practice of Natural History in the Eighteenth Century.* Chicago: University of Chicago Press, 2014.

———. *The Man Who Flattened the Earth: Maupertuis and the Sciences in the Enlightenment.* Chicago: University of Chicago Press, 2002.

Thévenot, Melchisédech. *Relations de divers voyages curieux.* Paris: Sébastien Cramoisy, 1664.

Thiout, Antoine. *Traité de l'horlogerie méchanique et pratique.* Paris: Moette, 1741.

Thomin, Marc Mitouflet. *Traité d'optique méchanique.* Paris: Coignard, 1749.

Three Letters Concerning the Forming of a Society: To Be Call'd, the Chamber of Arts. For the Preserving and Improvement of Operative Knowledge, the Mechanical Arts, Inventions, and Manufactures. London, 1722.

Thuillier, Guy, and Arthur Birembaut. "Une source inédite: Les cahiers du chimiste Jean Hellot (1685–1766)." *Annales: Histoire, Sciences Sociales* 21 (1966): 357–64.

Torlais, Jean. *Réaumur: Un esprit encyclopédique en dehors de l'Encyclopédie.* Paris: Desclée de Brouwer, 1936.

Toussaint, François-Vincent. *Observations périodiques sur la physique, l'histoire naturelle et les arts.* Paris: Pissot, Lambert, Cailleau, 1756.

Turner, Anthony J. "Grollier de Servière, the Brothers Monconys: Curiosity and Collecting in Seventeenth-Century Lyon." *Journal of the History of Collections* 20 (2008): 205–15.

———. "Melchisédech Thévenot, the Bubble Level, and the Artificial Horizon." *Nuncius* 7 (1992): 131–45.

Van Damme, Stéphane. *Paris, capitale philosophique de la Fronde à la Révolution.* Paris: Odile Jacob, 2005.

———. *Le Temple de la sagesse: Savoirs, écriture, et sociabilité urbaine (Lyon, 17–18e siècles).* Paris: Editions de l'EHESS, 2005.

———. *A toutes voiles vers la vérité: Une autre histoire de la philosophie au temps des Lumières.* Paris: Le Seuil, 2014.

Van Veen, Otto. *Amorum emblemata.* Antwerp: Swingen, 1609.

Vardi, Liana. *The Physiocrats and the World of the Enlightenment.* Cambridge: Cambridge University Press, 2012.

Vaucanson, Jacques de. "Construction d'un nouveau tour à filer la soie." *Histoire et mémoires de l'Académie Royale des Sciences pour l'année 1749* (1753): 142–54.

———. *Le mécanisme du fluteur automate: Presenté a messieurs de l'Académie Royale des Sciences.* Paris: Guerin, 1738.

———. "Second mémoire sur la filature des soies." *Histoire et mémoires de l'Académie Royale des Sciences pour l'année 1770* (1773): 437–58.

Velde, François. "The Life and Times of Nicolas Dutot." *Journal of the History of Economic Thought* 34 (2012): 67–107.

Venturi, Franco. *The End of the Old Regime in Europe, 1768–1776: The First Crisis.* Princeton, N.J.: Princeton University Press, 1989.

———. *Le origini dell'Enciclopedia.* Roma: U Edizioni, 1946.

Venturino, Diego. "L'historiographie révolutionnaire française et les lumières, de Paul Buchez à Albert Sorel." In *Historiographie et usages des lumières,* edited by Giuseppe Ricuperati, 21–58. Berlin: Berlin Verlag Arno Spitz, 2002.

Vérin, Hélène. "La technologie: Science autonome ou science intermédiaire?" *Documents pour l'histoire des techniques. Nouvelle série* 14 (2007): 134–43.

Vila, Anne, ed. *A Cultural History of the Senses in the Age of Enlightenment.* London: Bloomsbury, 2016.

Voltaire. *Le Mondain.* 1736. https://fr.wikisource.org/wiki/Le_Mondain.

———. *Lettres Philosophiques.* Amsterdam: Lucas, 1734.

Voskuhl, Adelheid. *Androids in the Enlightenment: Mechanics, Artisans, and Cultures of the Self.* Chicago: University of Chicago Press, 2013.

Watts, George B. "The *Encyclopédie* and the *Descriptions des arts et métiers*." *French Review* 25 (1952): 444–54.

Wellman, Kathleen Anne. *Making Science Social: The Conferences of Théophraste Renaudot, 1633–1642.* Norman: University of Oklahoma Press, 2003.

Wilson, Gillian, ed. *European Clocks in the J. Paul Getty Museum.* Los Angeles: Getty Museum, 1996.

Winner, Langdon. "Do Artifacts Have Politics?" *Daedalus* 109 (1980): 121–36.

Withers, Charles W. J. *Placing the Enlightenment: Thinking Geographically About the Age of Reason.* Chicago: University of Chicago Press, 2007.

Yeo, Richard R. *Encyclopaedic Visions: Scientific Dictionaries and Enlightenment Culture.* Cambridge: Cambridge University Press, 2001.

Zilsel, Edgar. "The Sociological Roots of Science." *American Journal of Sociology* 47 (1942): 544–62.

INDEX

Page numbers in *italics* refer to illustrations

Académie d'Architecture, 118, 119, 138
Académie de Chirurgie, 27, 107–8, 133–35, 138, 140, 141
Académie de Peinture et de Sculpture, 3, 12, 118, 119, 138, 163
Académie de Saint-Luc, 118, 155
Académie Française, 10, 18, 24, 40, 137, 180
Académie Royale des Sciences, 14, 25, 85, 93, 96, 110, 148–49, 150, 180; artisans disdained by, 2, 22; compensation to members of, 40; founding of, 23, 32; *History of the Arts* project of, 32–33, 36–37, 38, 39, 41–75, 79, 80, 109, 124, 148, 162; manufacturing standards enforced by, 109; mechanical arts of interest to, 17; reform of, 47–49; Société des Arts and, 27, 81–83, 92, 125, 129–33
Admirable Discourses on the Nature of Waters and Fountains (Palissy), 54
Advancement of Learning, The (Bacon), 31
agriculture, 46
Albertus Magnus, 263n78
alchemy, 71, 100; artisan tradition linked to, 4, 55; Du Tasta's work in, 65–66, 99; at Paris Mint, 64, 67; Philippe II's work in, 58, 65, 66; Réaumur's work in, 55

Alder, Ken, 163
Alembert, Jean Le Rond d', 209; *enchaînement* viewed by, 208; as *Encyclopédie* editor, 7, 18, 46, 152, 163, 194, 207, 211, 213; as *pensionnaire mécanicien*, 126, 204; Société des Arts failure viewed by, 139–40
Alary, Abbé, 180
Amorum emblemata (Veen), 16
anatomy, 8, 17, 36, 38, 135, 171, 194, 195, 212
anchors, 60–61, 63, 68–69, 70
Anisson, Jean, 49
Aphrodite (Greek deity), 1
Archimedes, 37
Arenberg, Léopold-Philippe, duc d', 83, 93, 107
Aristophanes, 1
armillary sphere, 105
Artificial Regulation of Time, The (Sully), 83–84, 85, 86, 107, 148
Art of Converting Wrought Iron into Steel (Réaumur), 25, 61–62, 63, 68, 70, 73
Art of Experiments, The (Nollet), 168, 170, 205
Art of Fire (Ferrand), 157–58

Art of Turning, The (Plumier), 164–65, 166, 167
Artois, Count of, 216
Ash, Eric, 21
astrolabe, 105
astronomy, 36, 38, 83, 84, 85, 112, 130
Aubert, Jean, 254n53, 256n88
automata, 105, 106, 160, 194–96, 201
Auvray, Louis, 150
Auzout, Adrien, 34, 36
Averlino, Antonio di Pietro (Filarete), 32
Aviat, Jean, 246n58

Bacon, Francis, 54, 83, 91, 124, 207, 208; encyclopedists inspired by, 31, 32, 36; history of the arts viewed by, 24, 31–32, 34; mechanical arts viewed by, 27, 32, 45–46, 151, 161, 185; types of knowledge distinguished by, 26, 81–82
Barrier, François-Julien, 137, 218, 257n94
Bassuel, Pierre, 133, 257n94
Baudouin, Benoît, 72
Beaumarchais, Pierre-Auguste Caron de, 109
beaux-arts, 18–19, 119, 215
Beckmann, Johann, 55
Belhoste, Bruno, 217
Belidor, Bernard Forest de, 249–50n34
Bénezet, Cyprien, 70
Beringhem, Marquis of, 116, 157
Bernoulli, Johann, 96
Berthoud, Ferdinand, 112, 140, 154, 155, 158, 215
Bertin, Henri Léonard Jean Baptiste, 195
Besson, Jacques, 43
Béthune, François-Annibal de, 68, 69–70
Bignon, Jean-Paul, 63, 71–72, 79, 92, 93–94, 125, 130, 133, 150; Académie reorganization planned by, 47, 48–49; Gallon viewed by, 131; *History of the Arts* committee formed by, 39–40, 49–50; Regent's Survey overseen by, 57–58, 60; Société des Arts leadership offered to, 121; Société des Arts proposal rebuffed by, 100
blacksmithing, 47
Blair, Ann, 145
Blakey, William, 85, 88, 89, 111–12
Boffrand, Germain, 128, 212, 254n153
Bon de Saint-Hilaire, François-Xavier, 56
Bonnier de La Mosson, Joseph, 113
Bottée, Claude, 137, 139
Boulduc, Gilles-François, 120
Bourbon, Louis de, comte de Clermont, 111, 123, 124, 141, 150, 181; absenteeism of, 136–37, 216; metallurgical activities of, 129, 130; military career of, 100, 151; Société des Arts sponsored by, 2, 23, 100–101, 102, 132, 136, 137, 138, 139, 153
Boyle, Robert, 34, 35, 39
Brief Description of a Clock of New Invention (Sully), 96, 97
Brunelleschi, Filippo, 32
Buache, Philippe, 123, 130
Buonanni, Filippo, 181
Buonarroti, Michelangelo, 5
Bureau of Commerce, 108, 110, 111
bureau typographique, 178–81, 184, 186
Butterfield horizontal dial, 113

Cabinet des beaux arts (Perrault), 18–19
Caffieri, Jean-Jacques, 112
cannons, 60, 63, 68–69, 70
cartography. *See* geography
Cassini de Thury, César-François, 216
Cassirer, Ernst, 6
cast iron, 25, 26, 60–62, 67, 68, 70, 71
Catherine de Medici, Queen, 54
Caus, Salomon de, 43
Century of Louis the Great, The (Perrault), 18
Chamber of Arts, 79–82, 83, 89
Chambers, Ephraim, 24, 209
Charles II, king of England, 105, 156
Chaulnes, Marie-Joseph-Louis d'Ailly, duke of, 216

chemistry, 4, 8
Chevotet, Jean-Michel, 119, 120
Clairaut, Alexis, 99, 119, 120, 125, 130, 131, 212
Clairaut, Jean-Baptiste, 88, 99, 119, 120, 126, 130, 179
clock- and watchmaking, 2, 4, 8, 27, 204–5, 214, 218; Académie des Sciences and, 130; Berthoud's writings on, 154; in England, 12, 156; French leadership in, 90; inventive process in, 156–57; Law's interest in, 26, 84, 87, 88, 90, 91–92; as liberal art, 17–18; longitude problem and, 85, 154, 174; as metaphor, 176–77, 189, 201–2, 205; pendulum clock, 84, 85, 93–94, 96–97, 98, 102, 105, 113, 116, 127, 151–52; practical importance of, 84, 104–5, 113, 215; regulation of, 110, 172–73; Société des Arts and, 103, 110, 111–24, 126, 138, 139, 151; Sully active in, 83–99, 107, 112, 127, 162, 171; theory vs. practice in, 127
Colbert, Jean-Baptiste, 18, 40, 42, 49, 89; Académie des Sciences co-founded by, 38, 63; encyclopedia of arts envisioned by, 39, 57, 207, 213–14; trade revitalization urged by, 24, 34–36, 43, 59, 87, 88, 93, 208
Commission des Arts, 40, 41, 42, 45, 47, 150, 163
Compagnie des Indies, 101–2
Condorcet, Jean-Antoine-Nicolas de Caritat, marquis de, 139, 202, 216, 217
Conversations on the Plurality of Worlds (Fontenelle), 39
Courcelle, Oliver, 24
Cour des Monnoyes, 64
Coustou, Guillaume, the Younger, 3, 12
Coyer, Gabriel-François, 93
Crestelet du Plessis (surgeons), 249–50n34
Critias (Plato), 1
Croissant de Garengeot, René-Jacques, 130, 138, 150, 256n74; as Académie de Chirurgie officer, 133; as guild administrator, 110; as surgeon, 126, 135, 137, 154, 156, 168, 171
Crozat, Antoine, marquis du Chatel, 120
Cyclopedia (Chambers), 24, 209, 210, 211

Dacier, Anne Le Fèvre, 2
Dandrieu, Jean-François, 249–50n34
Dante Alighieri, 5
Daston, Lorraine, 53–54
David, Michel-Antoine, 209
da Vinci, Leonardo, 5, 32
de la Tour, Pierre, 71
Delisle, Guillaume, 119, 132
Delisle, Joseph-Nicolas, 96
Demeulenaere-Douyère, Christiane, 74
Demon (investor), 246n58
Deparcieux, Antoine, 212
dérogeance, 93
Desaguliers, Jean Theophilus, 156, 183
des Billettes, Gilles Filleau, 58, 63, 72; aristocratic background of, 40; *History of the Arts* overseen by, 41, 42, 45–54, 60, 79, 162; ordering principles considered by, 46–47, 60, 73
Descartes, René, 54, 96, 105, 124, 176, 184, 186
Deschiseaux, Pierre, 126, 257n94
Deseuttre (engraver), 249–50n34
Desrosiers, F., 119
Deydier, Henri, 200
Dictionary of Commerce (Savary), 108
Dictionary of the French Academy, 10
Diderot, Denis, 48, 108, 152–53, 181, 194, 202, 208–9; on artisans vs. *artistes*, 4–5, 7, 8, 9, 12, 14, 18; artisans' workshops visited by, 46; Colbert praised by, 208; as freethinker, 211; history of arts viewed by, 31, 32, 37, 55–56, 175, 215; illustrations modified by, 163–64; intellectual forebears of, 33; "language of the arts" advocated by, 211–12; stocking machine explained by, 45; on theory vs. practice, 213
Discours admirables (Palissy), 102

Divine Comedy (Dante), 5
Du Châtelet, Gabrielle Emilie Le Tonnelier de Breteuil, marquise, 8, 171, 187
Dufay, Charles de Cisternay, 67, 109, 181–83, 187, 203
Duhamel du Monceau, Henri-Louis, 73–74
Du Marsais, César, 176–77, 195, 201
Dumas, Louis, 28, 178–79
Dumoustier, Hélène, 74
Du Quet (inventor), 125
Durand, Laurent, 209
Durey d'Harnoncourt, Pierre, 120
Duterte, Jean Baptiste II, 123, 130, 257n94
Dutot, Nicolas, 139
Du Verney, Jacques-François-Marie, 133
Duvivier, Jean, 254n53

École des Mines, 217
École des Ponts et Chaussées, 217
École Polytechnique, 217
École Royale du Génie, 217
Edelstein, Daniel, 10
edict of Nantes, 87
electricity, 187
enameling, 8, 54, 112, 154, 157, 158, 181–82, 193
enchaînement of the arts, 45–47, 48, 49, 51–52, 53, 63, 148–49, 207
Encyclopédie (Diderot and d'Alembert), 17, 31, 45, 49, 149, 176, 202, 217; artisans vs. *artistes* in, 4, 7–9, 12, 37, 56, 108, 215; Du Marsais's contributions to, 176; editorial policies of, 210; as Enlightenment project, 211, 212–13; frontispiece of, 208; illustrations in, 163–64; ordering principles of, 207–8; philosophy preeminent in, 208, 213; prospectus of, 7, 209, 211; Société des Arts linked to, 23–24, 33, 99, 209, 212; success of, 32, 73, 152–53; visual apparatus in, 33, 163–64, 190, 191, 192, 199
Enderlin, Henri, 88, 113, 127

engraving, 8, 12, 74–75, 119, 121, 126, 161, 164, 186
Enlightenment, 3–4, 6–7, 10, 21, 174, 177, 218; Industrial, 22–23
Enquête du Régent, 57–59
Erigena, John Scotus, 157
Essay on Horology (Berthoud), 154
Eugene of Savoy, prince of Savoy, 171
Evelyn, John, 34

Faget, Jean, 123, 133
Fayolle (engineer), 128
Félibien, André, 14–15, 16, 46, 124, 155, 191, 209
Ferrand, Jacques-Philippe, 155, 157–58, 159
Filarete (Antonio di Petro Averlino), 32
Filleau des Billettes, Gilles. *See* des Billettes, Gilles Filleau
Fleury, André-Hercule de, 101
Folard, Jean-Charles de, 148
Fontenelle, Bernard Le Bovier de, 39, 40, 52, 127, 140
Fousjean (laboratory assistant), 60
Fox, Celina, 9
Furetière, Antoine, 4, 46, 160

Gallon, Jean-Gaffin, 131, 137, 151, 160, 171, 212, 257n94
Game of the World (Jaugeon), 40
Gaudron, Pierre, 103, 110, 123, 138, 150, 157–58, 162
Gaultier, Pierre, 246n58
Gay, Peter, 6
Genlis, Stéphanie-Félicité Du Crest, comtesse de, 194
Geoffroy, Étienne François, 244n11
geography, 36, 117, 119; clockmaking linked to, 84, 94, 174; commerce and colonialism linked to, 88, 174; Société des Arts and, 2, 101, 115, 130
Germain, Thomas, 88, 112, 120, 123, 137, 218
Germany, 60

Giorgio Martini, Francesco di, 32
glassmaking, 35, 93, 122–23, 151, 161, 168, 182
goldsmithing, 8, 46, 64, 67, 71, 88, 112, 121
Gourdain, Nicolas, 110, 257n94
Graham, George, 97
Grandjean, Philippe, 49
Grandjean de Fouchy, Jean Paul, 99, 119–20, 126, 131, 132, 139, 175, 212
Great Britain, 87
Grignion, Thomas, 248–49n16
Grollier de Servière, Nicolas, 40, 160
Grosse, Jean, 120, 131
Gua de Malves, Jean-Paul de, 99, 126, 131, 149, 209, 210–12, 256n88
Guérard, Nicolas, 10–12, 17, 19, 81
Guichard, Charlotte, 118
gunpowder, 40–41, 161
Guyot, Edme, 130

Habert, Jean Charles, 131, 256n88
Harkness, Deborah, 9
Hellot, Jean, 203, 257n94
Helvétius, Jean Claude Adrien, 257n94
Hensing, Johann Thomas, 65
Hephaestus (Vulcan; Greek deity), 1, 2–3, 9, 12, 18, 111
Hera (Greek deity), 1
Hilaire-Pérez, Liliane, 9, 23, 216
Hillerin de Boistissandeau, Jean-Baptiste Laurent, 120, 257n94
History of Insects (Réaumur), 74
History of the Arts, 39, 79, 80, 124, 148, 162; Académie des Sciences reorganization and, 47–50; Baconian inspiration for, 36, 46; Colbert's view of, 38; objections to, 57; information gathering for, 51; ordering principles of, 45–47, 51; as political economy project, 63, 67; public purposes of, 37, 41–42, 52, 56–57, 62, 68, 72; Réaumur's leadership of, 50, 52–75, 109; unpublished precursor of, 32–33
Hobbes, Thomas, 142

Homberg, Guillaume, 58
Homer, 2, 5
Hooke, Robert, 34
Huygens, Christiaan, 20, 34, 36, 96, 127
Hynault (member of Parlement de Paris), 120, 150

Iliad (Homer), 2
Industrial Enlightenment, 22–23
Institutions of Physics (Du Châtelet), 171
intersemiotic translation, 146, 163, 165, 167–68
Israel, Jonathan, 7

Jakobson, Roman, 146
Jarosson, Pierre, 69, 70
Jaucourt, Louis de, 8–9, 18, 261n19
Jaugeon, Jacques, 40, 49, 52, 73
Jordan, Charles-Étienne, 73
Journal de Trévoux, 147–48
Jousse, Daniel, 120
Jouvenet, Jean, 20
Jullien, Nicolas, 112

Kant, Immanuel, 174
Kepler, Johannes, 112

La Condamine, Charles-Marie de, 131, 132, 195, 212
La Hire, Philippe de, 96, 113
Lambert (investor), 246n58
La Mettrie, Julien de, 176
Languet de Gergy, Jean-Joseph, 98–99, 100, 101, 120, 129
La Popelinière, Alexandre Jean Joseph Le Riche de, 194
La Peyronie, François Gigot de, 108, 252n95
Larmessin, Nicolas II de, 12, 13, 17, 19
Lassay, Armand-Léon de Madaillan de Lesparre, marquis de, 93
la Tour, Pierre de, 246n58
La Vallière, Duke of, 252n94

Lavoisier, Antoine, 216
Law, John, 123, 137, 139; financial schemes of, 61, 65, 66, 87, 100; French manufacturing promoted by, 87–88, 89, 91–92, 93, 112; watchmaking interests of, 26, 84, 87, 88, 90, 91–92
Law, William, 87–88
Laws (Plato), 1
Le Breton, André, 209
Le Cat, Claude-Nicolas, 194
Lectures on Experimental Physics (Nollet), 168, 169, 187, 189, 191, 193
Le Dran, Henri François, 133, 134, 150, 154, 156
Le Guay de Prémontval, Pierre, 212
Leibniz, Gottfried Wilhelm, 36, 83–84, 96
Le Maire, Jacques, 113, 115, 120, 126, 130
Lemoyne, François, 120, 123, 254n53
Lemoyne, Jean-Baptiste, 254n53
Le Normand (treasurer of Société des Arts), 99
Leonardo da Vinci, 5, 32
Lepaute, Jean-André, 153, 160
Le Pautre, Jean, 19
Lequin (goldsmith), 171
Le Roy, Julien, 90, 115, 125, 137, 156, 158, 167, 204–5, 215, 218, 257n94; Blakey's collaboration with, 112; as collector, 160; mechanical imperfections acknowledged by, 107, 113; reforms recommended by, 110; royal commission received by, 116–17, 157; as Société des Arts member, 88, 99, 103, 119, 120, 124, 130, 138; Voltaire rebutted by, 161; works consulted by, 171; workshop of, 111; writing stressed by, 172–73
Le Roy, Pierre, 99, 110, 119, 120
Le Sueur, Nicolas, 116
levée du roi, 116, 117
Liébaux, Henri, 88, 99, 119, 120, 121, 123, 130, 141
Lives of the Most Excellent Painters, Sculptors, and Architects (Vasari), 5

Locke, John, 36, 122, 124, 141–42, 178
locksmithing, 47, 71
Long, Pamela, 9, 27, 145, 155
Longitude Act, 97
Louis, Antoine, 140
Louis, duke of Burgundy, 40
Louis XIV, king of France, 18, 24, 34–35, 38, 41, 42, 58, 97, 105, 116, 156
Louis XV, king of France, 23, 91–92, 97, 101, 109, 116, 134, 204–5
Louvois, François-Michel Le Tellier, marquis de, 38
Louvre, Musée du, 2, 118
Lucotte, Jacques-Raymond, 8

Machiavelli, Niccolò, 148
Machines and Inventions Approved by the Académie des Sciences (Gallon), 132, 171, 212
Magalotti, Lorenzo, 36
Maillet, Benoît de, 141
Malebranche, Nicolas, 124, 159, 210
Mallet, Edmé-Francois, 261n19
Man, a Machine (La Mettrie), 176
mapmaking. *See* geography
Marmontel, Jean-François, 194
Maupertuis, Pierre-Louis Moreau de, 115
Maurepas, Jean-Frédéric Phélypeaux, 98, 126, 197
McClellan, James E., 101
Medalon, Daniel, 133, 135, 150, 179, 209–10, 252n95
Medici, Leopoldo de, 101
Meldecreutz, Jonas De, 121
mercantilism, 35, 87, 108
Mercure de France, 149, 173, 177, 179, 201
Merne (engraver), 123
metallurgy, 8, 26, 46–47, 58, 100, 101, 122–23, 151; Clermont active in, 129, 130; in Greek myth, 1; Nollet active in, 186; political economy of, 61–68; Réaumur active in, 25, 55, 59, 60–63, 67–72, 73, 90, 130
Michelangelo, 5

microscopy, 21
Minard, Philippe, 108, 109
Minerva (Greek deity), 10
mining, 93, 185
minting, of coins, 60, 64–67
Mississippi Company, 61, 66, 67, 87
Moitte, Pierre-Etienne, 125
Mokyr, Joel, 7, 22–23
Moncrif, François-Augustin de Paradis de, 137, 139, 150
Morand, Sauveur-François, 134
Morville, Charles-Jean-Baptiste, Fleuriau de, 136–37
Mukerji, Chandra, 41
Musschenbroek, Pieter van, 183, 257n94

Natural History (Pliny), 37
navigation, 101, 130; Académie des Sciences active in, 36, 55; commerce and colonialism dependent on, 84, 88; Law's proposals for, 87; longitude problem in, 26, 84–85, 87, 88, 94, 96, 97–98, 152, 154, 174, 215; watchmaking linked to, 26, 93, 94, 112, 115, 152, 174, 215
Netherlands, 84, 87
New Art of Making Steel and Cast Iron (Réaumur), 62
New Atlantis (Bacon), 91
New Discoveries on the Art of War (Folard), 148
Newton, Isaac, 85, 88, 96, 171, 183–86
Nine Years' War (1688–97), 49
Noailles, Adrien Maurice, duc de, 92
Noel, Nicolas, 171
Nollet, Jean-Antoine, 120, 131, 197; as enameller, 181–82; as instrument maker, 158, 160, 181, 182, 183, 218; as physicist and pedagogue, 28, 141, 168, 182, 183–91, 200, 205; scientific openness urged by, 156, 185; theological training of, 181; Vaucanson compared with, 193, 202–3
Norman, Larry, 20
Novum organum (Bacon), 32, 161, 208

Observations on Bloodletting (Quesnay), 194
Odyssey (Homer), 5
Oldenburg, Henry, 34, 35, 36
On Ancient Footwear (Baudouin), 72
optics, 19, 21, 122–23, 151, 161
Oudry, Jean-Baptiste, 120, 137, 218, 254n53

Paillasson, Charles, 9
Palissy, Bernard, 32, 54–55, 102
Panciroli, Guido, 37, 41, 51
paper money, 61, 91
Papillon, Jean-Michel, 120, 137, 164, 171, 212
Paracelsus, 65
Paré, Ambroise, 156
Parasceve ad historiam naturalem et experimentalem (Bacon), 32
Pelays, Jacques Louis, 99, 102
Pellegrini, Giovanni Antonio, 250n36
Perrault, Charles, 18–20, 40, 41
Petit, Pierre, 34, 36, 240n10
Petty, William, 34
Phélypeaux, Louis, 49
Philippe II, duc d'Orléans, 57–59, 60, 63, 68, 84, 123, 139
Philosopher, The (Du Marsais), 176, 178
Philosophical Letters (Voltaire), 161, 171, 184
physiocrats, 146, 204
Picon, Antoine, 216–17
Picot, Jean-Baptiste, 43, 45
Pigeon d'Osangis, Jean, 105, 106, 120
Pimaudan, abbé de, 120
Plato, 1
Plélo, Louis-Robert-Hippolyte de Bréhan, comte de, 132
Pliny, the Elder, 37
Pluche, Noël Antoine, 104, 185, 209
Plumier, Charles, 16, 164–65
Polinière, Pierre, 184
Pompadour, Jeanne Antoinette Poisson, marquise de, 141
Preliminary Discourse (Diderot), 7

Principe, Larry, 65
Principles of Architecture, Sculpture, Painting, and of the Arts That Depend on Them (Félibien), 14, 124, 155, 191
printing, 8, 40, 49
Privat de Molières, Joseph, 132
privilèges, 43
Procope, Michel, 134
"Projet d'une Compagnie des Sciences et Arts," 36–37, 38, 41, 50, 59, 74, 100
Prospectus (Diderot), 7, 209, 211
Proust, Jacques, 212
Puisieux, Jean-Baptiste de, 119, 120, 212
putti, 16–17, 20

Quarrel of the Ancients and the Moderns, 18, 20–21, 24, 40–41, 42, 46, 51
Quesnay, François, 131, 186, 211, 212, 256n74; as Académie de Chirurgie secretary, 133, 141; bloodletting viewed by, 194; theory of surgery expounded by, 146–47, 149, 153, 158
Quillau, Gabriel-François, 123

Raillard, Claude, 110, 151
Rameau, Jean-Philippe, 194, 218, 249–50n34, 257n94
Ramelli, Agostino, 43, 44
Raphael, 5
Raux, Jean, 120, 141, 181, 191, 193, 200
Réaumur, René Antoine Ferchault de, 52–75, 81, 89, 91, 97, 110, 182, 203; Dufay sponsored by, 182; entomological research of, 53; *History of the Arts* project overseen by, 50, 52–75, 109; metallurgical activities of, 25, 55, 59, 60–63, 67–72, 73, 90, 130; Regent's Survey overseen by, 57–61, 69; Société des Arts membership controlled by, 131–32, 133
Reflections on the Usefulness That the Académie des Sciences Could Have for the Kingdom (Réaumur), 63
Regent's Survey, 57–61, 69, 74, 153
Rémond, Nicolas-François, 84

Rémond de Sainte-Albine, Pierre, 127–28, 129, 149, 166, 173
Renard de Roussiac (investor), 246n58
Renard du Tasta, Mathieu, 65–68, 99, 139, 150, 249–50n34, 256n88
répondans, 123–24, 140, 141
Republic of Letters, 27, 37, 118, 121–22, 138, 149, 156, 161, 214
Restout, Jean II, 254n530
Richelet, Pierre, 4, 8, 9
Rigoud, François, 101
Riskin, Jessica, 195
Roettiers (engraver), 99, 246n58
Rohault, Jacques, 184
Roland de la Platière, Jean-Marie, 197
Rossi, Paolo, 236n16
Rousseau, Jean-Jacques, 181
Rousselot, Jean, 246n58
Royal Manufacture of English Watches, 26, 88, 89–92, 98
Royal Manufacture of Malleable Cast Iron and Steel, 25, 26, 67, 68–72, 75, 90, 99, 109
Royal Observatory, 131
Royal Press, 49
Royal Society of Arts, 79, 216
Royal Society of London, 14, 34, 81–82, 96
Royllet, Honoré Sébastien, 120, 140, 141

St. Bartholomew's Day massacre, 104
Saint-Lambert, Jean-François de, 8, 18
Saurin, Joseph, 125
Savary, Jacques, 108
Schaffer, Simon, 195
Scientific Revolution, 9, 21, 22–23
Sheridan, Geraldine, 163
Shiner, Larry, 5, 6
silk industry, 28, 56, 194–201, 203–4
Silva, Jean-Baptiste, 218
Simmoneau, Louis, 20, 49, 74–75
Slodtz, Michel-Ange, 120
Smith, Adam, 18
Smith, Pamela, 9, 54, 145, 147

Société des Arts, 3, 9, 24, 25, 93, 157, 182, 194, 216, 218; Académie de Chirurgie and, 133–35; Académie Royale des Sciences and, 27, 81–83, 92, 125, 129–33; *bureau typographique* and, 179–80, 186; clockmaking and, 103, 110, 111–24, 126, 138, 139, 151; dissolution of, 27, 107, 136–42, 172, 181, 215; first iteration of, 26, 50, 79–80, 88; founding of, 75, 79, 88, 92; frictions within, 13; Hephaestus myth invoked by, 2; membership roster of, 221–33; official role sought by, 22, 23, 110–11, 202, 214–15, 216; political project articulated by, 26, 107, 172; publications of, 147–53, 173, 209–15; second iteration of, 72, 97, 98–103; social microcosm, 118–24; surgeons belonging to, 2, 17–18, 99, 103, 110, 121, 123, 124, 135, 140, 150, 154, 168; theory vs. practice and, 124–29, 183, 184, 191, 202
Société Libre d'Emulation, 215–17
Souchet, Octavian, 246n58
Spain, 84
Spectacle of Nature, The (Pluche), 104, 185
standardization, 163
steelmaking, 25, 26, 60–63, 67, 68, 71, 87
Steno, Nicolas, 36
Stewart, Larry, 9, 236n16
Sturdy, David, 74
Sully, Henry, 79, 119, 120, 149, 153, 204; death of, 100; horology encyclopedia proposed by, 151–52; improvement stressed by, 89, 91, 96, 97, 98, 99, 102, 151; inventive process viewed by, 160, 161–62; Law's collaboration with, 87–88; Leibniz's sponsorship of, 83–84; Le Roy's collaborations with, 116, 130; longitude problem attacked by, 85, 88, 93–97, 98, 154, 215; natural vs. artificial division of time viewed by, 112–13; scientific collaboration urged by, 97–98, 102; senses and intellect viewed by, 159; watchmaking activities of, 83–99, 107, 112, 127, 162, 171; works consulted by, 171;
writings of, 83–84, 85, 86, 96, 97, 107, 148, 173–74
surgery, 8, 119, 121, 137, 156, 171, 173, 214; academy of, 27, 107–8, 133–35, 138, 140, 141; prejudices against, 109–10, 111, 124, 134, 147, 153; Quesnay's theory of, 146–47, 149, 153, 158; Société des Arts and, 2, 17–18, 99, 103, 110, 121, 123, 124, 135, 140, 150, 154, 168
Swammerdam, Jan, 36

Taglini, Carlo, 133
telescopes, 21, 98, 156
Telliamed (Maillet), 141
Teniers, David, 66
Terrall, Mary, 53, 74
"theaters of machines," 42–43
Thévenot, Melchisédech, 34, 35–36, 38
Thibault, Nicolas, 69
Thiout, Antoine, 147, 148–49, 215
Thomlin, Marc Mitouflet, 123, 159
Three Letters Concerning the Forming of a Society, 79–82, 83, 90, 92, 102
Toussaint, François-Vincent, 261n19
travel writing, 36, 37–38
Treatise on Mechanical and Practical Horology (Thiout), 147, 148–49
Treatise on the Varnish Commonly Called Chinese (Buonanni), 181
Trojan War, 1
Truchet, Sébastien, 40, 73
Trudaine, Daniel, 127, 203
Two Books on Memorable Things Lost and Found (Panciroli), 37, 51

Varignon, Pierre, 52
Vasari, Giorgio, 5
Vassé, François-Antoine, 254n53
Vaucanson, Jacques de, 28, 126–27, 131, 160, 193, 193–205, 217
Veen, Otto van, 16
Verdier, César, 123, 133, 212, 256n74
Vérin, Hélène, 21, 27, 82, 145
Versailles, château de, 26, 40, 90

Vigné de Vigny, Pierre, 120, 123, 254n53
Villiers, Jacques François de
Vitruvius Pollio, 8
Viviani, Vincenzo, 36
Voltaire, 9, 93, 136, 161, 171, 181, 183, 184, 218

War of the Polish Succession (1733–38), 136
War of the Spanish Succession (1701–14), 49

watchmaking. *See* clock- and watchmaking
Watelet, Claude-Henri, 9
Wisdom of the Ancients (Bacon), 45–46
Wolff, Christian, 210
Wordling, The (Voltaire), 218

Xenophon, 5

Yeo, Richard, 145